IFAE 2006

Incontri di Fisica delle Alte Energie
Italian Meeting on High Energy Physics

G. Montagna · O. Nicrosini · V. Vercesi (Eds.)

IFAE 2006

Incontri di Fisica delle Alte Energie
Italian Meeting on High Energy Physics

Pavia, 19–21 April 2006

 Springer

GUIDO MONTAGNA,
ORESTE NICROSINI,
VALERIO VERCESI

INFN – Sezione di Pavia and
Dipartimento di Fisica Nucleare
e Teorica – Università di Pavia
(Italy)

Library of Congress Control Number: 2006938540

ISBN 978-88-470-0529-7 Springer Berlin Heidelberg New York

Springer is a part of Springer Science+Business Media

springer.com

© Springer-Verlag Italia 2007

Cover concept: Simona Colombo, Milano

Typesetting: LE-TeX Jelonek, Schmidt, Vöckler GbR, Leipzig

Printing and binding: Grafiche Porpora, Segrate, Milano

Printed on acid-free paper   57/3141/NN - 5 4 3 2 1 0

Printed in Italy

Springer-Verlag Italia Srl, Via Decembrio 28, I-20137 Milano

Preface

The 2006 edition of the IFAE ("Incontri di Fisica delle Alte Energie") Workshop reviews the recent and most important advancements in High-Energy Physics and Astroparticle Physics, including reports as well on multi-disciplinary applications of detector developments and on Grid computing. The Workshop (http://www.pv.infn.it/ifae2006/) was held in Pavia from April 19th to 21st, in the beautiful medieval frame of the "San Tommaso" Congress Centre and saw the participation of more than 150 researchers.

Presentations, both theoretical and experimental, addressed the status of Standard Model and Flavour physics, Neutrino and Cosmological topics, new insights beyond the present understanding of fundamental particle physics and cross-fertilization in areas such as medicine, biology, technological spin-offs and computing. Special emphasis was given to the expectations of the forthcoming Large Hadron Collider, due in operation next year, the status of its experiments and the possibilities offered by this new energy frontier. The venue of plenary sessions interleaved with parallel ones allowed for a rich exchange of ideas, presented in these Proceedings, that form a coherent picture of the findings and of the open questions in our field.

We are happy to have had the opportunity of organizing such an event in Pavia, and we are pleased of the enthusiastic response of the community.

We acknowledge the financial contributions of STMicroelectronics, IBM, CAEN, Banca Intermobiliare and the sponsorship of Comune di Pavia. We are deeply indebted to INFN and University of Pavia, whose support has been fundamental to accomplish the high standards of the meeting. We are grateful to Diana A. Scannicchio and Carlo M. Carloni Calame for their extensive and generous help. And we warmly thank all the speakers and participants, who have been the real engine behind the success of this Workshop.

Pavia, November 2006   *The Editors*

IFAE 2006: the participants

Contents

**PARALLEL SESSION: Standard Model Physics
(P. Azzi and F. Piccinini, conveners)**

PARALLEL SESSION: New Physics (A. Perrotta and A. Strumia, conveners)

**PARALLEL SESSION: Neutrinos and Cosmic Rays
(E. Lisi and L. Patrizii, conveners)**

**PARALLEL SESSION: Detectors and New Technologies
(A. Cardini, M. Michelotto and V. Rosso, conveners)**

Monolithic Active Pixel Sensors in a 130 nm Triple Well CMOS Technology

The external scanning proton microprobe in Florence: set-up and examples of applications

Infrastructure of the ATLAS Event Filter

The CMS High-Level Trigger

WLCG Service Challenges and Tiered architecture in the LHC era

List of Participants

Abbiendi Giovanni
INFN, Sezione di Bologna

Altarelli Guido
Università di Roma Tre

Ambroglini Filippo
Università di Perugia

Andreazza Attilio
Università di Milano

Annovi Alberto
Laboratori Nazionali di Frascati

Antinori Federico
INFN, Sezione di Padova

Antonelli Vito
Università di Milano

Azzi Patrizia
INFN, Sezione di Padova

Balossini Giovanni
Università di Pavia

Banfi Andrea
Università di Milano Bicocca

Baracchini Elisabetta
Università di Roma "La Sapienza"

Barbieri Riccardo
Scuola Normale Superiore, Pisa

Bellomo Massimiliano
INFN, Sezione di Pavia

Bocci Andrea
Scuola Normale Superiore, Pisa

Boffi Sigfrido
Università di Pavia

Bolognesi Sara
Università di Torino

Bolzoni Paolo
Università di Milano

Bonacorsi Daniele
INFN-CNAF, Sezione di Bologna

Bozzi Concezio
INFN, Sezione di Ferrara

Bozzi Giuseppe
LPSC, Grenoble

Brambilla Massimo
Università di Pavia

Brigliadori Luca
INFN, Sezione di Bologna

Cadeddu Sandro
INFN, Sezione di Cagliari

Caffo Michele
INFN, Sezione di Bologna

Cardelli Edoardo
INFN, Sezione di Cagliari

Cardini Alessandro
INFN, Sezione di Cagliari

Carloni Calame Carlo Michel
INFN, Sezione di Pavia

Cartaro Concetta
Università di Trieste

Casadio Roberto
Università di Bologna

Cei Fabrizio
Università di Pisa

Cherubini Roberto
Laboratori Nazionali di Legnaro

Chiavassa Andrea
Università di Torino

Ciafaloni Paolo
INFN, Sezione di Lecce

Cirelli Marco
Yale University

Cobal Marina
Università di Udine

Conta Claudio
INFN, Sezione di Pavia

Contino Roberto
INFN, Sezione di Roma 1

Cozzi Michela
INFN, Sezione di Bologna

Cremonesi Oliviero
INFN, Sezione di Milano

Crosetti Giovanni
Università della Calabria

De Marco Daniel
University of Delaware

De Sanctis Umberto
Università di Milano

Devecchi Federica
Università di Pavia

Di Pompeo Francesco
Laboratori Nazionali del Gran Sasso

Dolce Donatello
INFN, Sezione di Firenze

Dosselli Umberto
INFN, Sezione di Padova

Dotti Andrea
INFN, Sezione di Pisa

Dreucci Marco
Laboratori Nazionali di Frascati

Eulisse Giulio
Northeastern University, Boston

Fabbri Fabrizio
INFN, Sezione di Bologna

Ferrandes Rossella
Università di Bari

Ferrari Roberto
INFN, Sezione di Pavia

Ferri Federico
INFN, Sezione di Milano

Ferroni Fernando
INFN, Sezione di Roma 1

Franchino Silvia
INFN, Sezione di Pavia

Frigerio Michele
SPhT, CEA/Saclay

Furcas Sara
INFN, Sezione di Cagliari

Gambino Paolo
INFN, Sezione di Torino

Gatti Flavio
INFN, Sezione di Genova

Gaudio Gabriella
INFN, Sezione di Pavia

Giagu Stefano
Università di Roma "La Sapienza"

Gianotti Fabiola
CERN, Ginevra

Giuntini Lorenzo
INFN, Sezione di Firenze

Goggi Giorgio
Università di Pavia

Govoni Pietro
INFN, Sezione di Milano

Gresele Ambra
Università di Trento

Ianni Aldo
Laboratori Nazionali del Gran Sasso

Introzzi Gianluca
INFN, Sezione di Pavia

Lanza Agostino
INFN, Sezione di Pavia

Lari Tommaso
INFN, Sezione di Milano

Lisi Eligio
INFN, Sezione di Bari

Livan Michele
Università di Pavia

Lombardo Maria Paola
Laboratori Nazionali di Frascati

Maltoni Fabio
Université de Louvain

Maltoni Michele
ICTP, Trieste

Marconi Umberto
INFN, Sezione di Bologna

Marzuoli Annalisa
Università di Pavia

Massarotti Paolo
INFN, Sezione di Napoli

Mastrolia Pierpaolo
University of Zurich

Meloni Davide
INFN, Sezione di Roma 1

Menici Lorenzo
Università di Roma "La Sapienza"

Mescia Federico
Laboratori Nazionali di Frascati

Mezzetto Mauro
INFN, Sezione di Padova

Michelotto Michele
INFN, Sezione di Padova

Mila Giorgia
INFN, Sezione di Torino

Mirizzi Alessandro
Università di Bari

Montagna Guido
Università di Pavia

Montanari Claudio
INFN, Sezione di Pavia

Morello Michael
Scuola Normale Superiore, Pisa

Morisi Stefano
Università di Milano

Navarria Francesco L.
Università di Bologna

Negri Andrea
University of California, Irvine

Neri Nicola
INFN, Sezione di Pisa

Nervo Marco
Università di Torino

Nguyen Federico
Università di Roma Tre

Nicrosini Oreste
INFN, Sezione di Pavia

Paganoni Marco
Università di Milano Bicocca

Paradisi Paride
Università di Roma Tor Vergata

Parenti Andrea
Università di Padova

Passaleva Giovanni
INFN, Sezione di Firenze

Passera Massimo
INFN, Sezione di Padova

Patrizii Laura
INFN, Sezione di Bologna

Pavan Maura
Università di Milano Bicocca

Perrotta Andrea
INFN, Sezione di Bologna

Piai Maurizio
University of Washington

Piazzoli Adalberto
Università di Pavia

Piccinini Fulvio
INFN, Sezione di Pavia

Piemonte Claudio
ITC-IRST, Trento

Polesello Giacomo
INFN, Sezione di Pavia

Polosa Antonio Davide
INFN, Sezione di Roma 1

Porretti Valentina
Università di Roma Tre

Pullia Antonino
Università di Milano Bicocca

Rappoldi Andrea
INFN, Sezione di Pavia

Raselli Gian Luca
INFN, Sezione di Pavia

Ratti Sergio
Università di Pavia

Rebuzzi Daniela
INFN, Sezione di Pavia

Re Valerio
Università di Bergamo

Riccardi Cristina
Università di Pavia

Riccobene Giorgio
Laboratori Nazionali del Sud

Rimoldi Adele
Università di Pavia

Rolli Simona
Tufts University, Naperville

Roncadelli Marco
INFN, Sezione di Pavia

Rosati Stefano
INFN, Sezione di Roma 1

Rossi Sandro
Fondazione CNAO, Milano

Rosso Valeria
Università di Pisa

Salamanna Giuseppe
Università di Roma "La Sapienza"

Salvatore Daniela
Università della Calabria

Salvini Paola
INFN, Sezione di Pavia

Sannino Francesco
NBI, Copenhagen

Santorelli Pietro
Università di Napoli

Scannicchio Diana
INFN, Sezione di Pavia

Scannicchio Domenico
Università di Pavia

Silvestrini Luca
INFN, Sezione di Roma 1

Sioli Maximiliano
Università di Bologna

Slavich Pietro
LAPTH, Annecy

Strumia Alessandro
Università di Pisa

Tancini Valentina
Università di Milano Bicocca

Torre Paola
Università di Pavia

Trentadue Luca
Università di Parma

Tripiccione Raffaele
Università di Ferrara

Ubiali Maria
Università di Genova

Uccirati Sandro
INFN, Sezione di Torino

Ullio Piero
SISSA, Trieste

Vagnoni Vincenzo
INFN, Sezione di Bologna

Vallecorsa Sofia
Università di Ginevra

Vecchi Stefania
INFN, Sezione di Bologna

Vercesi Valerio
INFN, Sezione di Pavia

Verducci Monica
CERN, Ginevra

Vicini Alessandro
Università di Milano

Viganò Sara
Università di Milano Bicocca

Vitale Vincenzo
Università di Udine

List of Not Received Contributions

The transparencies of the related talks are available at the web page
http://www.pv.infn.it/ifae2006/talks/

F. Antinori
Status and perspectives
of research in ultrarelativistic
nucleus-nucleus collisions

R. Barbieri
Future perspectives of elementary
particles phenomenology

M.P. Lombardo
Phase transitions
in the Standard Model

P. Ciafaloni
Electroweak corrections
at the TeV scale

F. Ambroglini
Minimum Bias and Underlying
Event studies at the LHC

A. Gresele
Top mass and cross section
at the Tevatron

A. Bocci
b-tagging for the Higgs search
at the LHC

F. Maltoni
MonteCarlo for new physics at LHC

P. Slavich
Split Susy and the LHC

P. Ullio
Dark matter and LHC

F.R. Navarria
Searches for extra dimensions
from LEP to LHC to NLC

F. Sannino
Dynamical EW breaking: a classic

M. Piai
Little(st) Higgs and LHC

P. Gambino
Inclusive radiative B decays:
an update

M. Morello
Rare and charmless decays at CDFII

F. Gatti
The neutrino mass
from β-decay: M_β

M. Maltoni
Synergies between accelerator
and atmospheric neutrinos searches

A. Chiavassa
High and ultra-high
energy cosmic rays

S. Rossi
The CNAO project

R. Cherubini
Hadron radiobiology and its
implications in hadrotherapy
and radioprotection

F. Di Pompeo
WArP: a double phase argon
detector for direct search
of dark matter

S. Cadeddu
Microelectronics for time calibration
of the muon system in LHCb

C. Piemonte
Development of 3D silicon detectors
at ITC-irst

M. Briscolini
High-performance computing
and data management
architectures in HEP

G. Eulisse
Interactive Web-based Analysis
Clients using AJAX: examples
for CMS, ROOT and GEANT4

INVITED TALKS

Particle Physics: a Progress Report

Guido Altarelli

Dipartimento di Fisica 'E. Amaldi', Università di Roma Tre and INFN, Sezione di Roma Tre, I-00146 Rome, Italy and
CERN, Department of Physics, Theory Division, CH-1211 Geneva 23, Switzerland
guido.altarelli@cern.ch

1 Introduction

I would like to present a concise review of where we stand in particle physics today. First I will discuss QCD, then the electroweak sector and finally the motivations and the avenues for new physics beyond the Standard Model.

2 QCD

QCD stands as a main building block of the Standard Model (SM) of particle physics. For many years the relativistic quantum field theory of reference was QED, but at present QCD offers a much more complex and intriguing theoretical laboratory. Indeed, due to asymptotic freedom, QCD can be considered as a better defined theory than QED. The statement that QCD is an unbroken renormalisable gauge theory with six kinds of triplets quarks with given masses completely specifies the form of the Lagrangian in terms of quark and gluon fields. From the compact form of its Lagrangian one might be led to think that QCD is a "simple" theory. But actually this simple theory has an extremely rich dynamical content, including the property of confinement, the complexity of the observed hadronic spectrum (with light and heavy quarks), the spontaneous breaking of (approximate) chiral symmetry, a complicated phase transition structure (deconfinement, chiral symmetry restauration, colour superconductivity), a highly non trivial vacuum topology (instantons, $U(1)_A$ symmetry breaking, strong CP violation, ...), the property of asymptotic freedom and so on.

How do we get predictions from QCD? There are non perturbative methods: lattice simulations (in great continuous progress), effective lagrangians valid in restricted specified domains [chiral lagrangians, heavy quark effective theories, Soft Collinear Effective Theories (SCET), Non Relativistic QCD...] and also QCD sum rules, potential models (for quarkonium) and so on. But the perturbative approach, based on asymptotic freedom and valid for hard processes, still remains the main quantitative connection to experiment.

Due to confinement no free coloured particles are observed but only colour singlet hadrons. In high energy collisions the produced quarks and gluons materialize as narrow jets of hadrons. Our understanding of the confinement mechanism has much improved thanks to lattice simulations of QCD at finite temperatures and densities [1]. The potential between two colour charges clearly shows a linear slope at large distances (linearly rising potential). The slope decreases with increasing temperature until it vanishes at a critical temperature T_C. Above T_C the slope remains zero. The phase transitions of colour deconfinement and of chiral restauration appear to happen together on the lattice. Near the critical temperature for both deconfinement and chiral restauration a rapid transition is observed in lattice simulations. In particular the energy density $\epsilon(T)$ is seen to sharply increase. The critical parameters and the nature of the phase transition depend on the number of quark flavours N_f and on their masses. For example, for $N_f = 2$ or $2 + 1$ (i. e. 2 light u and d quarks and 1 heavier s-quarks), $T_C \sim 175\,\mathrm{MeV}$ and $\epsilon(T_C) \sim 0.5-1.0\,\mathrm{GeV/fm^3}$. For realistic values of the masses m_s and $m_{u,d}$ the phase transition appears to be a second order one, while it becomes first order for very small or very large $m_{u,d,s}$. The hadronic phase and the deconfined phase are separated by a crossover line at small densities and by a critical line at high densities. Determining the exact location of the critical point in T and μ_B is an important challenge for theory which is also important for the interpretation of heavy ion collision experiments. At high densities the colour superconducting phase is probably also present with bosonic diquarks acting as Cooper pairs.

A large investment is being done in experiments of heavy ion collisions with the aim of finding some evidence of the quark gluon plasma phase. Many exciting results have been found at the CERN SPS in the past years and more recently at RHIC. At the CERN SPS some experimental hints of variation with the energy density were found in the form, for example, of J/Ψ production suppression or of strangeness enhancement when going from p–A to Pb–Pb collisions. Indeed a posteriori the CERN SPS appears well positioned in energy to probe the transition region, in that a marked variation of different observables was observed. The most impressive effect detected at RHIC, interpreted as due to the formation of a hot and dense bubble of matter, is the observation of a strong suppression of back-to-back correlations in jets from central collisions in Au–Au, showing that the jet that crosses the bulk of the dense region is absorbed. The produced hot matter shows a high degree of collectivity, as shown by the observation of elliptic flow (produced hadrons show an elliptic distribution while it would be spherical for a gas) and resembles a perfect liquid with small or no viscosity. However, for quark gluon plasma, it is fair to say that the significance of each single piece of evidence can be questioned and one is still far from an experimental confirmation of a phase transition. The experimental programme on heavy ion collisions will continue at RHIC and then at the LHC where ALICE, a dedicated heavy ion collision experiment, is in preparation.

As we have seen, a main approach to non perturbative problems in QCD is by simulations of the theory on the lattice, a technique started by K. Wilson in 1974 which has shown continuous progress over the last decades. One recent big step, made possible by the availability of more powerful dedicated computers, is the evolution from quenched (i. e. with no dynamical fermions) to unquenched calculations. In doing so an evident improvement in the agreement of predictions with the data is obtained. For example [2], modern unquenched simulations reproduce the hadron spectrum quite well. Calculations with dynamical fermions (which take into account the effects of virtual quark loops) imply the evaluation of the quark determinant which is a difficult task. How difficult depends on the particular calculation method. There are several approaches (Wilson, twisted mass, Kogut-Susskind staggered, Ginsparg–Wilson fermions), each with its own advantages and disadvantages (including the time it takes to run the simulation on a computer). Another area of progress is the implementation of chiral extrapolations: lattice simulation is limited to large enough masses of light quarks. To extrapolate the results down to the physical pion mass one can take advantage of the chiral effective theory in order to control the chiral logs: $\log(m_{\mathrm{q}}/4\pi f_\pi)$. For lattice QCD one is now in an epoch of pre-dictivity as opposed to the post-dictivity of the past. And in fact the range of precise lattice results currently includes many domains: the QCD coupling constant (the value $\alpha_{\mathrm{s}}(m_Z) = 0.1170(12)$ has been recently quoted [3]: the central value is in agreement with other determinations but one would not trust the stated error as the total uncertainty), the quark masses, the form factors for K and D decay, the B parameter for kaons, the decay constants f_K, f_D, f_{Ds}, the B_{c} mass, the nucleon axial charge g_{A} (the lattice result [4] is close to the experimental value $g_{\mathrm{A}} \sim 1.25$ and well separated from the $SU(6)$ value $g_{\mathrm{A}} = 5/3$) and many more.

Recently some surprising developments in hadron spectroscopy have attracted the general attention. Ordinary hadrons are baryons, $B \sim qqq$ and mesons $M \sim q\bar{q}$. For a long time the search for exotic states was concentrated on glueballs, gg bound states, predicted at $M \gtrsim 1.5\,\mathrm{GeV}$ by the lattice. As well known, experimentally glueballs were never clearly identified, probably because they are largely mixed with states made up of quark-antiquark pairs. Hybrid states ($q\bar{q}g$ or $qqqg$) have also escaped detection. Recently a number of unexpected results have revamped the interest for hadron spectroscopy. Several experiments have reported new narrow states, with widths below a few MeV(!!): $\Theta^+(1540)$ with the quantum numbers of nK^+ or pK^0_S or, in terms of quarks, of $uudd\bar{s}$; $D^+_{sJ}(2317) \sim D_s\pi$, $D^+_{sJ}(2460) \sim D^*_s\pi$, ... and $X^0(3872) \sim \pi\pi J/\Psi$. The interpretations proposed are in terms of pentaquarks ($[ud][ud]\bar{s}$ for Θ^+ for example), tetraquarks ($[qq][\bar{q}\bar{q}]$) vs. meson-meson molecules for low lying scalar mesons or for X^0 and also in terms of chiral solitons. Tetraquarks and pentaquarks are based on diquarks: $[qq]$ of spin 0, antisymmetric in colour, $\bar{3}$ of $SU(3)_{\mathrm{colour}}$, and antisymmetric in flavour, $\bar{3}$ of $SU(3)_{\text{flavour}}$. Tetraquarks were originally proposed for scalar mesons by Jaffe [5]. It is well known that there are two clusters of scalar mesons: one possible nonet at high mass, around $1.5\,\text{GeV}$, and a low lying nonet below $1\,\text{GeV}$. The light nonet presents an inversion in the spectrum: the mesons that would contain s-quarks in the conventional $q\bar{q}$ picture and would hence be heavier are actually lighter. In the tetraquark interpretation this becomes clear because the s-quarks with index "3" of the conventional picture is now replaced be the diquark $[ud]$. However, one can still formulate doubts about the existence of so many scalar states [6]. The tetraquark interpretation for the doubly charmed $X^0(3872)$ has been proposed recently by Maiani et al. [7] as opposed to that in terms of a D–$D*$ molecule by Braaten and Kusunoki [8]. Both models appear to face difficulties with the data. For putative pentaquark states like the Θ^+ doubts on their existence have much increased recently. Not only there are mass inconsistencies among different experiments, evident tension between a small width and large production rates and the need of an exotic production mechanism to explain the lack of evidence at larger energies. But the most disturbing fact is the absence of the signal in some specific experiments where it is difficult to imagine a reason for not seeing it [9].

We now discuss perturbative QCD [10]. In the QCD Lagrangian quark masses are the only parameters with dimensions. Naively (or classically) one would expect massless QCD to be scale invariant so that dimensionless observables would not depend on the absolute energy scale but only on ratios of energy variables. While massless QCD in the quantum version, after regularisation and renormalisation, is finally not scale invariant, the theory is asymptotically free and all the departures from scaling are asymptotically small, logarithmic and computable in terms of the running coupling $\alpha_s(Q^2)$. Mass corrections, present in the realistic case together with hadronisation effects, are suppressed by powers. The QCD beta function that fixes the running coupling is known in QCD up to 4 loops in the MS or $\overline{MS}$ definitions and the expansion is well behaved. The 4-loop calculation by van Ritbergen, Vermaseren and Larin [11] involving about 50.000 4-loop diagrams is a great piece of work. The running coupling is a function of $Q^2/\Lambda_{\text{QCD}}^2$, where Λ_{QCD} is the scale that breaks scale invariance in massless QCD. Its value in $\overline{MS}$, for 5 flavours of quarks, from the PDG'06 is $\Lambda_{\text{QCD}} \sim 222(25)\,\text{MeV}$. This fundamental constant of nature, which determines the masses of hadrons, is a subtle effect arising from defining the theory at the quantum level. There is no hierarchy problem in QCD, in that the logarithmic evolution of the running makes the smallness of Λ_{QCD} with respect to the Planck mass M_{Pl} natural: $\Lambda_{\text{QCD}} \sim M_{\text{Pl}} \exp\left[-1/2b\alpha_s(M_{\text{Pl}}^2)\right]$.

The measurements of $\alpha_s(Q^2)$ are among the main quantitative tests of the theory. The most precise and reliable determinations are from e^+e^- colliders (mainly at LEP: inclusive hadronic Z decay, inclusive hadronic τ decay, event shapes and jet rates) and from scaling violations in Deep Inelastic Scattering (DIS). Z decay widths are very clean: the perturbative expansion is known to 3-loops, power corrections are controlled by the light-cone operator expansion and are very suppressed due to m_Z being very large. For measuring $\alpha_{\rm s}(Q^2)$ [12] the basic quantity is $\Gamma_{\rm h}$ the Z hadronic partial width. It enters in $R_{\rm l}$, $\sigma_{\rm h}$, $\sigma_{\rm l}$ and Γ_Z (the width ratio of hadrons to leptons, the hadron cross section at the peak, the charged lepton cross section at the peak and the total width, respectively) which are separately measured with largely independent systematics. From combining all these measurements one obtains $\alpha_{\rm s}(m_Z^2) = 0.1186(27)$ [13]. The error is predominantly theoretical and is dominated by our ignorance on m_H and from higher orders in the QCD expansion (the possible impact of new physics is very limited, given the results of precision tests of the SM at LEP). The measurement of $\alpha_{\rm s}(m_Z)$ from τ decay is based on R_τ, the ratio of the hadronic to leptonic widths. R_τ has a number of advantages that, at least in part, tend to compensate for the smallness of m_τ. First, R_τ is maximally inclusive, more than $R_{e+e-}(s)$, because one also integrates over all values of the invariant hadronic squared mass. Analyticity is used to transform the integral into one on the circle at $|s| = m_\tau^2$. Also, a factor $(1 - \frac{s}{m_\tau^2})^2$ that appears in the integral kills the sensitivity of the region $\mathrm{Re}(s) = m_\tau^2$ where the physical cut and the associated thresholds are located. Still the quoted result (PDG'06) looks a bit too precise: $\alpha_{\rm s}(m_Z^2) = 0.120(3)$. This precision is obtained by taking for granted that corrections suppressed by $1/m_\tau^2$ are negligible. This is because, in the massless theory, no dim-2 Lorentz and gauge invariant operators exist that can appear in the light cone expansion. In the massive theory, the coefficient of $1/m_\tau^2$ does not vanish but is proportional to light quark mass-squared m^2. This is still negligible if m is taken as a Lagrangian mass of a few MeV. But would not at all be negligible, actually would much increase the theoretical error, if it is taken as a constituent mass of order $m \sim \Lambda_{\rm QCD}$. Most people believe the optimistic version. I am not convinced that the gap is not filled up by ambiguities of $O(\Lambda_{\rm QCD}^2/m_\tau^2)$ e. g. from ultraviolet renormalons. In any case, one can discuss the error, but it is true and remarkable, that the central value from τ decay, obtained at very small Q^2, when evolved at $Q^2 = m_Z^2$, is in perfect agreement with all other precise determinations of $\alpha_{\rm s}(m_Z^2)$ at more typical LEP values of Q^2. The measurements of $\alpha_{\rm s}$ from event shapes and jet rates are affected by non perturbative hadronic corrections which are difficult to precisely assess. The combined result gives $\alpha_{\rm s}(m_Z^2) = 0.120(5)$ (PDG'06). By measuring event shapes at different energies in the LEP1 and LEP2 ranges one also directly sees the running of $\alpha_{\rm s}$.

In DIS QCD predicts the Q^2 dependence of a generic structure function $F(x, Q^2)$ at each fixed x, not the x shape. But the Q^2 dependence is related to the x shape by the QCD evolution equations. For each x-bin the data allow to extract the slope of an approximately straight line, the log slope: $\mathrm{d}\log F(x, Q^2)/\mathrm{d}\log Q^2$. For most x values the Q^2 span and the precision of the data are not much sensitive to the curvature. A single value of $\Lambda_{\rm QCD}$ must be fitted to reproduce the collection of the log slopes. The QCD theory of scaling violations, based on the renormalization group and the light-cone operator expansion, is crystal clear. Recently ('04) the formidable task of computing the splitting functions at NNLO accuracy has been completed by Moch, Vermaseren and Vogt, a really monumental, fully analytic calculation [14]. For the determination of α_s the scaling violations of non-singlet structure functions would be ideal, because of the minimal impact of the choice of input parton densities. Unfortunately the data on non-singlet structure functions are not very accurate. For example, NNLO determinations of α_s from the CCFR data on $F_{3\nu N}$ with different techniques have led to the central values $\alpha_s(m_Z^2) = 0.1153$ [15]), $\alpha_s(m_Z^2) = 0.1174$ [16], $\alpha_s(m_Z^2) = 0.1190$ [17], with average and common estimated error of $\alpha_s(m_Z^2) = 0.117(6)$ which I will use later. When one measures α_s from scaling violations on F_2 from e or μ beams, the data are abundant, the errors small but there is an increased dependence on input parton densities and especially a strong correlation between the result on α_s and the input on the gluon density. There are several most complete and accurate derivations of α_s from scaling violations in F_2 with different, sophisticated methods (Mellin moments, Bernstein moments, truncated moments...). We quote here the result at NNLO accuracy from MRST'04 (see PDG'06): $\alpha_s(m_Z^2) = 0.1167(40)$.

More measurements of α_s could be listed: I just reproduced those which I think are most significant and reliable. There is a remarkable agreement among the different determinations. If I directly average the five values listed above from inclusive Z decay, from R_τ, from event shapes and jet rates in e^+e^-, from F_3 and from F_2 in DIS I obtain $\alpha_s(m_Z^2) = 0.1187(16)$ in good agreement with the PDG'06 average $\alpha_s(m_Z^2) = 0.1176(20)$.

The importance of DIS for QCD goes well beyond the measurement of α_s. In the past it played a crucial role in establishing the reality of quarks and gluons as partons and in promoting QCD as the theory of strong interactions. Nowadays it still generates challenges to QCD as, for example, in the domain of structure functions at small x or of polarized structure functions or of generalized parton densities and so on.

The problem of constructing a convergent procedure to include the BFKL corrections at small x in the singlet splitting functions, in agreement with the small-x behaviour observed at HERA, has been a long standing puzzle which has now been essentially solved. The naive BFKL rise of splitting functions is tamed by resummation of collinear singularities and by running coupling effects. The resummed expansion is well behaved and the result is close to the perturbative NLO splitting function in the region of HERA data at small x [18,19].

In polarized DIS one main question is how the proton helicity is distributed among quarks, gluons and orbital angular momentum: $1/2\Delta\Sigma + \Delta g + L_z = 1/2$ [20]. The quark moment $\Delta\Sigma$ was found to be small: typically, at $Q^2 \sim 1\,\mathrm{GeV}^2$, $\Delta\Sigma_{\exp} \sim 0.2$ (the "spin crisis"). Either $\Delta g + L_z$ is large or there are contributions to $\Delta\Sigma$ at very small x outside of the measured region. Δg evolves like $\Delta g \sim \log Q^2$, so that eventually should become large (while $\Delta\Sigma$ and $\Delta g + L_z$ are Q^2 independent in LO). It will take long before this log growth of Δg will be confirmed by experiment! Δg can be measured indirectly by scaling violations and directly from asymmetries, e. g. in $c\bar{c}$ production. Existing direct measurements by Hermes, Compass, and at RHIC are still very crude and show no hint of a large Δg. The perspectives of better measurements are good at Compass and RHIC in the near future.

Another important role of DIS is to provide information on parton density functions (PDF) which are instrumental for computing cross-sections of hard processes at hadron colliders via the factorisation formula. The predictions for cross sections and distributions at pp or $p\bar{p}$ colliders for large p_T jets or photons, for heavy quark production, for Drell–Yan, W and Z production are all in very good agreement with experiment. There was an apparent problem for b quark production at the Tevatron, but the problem appears now to be solved by a combination of refinements (log resummation, B hadrons instead of b quarks, better fragmentation functions...) [21]. The QCD predictions are so solid that W and Z production are actually considered as possible luminosity monitors for the LHC.

A great effort is being devoted to the preparation to the LHC. Calculations for specific processes are being completed. A very important example is Higgs production via $g+g \to H$. The amplitude is dominated by the top quark loop. Higher order corrections can be computed either in the effective lagrangian approach, where the heavy top is integrated away and the loop is shrunk down to a point [the coefficient of the effective vertex is known to α_s^4 accuracy [22]], or in the full theory. At the NLO [23] the two approaches agree very well for the rate as a function of m_H. Rapidity and p_T distributions have also been evaluated at NLO [23]. The $[\log(p_T/m_H)]^n$ have been resummed in analogy with what was done long ago for W and Z production. Recently the NNLO analytic calculation for the rate has been completed in the effective lagrangian formalism [23, 24].

The activity on event simulation also received a big boost from the LHC preparation. General algorithms for performing NLO calculations numerically (requiring techniques for the cancellation of singularities between real and virtual diagrams), for example the dipole formalism by Catani, Seymour et al. [25], have been developed. The matching of matrix element calculation of rates together with the modeling of parton showers has been realised in packages, as for example in the MC@NLO based on HERWIG. The matrix element calculation, improved by resummation of large logs, provides the hard skeleton (with large p_T branchings) while the parton shower is constructed by a sequence of factorized collinear emissions fixed by the QCD splitting functions. In addition, at low scales a model of hadronisation completes the simulation. The importance of all the components, matrix element, parton shower and hadronisation can be appreciated in simulations of hard events compared with the Tevatron data.

Before closing I would like to mention some very interesting developments at the interface between string theory and QCD, twistor calculus. A precursor work was the Parke-Taylor result in 1986 [27] on the amplitudes for n

gluons (all taken as incoming) with given helicities. Inspired by dual models, they derived a compact formula for the maximum non vanishing helicity violating amplitude (with $n - 2$ plus and 2 minus helicities) in terms of spinor products. Using the relation between strings and gauge theories in twistor space Witten [28] developed in '03 a formalism in terms of effective vertices and propagators that allows to compute all helicity amplitudes. The method, much faster than Feynman diagrams, leads to very compact results. Since then rapid progress followed [23]: for tree level processes powerful recurrence relations were established (Britto, Cachazo, Feng; Witten), the method was extended to include massless fermions (Georgiu, Khoze) and also external EW vector bosons (Bern et al.) and Higgs particles (Dixon, Glover, Khoze, Badger et al.). The level already attained is already important for multijet events at the LHC. And the study of loop diagrams has been started. In summary, this road looks very promising.

A different string connection is the attempt at obtaining results on QCD from the AdS correspondence, pioneered by Maldacena [29]. The starting point is the holographic correspondence between $D = 10$ string theory and the $N = 4$ SUSY Yang-Mills in four dimensions at large N_c. From there to get to real life QCD the way looks impervious, but a number of results for actual processes have been advocated and the perspective is exciting [30].

In conclusion, I think that the domain of QCD appears as one of great maturity but also of robust vitality with many rich branches and plenty of new blossoms. The physics content of QCD is very large and our knowledge, especially in the non perturbative domain, is still very limited but progress both from experiment (LEP, HERA, Tevatron, RHIC, LHC...) and from theory is continuing at a healthy rate. And all the QCD predictions that we were able to formulate and to test are in very good agreement with experiment.

3 The Physics of Flavour

In the last decade great progress in different areas of flavour physics has been achieved. In the quark sector, the amazing results of a generation of frontier experiments, obtained at B factories and at accelerators, have become available. QCD has been playing a crucial role in the interpretation of experiments by a combination of effective theory methods (heavy quark effective theory, NRQCD, SCET), lattice simulations and perturbative calculations. The hope of these experiments was to detect departures from the CKM picture of mixing and CP violation as signals of new physics. Finally, B mixing and CP violation agree very well with the SM predictions based on the CKM matrix [31]. The recent measurement of Δm_s by CDF and D0, in fair agreement with the SM expectation, has closed another door for new physics. It is only in channels that are forbidden at tree level and occur through penguin loops (as is the case for $B \to \pi K$ modes) that some deviation could be hidden. The amazing performance of the SM in flavour changing transitions and for

CP violation in K and B decays poses a strong constraint on all proposed models of new physics.

In the leptonic sector the study of neutrino oscillations has led to the discovery that at least two neutrinos are not massless and to the determination of the mixing matrix [32]. Neutrinos are not all massless but their masses are very small. Probably masses are small because νs are Majorana particles, and, by the see-saw mechanism, their masses are inversely proportional to the large scale M where lepton number (L) violation occurs (as expected in GUT's). Indeed the value of $M \sim m_{\nu R}$ from experiment is compatible with being close to $M_{\mathrm{GUT}} \sim 10^{14} - 10^{15}\,\mathrm{GeV}$, so that neutrino masses fit well in the GUT picture and actually support it. It was realized that decays of heavy ν_{R} with CP and L violation can produce a B–L asymmetry. The range of neutrino masses indicated by neutrino phenomenology turns out to be perfectly compatible with the idea of baryogenesis via leptogenesis [33]. This elegant model for baryogenesis has by now replaced the idea of baryogenesis near the weak scale, which has been strongly disfavoured by LEP. It is remarkable that we now know the neutrino mixing matrix with good accuracy. Two mixing angles are large and one is small. The atmospheric angle θ_{23} is large, actually compatible with maximal but not necessarily so: at 3σ: $0.31 \leq \sin^2 \theta_{23} \leq 0.72$ with central value around 0.5. The solar angle θ_{12} is large, $\sin^2 \theta_{12} \sim 0.3$, but certainly not maximal (by more than 5σ). The third angle θ_{13}, strongly limited mainly by the CHOOZ experiment, has at present a 3σ upper limit given by about $\sin^2 \theta_{13} \leq 0.08$. While these discoveries are truly remarkable, it is somewhat depressing that the detailed knowledge of both the quark and the neutrino mixings has not led so far to a compelling solution of the dynamics of fermion masses and mixings: our models can reproduce, actually along different ways, the observed values, but we do not really understand their mysterious pattern.

4 Precision Tests of the Standard Electroweak Theory

The results of the electroweak precision tests as well as of the searches for the Higgs boson and for new particles performed at LEP and SLC are now available in final form. Taken together with the measurements of m_{t}, m_W and the searches for new physics at the Tevatron, and with some other data from low energy experiments, they form a very stringent set of precise constraints [13] to compare with the Standard Model (SM) or with any of its conceivable extensions. When confronted with these results, on the whole the SM performs rather well, so that it is fair to say that no clear indication for new physics emerges from the data [34]. The main lesson of precision tests of the standard electroweak theory can be summarised as follows. The couplings of quark and leptons to the weak gauge bosons $W^{\pm}$ and Z are indeed precisely those prescribed by the gauge symmetry. The accuracy of a few per-mille for these tests implies that, not only the tree level, but also the structure of quantum corrections has been verified. To a lesser accuracy the triple gauge vertices γWW and ZWW have also been found in agreement with the specific prediction of the $SU(2) \otimes U(1)$ gauge theory. This means that it has been verified that the gauge symmetry is unbroken in the vertices of the theory: all currents and charges are indeed symmetric. Yet there is obvious evidence that the symmetry is otherwise badly broken in the masses. This is a clear signal of spontaneous symmetry breaking. The practical implementation of spontaneous symmetry breaking in a gauge theory is via the Higgs mechanism. The Higgs sector of the SM is still very much untested. What has been tested is the relation $M_W^2 = M_Z^2 \cos^2 \theta_W$, modified by small, computable radiative corrections. This relation means that the effective Higgs (be it fundamental or composite) is indeed a weak isospin doublet. The Higgs particle has not been found but in the SM its mass can well be larger than the present direct lower limit $m_H \gtrsim 114\,\text{GeV}$ obtained from direct searches at LEP-2. The radiative corrections computed in the SM when compared to the data on precision electroweak tests lead to a clear indication for a light Higgs, not too far from the present lower bound. The exact upper limit for m_H in the SM depends on the value of the top quark mass m_t (the one-loop radiative corrections are quadratic in m_t and logarithmic in m_H). The measured value of m_t went down recently (as well as the associated error) according to the results of Run II at the Tevatron. The CDF and D0 combined value is at present [35] $m_t = 171.4 \pm 2.1\,\text{GeV}$ (it went slightly down with respect to the value from Run I). As a consequence the present limit on m_H is more stringent [36]: $m_H < 199\,\text{GeV}$ (at 95% c.l., after including the information from the 114 GeV direct bound).

5 Outlook on Avenues beyond the Standard Model

No signal of new physics has been found neither in electroweak precision tests nor in flavour physics. Given the success of the SM why are we not satisfied with that theory? Why not just find the Higgs particle, for completeness, and declare that particle physics is closed? The reason is that there are both conceptual problems and phenomenological indications for physics beyond the SM. On the conceptual side the most obvious problems are that quantum gravity is not included in the SM and the related hierarchy problem. Among the main phenomenological hints for new physics we can list coupling unification, dark matter, neutrino masses, baryogenesis and the cosmological vacuum energy.

The computed evolution with energy of the effective SM gauge couplings clearly points towards the unification of the electro-weak and strong forces (Grand Unified Theories: GUT's) at scales of energy $M_{\text{GUT}} \sim 10^{15}-10^{16}\,\text{GeV}$ which are close to the scale of quantum gravity, $M_{\text{Pl}} \sim 10^{19}\,\text{GeV}$. One is led to imagine a unified theory of all interactions also including gravity (at present superstrings provide the best attempt at such a theory). Thus GUT's and the realm of quantum gravity set a very distant energy horizon that modern particle theory cannot ignore. Can the SM without new physics be valid up to such large energies? One can imagine that some obvious problems could be postponed to the more fundamental theory at the Planck mass. For example, the explanation of the three generations of fermions and the understanding of fermion masses and mixing angles can be postponed. But other problems must find their solution in the low energy theory. In particular, the structure of the SM could not naturally explain the relative smallness of the weak scale of mass, set by the Higgs mechanism at $\mu \sim 1/\sqrt{G_F} \sim 250\,\mathrm{GeV}$ with G_F being the Fermi coupling constant. This so-called hierarchy problem is due to the instability of the SM with respect to quantum corrections. This is related to the presence of fundamental scalar fields in the theory with quadratic mass divergences and no protective extra symmetry at $\mu = 0$. For fermion masses, first, the divergences are logarithmic and, second, they are forbidden by the $SU(2) \otimes U(1)$ gauge symmetry plus the fact that at $m = 0$ an additional symmetry, i.e. chiral symmetry, is restored. Here, when talking of divergences, we are not worried of actual infinities. The theory is renormalisable and finite once the dependence on the cut off Λ is absorbed in a redefinition of masses and couplings. Rather the hierarchy problem is one of naturalness. We can look at the cut off as a parameterization of our ignorance on the new physics that will modify the theory at large energy scales. Then it is relevant to look at the dependence of physical quantities on the cut off and to demand that no unexplained enormously accurate cancellations arise.

The hierarchy problem can be put in very practical terms: loop corrections to the higgs mass squared are quadratic in Λ. The most pressing problem is from the top loop. With $m_{\mathrm{h}}^2 = m_{\mathrm{bare}}^2 + \delta m_{\mathrm{h}}^2$ the top loop gives

$$\delta m_{\mathrm{h}|\mathrm{top}}^2 \sim -\frac{3G_F}{2\sqrt{2}\pi^2} m_{\mathrm{t}}^2 \Lambda^2 \sim -(0.2\Lambda)^2\,. \tag{1}$$

If we demand that the correction does not exceed the light Higgs mass indicated by the precision tests, Λ must be close, $\Lambda \sim o(1\,\mathrm{TeV})$. Similar constraints arise from the quadratic Λ dependence of loops with gauge bosons and scalars, which, however, lead to less pressing bounds. So the hierarchy problem demands new physics to be very close (in particular the mechanism that quenches the top loop). Actually, this new physics must be rather special, because it must be very close, yet its effects are not clearly visible (the "LEP Paradox" [37]). Examples of proposed classes of solutions for the hierarchy problem are:

Supersymmetry. In the limit of exact boson-fermion symmetry the quadratic divergences of bosons cancel so that only log divergences remain. However, exact SUSY is clearly unrealistic. For approximate SUSY (with soft breaking terms), which is the basis for all practical models, Λ is replaced by the splitting of SUSY multiplets, $\Lambda \sim m_{\mathrm{SUSY}} - m_{\mathrm{ord}}$. In particular, the top loop is quenched by partial cancellation with s-top exchange, so the s-top cannot be too heavy.

Technicolor. The Higgs system is a condensate of new fermions. There are no fundamental scalar Higgs sector, hence no quadratic devergences associated to the μ^2 mass in the scalar potential. This mechanism needs a very strong binding force, $\Lambda_{\rm TC} \sim 10^3 \, \Lambda_{\rm QCD}$. It is difficult to arrange that such nearby strong force is not showing up in precision tests. Hence this class of models has been disfavoured by LEP, although some special class of models have been devised aposteriori, like walking TC, top-color assisted TC etc (for recent reviews, see, for example, [40]).

Large extra dimensions. The idea is that $M_{\rm Pl}$ appears very large, or equivalently that gravity appears very weak, because we are fooled by hidden extra dimensions so that the real gravity scale is reduced down to a lower scale, even possibly down to $o(1 \, {\rm TeV})$. This possibility is very exciting in itself and it is really remarkable that it is compatible with experiment.

"Little Higgs" models. In these models extra symmetries allow $m_{\rm h} \neq 0$ only at two-loop level, so that Λ can be as large as $o(10 \, {\rm TeV})$ with the Higgs within present bounds (the top loop is quenched by exchange of heavy vectorlike new quarks with charge $2/3$).

We now briefly comment in turn on these possibilities.

SUSY models are the most developed and most widely accepted. Many theorists consider SUSY as established at the Planck scale $M_{\rm Pl}$. So why not to use it also at low energy to fix the hierarchy problem, if at all possible? It is interesting that viable models exist. The necessary SUSY breaking can be introduced through soft terms that do not spoil the good convergence properties of the theory. Precisely those terms arise from supergravity when it is spontaneoulsly broken in a hidden sector. This is the case of the MSSM [41]. Of course, minimality is only a simplicity assumption that could possibly be relaxed. For example, adding an additional Higgs singlet S considerably helps in addressing naturalness constraints [38,39]. Minimal versions or, even more, very constrained versions like the CMSSM (where simple conditions at the GUT scale are in addition assumed) are economic in terms of new parameters but could be to some extent misleading. Still, the MSSM is a completely specified, consistent and computable theory which is compatible with all precision electroweak tests. In this most traditional approach SUSY is broken in a hidden sector and the scale of SUSY breaking is very large of order $\Lambda \sim \sqrt{G_F^{-1/2} M_{\rm Pl}}$. But since the hidden sector only communicates with the visible sector through gravitational interactions the splitting of the SUSY multiplets is much smaller, in the TeV energy domain, and the Goldstino is practically decoupled. But alternative mechanisms of SUSY breaking are also being considered. In one alternative scenario [42] the (not so much) hidden sector is connected to the visible one by ordinary gauge interactions. As these are much stronger than the gravitational interactions, Λ can be much smaller, as low as 10–100 TeV. It follows that the Goldstino is very light in these models (with mass of order or below 1 eV typically) and is the lightest, stable SUSY particle, but its couplings are observably large. The radiative decay of the lightest neutralino into the Goldstino leads to detectable photons. The signature of photons comes out naturally in this SUSY breaking pattern: with respect to the MSSM, in the gauge mediated model there are typically more photons and less missing energy. The main appeal of gauge mediated models is a better protection against flavour changing neutral currents but naturality problems tend to increase. As another possibility it has been pointed out that there are pure gravity contributions to soft masses that arise from gravity theory anomalies [43]. In the assumption that these terms are dominant the associated spectrum and phenomenology have been studied. In this case gaugino masses are proportional to gauge coupling beta functions, so that the gluino is much heavier than the electroweak gauginos, and the wino is most often the lightest SUSY particle.

What is really unique to SUSY with respect to all other extensions of the SM listed above is that the MSSM or other non minimal SUSY models are well defined and computable up to M_{Pl} and, moreover, are not only compatible but actually quantitatively supported by coupling unification and GUT's. At present the most direct phenomenological evidence in favour of supersymmetry is obtained from the unification of couplings in GUTs. Precise LEP data on $\alpha_{\mathrm{s}}(m_Z)$ and $\sin^2\theta_W$ show that standard one-scale GUTs fail in predicting $\sin^2\theta_W$ given $\alpha_{\mathrm{s}}(m_Z)$ (and $\alpha(m_Z)$) while SUSY GUTs are in agreement with the present, very precise, experimental results. If one starts from the known values of $\sin^2\theta_W$ and $\alpha(m_Z)$, one finds [45] for $\alpha_{\mathrm{s}}(m_Z)$ the results: $\alpha_{\mathrm{s}}(m_Z) = 0.073 \pm 0.002$ for Standard GUTs and $\alpha_{\mathrm{s}}(m_Z) = 0.129 \pm 0.010$ for SUSY GUTs to be compared with the world average experimental value $\alpha_{\mathrm{s}}(m_Z) = 0.118 \pm 0.002$. Another great asset of SUSY GUT's is that proton decay is much slowed down with respect to the non SUSY case. First, the unification mass $M_{\mathrm{GUT}} \sim$ few 10^{16} GeV, in typical SUSY GUT's, is about 20-30 times larger than for ordinary GUT's. This makes p decay via gauge boson exchange negligible and the main decay amplitude arises from dim-5 operators with higgsino exchange, leading to a rate close but still compatible with existing bounds (see, for example, [44]). It is also important that SUSY provides an excellent dark matter candidate, the neutralino. We finally recall that the range of neutrino masses as indicated by oscillation experiments, when interpreted in the see-saw mechanism, point to M_{GUT} and give additional support to GUTs [32].

In spite of all these virtues it is true that the lack of SUSY signals at LEP and the lower limit on m_H pose problems for the MSSM. The lightest Higgs particle is predicted in the MSSM to be below $m_{\mathrm{h}} \lesssim 135$ GeV. The limit on the SM Higgs $m_H \gtrsim 114$ GeV considerably restricts the available parameter space of the MSSM requiring relatively large $\tan\beta$ ($\tan\beta \gtrsim 2-3$: at tree level $m_h^2 = m_Z^2 \cos^2 2\beta$) and rather heavy s-top (the loop corrections increase with $\log m_{\tilde{t}}^2$). But we have seen that a heavy s-top is unnatural, because it enters quadratically in the radiative corrections to $\delta m_{\mathrm{h|top}}^2$. Stringent naturality constraints also follow from imposing that the electroweak symmetry breaking occurs at the right place: in SUSY models the breaking is induced by the running of the H_u mass starting from a common scalar mass m_0 at M_GUT. The squared Z mass m_Z^2 can be expressed as a linear combination of the SUSY parameters m_0^2, $m_{1/2}^2$, A_t^2, μ^2, ... with known coefficients. Barring cancellations that need fine tuning, the SUSY parameters, hence the SUSY s-partners cannot be too heavy. The LEP limits, in particular the chargino lower bound $m_{\chi+} \gtrsim 100\,\mathrm{GeV}$, are sufficient to eliminate an important region of the parameter space, depending on the amount of allowed fine tuning. For example, models based on gaugino universality at the GUT scale are discarded unless a fine tuning by at least a factor of 20 is not allowed. Without gaugino universality [46] the strongest limit remains on the gluino mass: $m_Z^2 \sim 0.7\, m_\mathrm{gluino}^2 + \ldots$ which is still compatible with the present limit $m_\mathrm{gluino} \gtrsim 200\,\mathrm{GeV}$.

The non discovery of SUSY at LEP has given further impulse to the quest for new ideas on physics beyond the SM. Large extra dimensions [47] and "little Higgs" [48] models are among the most interesting new directions in model building. Large extra dimension models propose to solve the hierarchy problem by bringing gravity down from M_Pl to $m \sim o(1\,\mathrm{TeV})$ where m is the string scale. Inspired by string theory one assumes that some compactified extra dimensions are sufficiently large and that the SM fields are confined to a 4-dimensional brane immersed in a d-dimensional bulk while gravity, which feels the whole geometry, propagates in the bulk. We know that the Planck mass is large because gravity is weak: in fact $G_\mathrm{N} \sim 1/M_\mathrm{Pl}^2$, where G_N is Newton constant. The idea is that gravity appears so weak because a lot of lines of force escape in extra dimensions. Assume you have $n = d-4$ extra dimensions with compactification radius R. For large distances, $r >> R$, the ordinary Newton law applies for gravity: in natural units $F \sim G_\mathrm{N}/r^2 \sim 1/(M_\mathrm{Pl}^2 r^2)$. At short distances, $r \lesssim R$, the flow of lines of force in extra dimensions modifies Gauss law and $F^{-1} \sim m^2 (mr)^{d-4} r^2$. By matching the two formulas at $r = R$ one obtains $(M_\mathrm{Pl}/m)^2 = (Rm)^{d-4}$. For $m \sim 1\,\mathrm{TeV}$ and $n = d - 4$ one finds that $n = 1$ is excluded ($R \sim 10^{15}\,\mathrm{cm}$), for $n = 2$ R is at the edge of present bounds $R \sim 1\,\mathrm{mm}$, while for $n = 4, 6$, $R \sim 10^{-9}, 10^{-12}\,\mathrm{cm}$. In all these models a generic feature is the occurrence of Kaluza–Klein (KK) modes. Compactified dimensions with periodic boundary conditions, as for quantization in a box, imply a discrete spectrum with momentum $p = n/R$ and mass squared $m^2 = n^2/R^2$. There are many versions of these models. The SM brane can itself have a thickness r with $r <\sim 10^{-17}\,\mathrm{cm}$ or $1/r >\sim 1\,\mathrm{TeV}$, because we know that quarks and leptons are pointlike down to these distances, while for gravity there is no experimental counter-evidence down to $R <\sim 0.1\,\mathrm{mm}$ or $1/R >\sim 10^{-3}\,\mathrm{eV}$. In case of a thickness for the SM brane there would be KK recurrences for SM fields, like W_n, Z_n and so on in the TeV region and above. There are models with factorized metric ($\mathrm{d}s^2 = \eta_{\mu\nu}\,\mathrm{d}x^\mu\,\mathrm{d}x^\nu + h_{ij}(y)\,\mathrm{d}y^i\,\mathrm{d}y^j$, where y (i,j) denotes the extra dimension coordinates (and indices), or models with warped metric ($\mathrm{d}s^2 = \mathrm{e}^{-2kR|\phi|}\eta_{\mu\nu}\,\mathrm{d}x^\mu\,\mathrm{d}x^\nu - R^2\phi^2$ [49]. In any case there are the towers of KK recurrences of the graviton. They are gravitationally coupled but there are a lot of them that sizably couple, so that the net result is a modification of cross-sections and the presence of missing energy.

Large extra dimensions provide a very exciting scenario [50]. Already it is remarkable that this possibility is compatible with experiment. However, there are a number of criticisms that can be brought up. First, the hierarchy problem is more translated in new terms rather than solved. In fact the basic relation $Rm = (M_{\mathrm{Pl}}/m)^{2/n}$ shows that Rm, which one would apriori expect to be $0(1)$, is instead ad hoc related to the large ratio M_{Pl}/m. In this respect the Randall-Sundrum variety is more appealing because the hierarchy suppression m_W/M_{Pl} could arise from the warping factor $e^{-2kR|\phi|}$, with not too large values of kR. The question of whether these values of kR are reasonable has been discussed in [51], which offer the best support to the solution of the hierarchy problem in this context. Also it is not clear how extra dimensions can by themselves solve the LEP paradox (the large top loop corrections should be controlled by the opening of the new dimensions and the onset of gravity): since m_H is light $\Lambda \sim 1/R$ must be relatively close. But precision tests put very strong limits on Λ. In fact in typical models of this class there is no mechanism to sufficiently quench the corrections. While no simple, realistic model has yet emerged as a benchmark, it is attractive to imagine that large extra dimensions could be a part of the truth, perhaps coupled with some additional symmetry or even SUSY. The Randall-Sundrum warped geometry has become the common framework for many attempts in this direction.

In the general context of extra dimensions an interesting direction of development is the study of symmetry breaking by orbifolding and/or boundary conditions. These are models where a larger gauge symmetry (with or without SUSY) holds in the bulk. The symmetry is reduced in the 4 dimensional brane, where the physics that we observe is located, as an effect of symmetry breaking induced geometrically by suitable boundary conditions. There are models where SUSY, valid in $n > 4$ dimensions is broken by boundary conditions [52], in particular the model of [53], where the mass of the Higgs is computable and can be estimated with good accuracy. Then there are "Higgsless models" where it is the SM electroweak gauge symmetry which is broken at the boundaries [54]. Or models where the Higgs is the 5th component of a gauge boson of an extended symmetry valid in $n > 4$ [55]. In general all these alternative models for the Higgs mechanism face severe problems and constraints from electroweak precision tests [56]. At the GUT scale, symmetry breaking by orbifolding can be applied to obtain a reformulation of SUSY GUT's where many problematic features of ordinary GUT's (e. g. a baroque Higgs sector, the doublet-triplet splitting problem, fast proton decay etc) are improved [50, 57].

In "little Higgs" models the symmetry of the SM is extended to a suitable global group G that also contains some gauge enlargement of $SU(2) \otimes U(1)$, for example $G \supset [SU(2) \otimes U(1)]^2 \supset SU(2) \otimes U(1)$. The Higgs particle is a pseudo-Goldstone boson of G that only takes mass at 2-loop level, because two distinct symmetries must be simultaneously broken for it to take mass, which requires the action of two different couplings in the same diagram. Then in the relation between $\delta m_{\rm h}^2$ and Λ^2 there is an additional coupling and an additional loop factor that allow for a bigger separation between the Higgs mass and the cut-off. Typically, in these models one has one or more Higgs doublets at $m_{\rm h} \sim 0.2\,{\rm TeV}$, and a cut-off at $\Lambda \sim 10\,{\rm TeV}$. The top loop quadratic cut-off dependence is partially canceled, in a natural way guaranteed by the symmetries of the model, by a new coloured, charge-2/3, vectorial quark χ of mass around $1\,{\rm TeV}$ (a fermion not a scalar like the s-top of SUSY models). Certainly these models involve a remarkable level of group theoretic virtuosity. However, in the simplest versions one is faced with problems with precision tests of the SM [58]. Even with vectorlike new fermions large corrections to the epsilon parameters arise from exchanges of the new gauge bosons W' and Z' (due to lack of custodial $SU(2)$ symmetry). In order to comply with these constraints the cut-off must be pushed towards large energy and the amount of fine tuning needed to keep the Higgs light is still quite large. Probably these bad features can be fixed by some suitable complication of the model (see for example, [59]). But, in my opinion, the real limit of this approach is that it only offers a postponement of the main problem by a few TeV, paid by a complete loss of predictivity at higher energies. In particular all connections to GUT's are lost. An interesting model that combines the idea of the Higgs as a Goldstone boson and warped extra dimensions was proposed and studied in refs. [60].

Finally, we stress the importance of the dark matter and of the cosmological constant or vacuum energy problem [61]. In fact, we know by now [62] that the Universe is flat and most of it is not made up of known forms of matter: $\Omega_{\rm tot} \sim 1$, $\Omega_{\rm baryonic} \sim 0.044$, $\Omega_{\rm matter} \sim 0.27$, where Ω is the ratio of the density to the critical density. Most is Dark Matter (DM) and Dark Energy (DE). We also know that most of DM must be cold (non relativistic at freeze-out) and that significant fractions of hot DM are excluded. Neutrinos are hot DM (because they are ultrarelativistic at freeze-out) and indeed are not much cosmo-relevant: $\Omega_\nu \lesssim 0.015$. Identification of DM is a task of enormous importance for both particle physics and cosmology. If really neutralinos are the main component of DM they will be discovered at the LHC and this will be a great service of particle physics to cosmology. More in general, the LHC is sensitive to a large variety of WIMP's (Weekly Interacting Massive Particles). WIMP's with masses in the $10\,{\rm GeV}$–$1\,{\rm TeV}$ range with typical electroweak crosssections contribute to Ω terms of $o(1)$. Also, these results on cosmological parameters have shown that vacuum energy accounts for about 2/3 of the critical density: $\Omega_\Lambda \sim 0.65$, Translated into familiar units this means for the energy density $\rho_\Lambda \sim (2\ 10^{-3}\,{\rm eV})^4$ or $(0.1\,{\rm mm})^{-4}$. It is really interesting (and not at all understood) that $\rho_\Lambda^{1/4} \sim \Lambda_{EW}^2/M_{\rm Pl}$ (close to the range of neutrino masses). It is well known that in field theory we expect $\rho_\Lambda \sim \Lambda_{\rm cutoff}^4$. If the cut off is set at $M_{\rm Pl}$ or even at $0(1\,{\rm TeV})$ there would an enormous mismatch. In exact SUSY $\rho_\Lambda = 0$, but SUSY is broken and in presence of breaking $\rho_\Lambda^{1/4}$ is in general not smaller than the typical SUSY multiplet splitting. Another closely related problem is "why now?": the time evolution of the matter or radiation density is quite rapid, while the density for a cosmological constant term would be flat. If so, then how comes that precisely now the two density sources are comparable? This suggests that the vacuum energy is not a cosmological constant term, buth rather the vacuum expectation value of some field (quintessence) and that the "why now?" problem is solved by some dynamical coupling of the quintessence field with gauge singlet fields (perhaps RH neutrinos).

Clearly the cosmological constant problem poses a big question mark on the relevance of naturalness as a relevant criterion also for the hierarchy problem: how we can trust that we need new physics close to the weak scale out of naturalness if we have no idea on the solution of the cosmological constant huge naturalness problem? The common answer is that the hierarchy problem is formulated within a well defined field theory context while the cosmological constant problem makes only sense within a theory of quantum gravity, that there could be modification of gravity at the sub-eV scale, that the vacuum energy could flow in extra dimensions or in different Universes and so on. At the other extreme is the possibility that naturalness is misleading. Weinberg [63] has pointed out that the observed order of magnitude of Λ can be successfully reproduced as the one necessary to allow galaxy formation in the Universe. In a scenario where new Universes are continuously produced we might be living in a very special one (largely fine-tuned) but the only one to allow the development of an observer (anthropic principle). One might then argue that the same could in principle be true also for the Higgs sector. Recently it was suggested [64] to abandon the no-fine-tuning assumption for the electro-weak theory, but require correct coupling unification, presence of dark matter with weak couplings and a single scale of evolution from the EW to the GUT scale. A "split SUSY" model arises as a solution with a fine-tuned light Higgs and all SUSY particles heavy except for gauginos, higgsinos and neutralinos, protected by chiral symmetry. But, then, we could also have a two-scale non-SUSY GUT with axions as dark matter. In conclusion, it is clear that naturalness can be a good heuristic principle but you cannot prove its necessity. The anthropic approach to the hierarchy problem is discussed in [65].

6 Summary and Conclusion

Supersymmetry remains the standard way beyond the SM. What is unique to SUSY, beyond leading to a set of consistent and completely formulated models, as, for example, the MSSM, is that this theory can potentially work up to the GUT energy scale. In this respect it is the most ambitious model because it describes a computable framework that could be valid all the way up to the vicinity of the Planck mass. The SUSY models are perfectly compatible with

GUT's and are actually quantitatively supported by coupling unification and also by what we have recently learned on neutrino masses. All other main ideas for going beyond the SM do not share this synthesis with GUT's. The SUSY way is testable, for example at the LHC, and the issue of its validity will be decided by experiment. It is true that we could have expected the first signals of SUSY already at LEP, based on naturality arguments applied to the most minimal models (for example, those with gaugino universality at asymptotic scales). The absence of signals has stimulated the development of new ideas like those of large extra dimensions and "little Higgs" models. These ideas are very interesting and provide an important reference for the preparation of LHC experiments. Models along these new ideas are not so completely formulated and studied as for SUSY and no well defined and realistic baseline has sofar emerged. But it is well possible that they might represent at least a part of the truth and it is very important to continue the exploration of new ways beyond the SM. New input from experiment is badly needed, so we all look forward to the start of the LHC.

I conclude by thanking the Organisers of this very inspiring Meeting: Guido Montagna, Oreste Nicrosini and Valerio Vercesi, for their kind invitation and great hospitality in Pavia.

References

1. See, for example, S. Ejiri, Nucl. Phys. B (Proc. Suppl.) **94**, 19 (2001); S. Aoki, Int. Journal of Mod. Phys. **A21**, 682 (2006), hep-lat 0509068 and references therein.
2. See, for example, A.S. Kronfeld *et al.*, Int. J. Mod. Phys. **A21**, 713 (2006) hep-lat 0509169] and references therein.
3. HPQCD Collaboration and UKQCD Collaboration (Q. Mason *et al.*). Phys. Rev. Lett. **95**, 052002 (2005), hep-lat 0503005.
4. J.W. Negele *et al.*, Int. J. Mod. Phys. **A21**, 720 (2006) hep-lat 0509101.
5. R.L. Jaffe, Phys. Rev. **D15**, 281 (1977); Phys. Rep. **409**, 1 (2005).
6. M.R. Pennington, Int. J. Mod. Phys. **A21**, 747 (2006), hep-ph 0509265
7. L. Maiani *et al.*, PoS HEP2005 105 (2006) hep-ph 0603021 and references therein.
8. E. Braaten, M. Kusunoki, Phys. Rev. **D72**, 054022 (2005), hep-ph 0507163 and references therein.
9. For a review, see, for example, E. Klempt, AIP Conf. Proc. **814**, 723 (2006).
10. For an introduction, see, for example, G Altarelli, hep-ph 0204179.
11. T. van Ritbergen, J.A.M. Vermaseren, S.A. Larin, Phys. Lett. **B400**, 379 (1997) hep-ph 9701390.
12. H. Stenzel, hep-ph 0501245].
13. http://lepewwg.web.cern.ch/LEPEWWG/
14. A. Vogt, S. Moch, J.A.M. Vermaseren, Nucl. Phys. **B691**, 129 (2004), hep-ph 0404111]; S. Moch, J.A.M. Vermaseren, A. Vogt, Nucl. Phys. **B688**101 (2004), hep-ph 0403192.
15. J. Santiago, F.J. Yndurain, Nucl. Phys. **B611**447 (2001), hep-ph 0102247.

16. C.J. Maxwell, A. Mirjalili, Nucl. Phys. **B645**, 298 (2000), hep-ph 0207069.

17. A.L. Katacv, G. Parcntc, Λ.V. Sidorov, J. Phys. **G29**, 1985 (2003), hcp-ph 0209024.

18. G. Altarelli, R.D. Ball, S. Forte, Nucl. Phys. **B742**, 1 (2006), hep-ph 0512237 and references therein.

19. M. Ciafaloni, D. Colferai, G.P. Salam, A.M. Stasto, Phys. Rev. **D68**, 114003 (2003), hep-ph 0307188 and references therein.

20. See the Proceedings of Spin '04, eds. K. Aulenbacher *et al.*, World Sci., Trieste, 2004.

21. M. Cacciari, S. Frixione, M.L. Mangano *et al.*, JHEP **0407**, 033 (2004), hep-ph 0312132; M. Cacciari, P. Nason, Phys. Rev. Lett. **89**, 122003 (2002), hep-ph 0204025.

22. K.G. Chetyrkin, B. Kniehl, M. Steinhauser, Phys. Rev. Lett. **79**, 353 (1997), hep-ph 9705240.

23. For a set of references, see, for example, J. Stirling, Proceedings of ICHEP, Beijing, China, 2004, hep-ph 0411372.

24. C. Anastasiou, K. Melnikov and F Petriello, Nucl. Phys. **B724**, 197 (2005), hep-ph 0501130.

25. S. Catani, M.H. Seymour, Nucl. Phys. **B485**, 291 (1997), Erratum-ibid. **B510**, 503 (1997), hep-ph 9605323

26. S. Frixione, B. Webber, hep-ph 0506182 and ref.s therein.

27. S. Parke, T.R. Taylor, Phys. Rev. Lett. **56**, 2459 (1986).

28. E Witten, Commun. Math. Phys. **252**, 189 (2004), hep-th 0312171

29. J.M. Maldacena, Adv. Theor. Math. Phys. **2**, 231 (1998).

30. For a review, see the talk by V. Shomerus at ICHEP'06, Moscow.

31. For a recent review, see, for example, P. Ball, R. Fleischer, hep-ph 0604249.

32. For a review see, for example, G. Altarelli and F. Feruglio, hep-ph 0405048.

33. For a recent review, see, for example, W. Buchmuller, R.D. Peccei and T. Yanagida, Ann. Rev. Nucl. Part. Sci. **55**, 311 (2005), hep-ph 0502169.

34. G. Altarelli and M. Grunewald, hep-ph 0404165.

35. D. Glenzinski, ICHEP'06, Moscow.

36. D. Wood, ICHEP'06, Moscow.

37. R. Barbieri and A. Strumia, hep-ph 0007265.

38. H.P. Nilles, M. Srednicki and D. Wyler, Phys. Lett. B **120**, 346 (1983); J.P. Derendinger and C.A. Savoy, Nucl. Phys. B **237**, 307 (1984); M. Drees, Int. J. Mod. Phys. A4, 3635 (1989); J.R. Ellis, J.F. Gunion, H.E. Haber, L. Roszkowski and F. Zwirner, Phys. Rev. D **39**, 844 (1989); T. Elliott, S.F. King and P.L. White, Phys. Lett. B **314**, 56 (1993) hep-ph 9305282; Phys. Rev. D **49**, 2435 (1994) hep-ph 9308309; U. Ellwanger, M. Rausch de Traubenberg and C.A. Savoy, Phys. Lett. B **315**, 331 (1993) hep-ph 9307322; B.R. Kim, A. Stephan and S.K. Oh, Phys. Lett. B **336** 200 (1994).

39. R. Barbieri *et al.*, hep-ph 0607332.

40. K. Lane, hep-ph 0202255; R.S. Chivukula, hep-ph 0011264.

41. For a recent introduction see, for example, S.P. Martin, hep-ph 9709356.

42. M. Dine and A.E. Nelson,Phys. Rev. **D48**, 1277 (1993); M. Dine, A.E. Nelson and Y. Shirman,Phys. Rev. **D51**, 1362 (1995); G.F. Giudice and R. Rattazzi, Phys. Rept. **322**, 419 (1999).

43. L. Randall and R. Sundrum, Nucl. Phys. **B557**, 79 (1999); G.F. Giudice *et al.*, JHEP**9812**, 027 (1998).

44. G. Altarelli, F. Feruglio and I. Masina, JHEP **0011**, 040 (2000).

45. P. Langacker and N. Polonsky, Phys. Rev. **D52**, 3081 (1995).

46. G. Kane *et al.*, Phys. Lett. **B551**, 146 (2003).

47. For a review and a list of refs., see, for example, J. Hewett and M. Spiropulu, hep-ph 0205196.

48. For a review and a list of refs., see, for example, M. Schmaltz, hep-ph 0210415.

49. L. Randall and R. Sundrum, Phys. Rev. Lett. **83**, 3370 (1999), **83**, 4690 (1999).

50. For a recent review, see for example, R. Rattazzi, hep-ph 0607055.

51. W.D. Goldberger and M.B. Wise, Phys. Rev. Letters **83**, 4922 (1999), hep-ph 9907447.

52. I. Antoniadis, C. Munoz and M. Quiros, Nucl. Phys. **B397**, 515 (1993); A. Pomarol and M. Quiros, Phys. Lett. **B438**, 255 (1998).

53. R. Barbieri, L. Hall and Y. Nomura, Nucl. Phys. **B624**, 63 (2002); R. Barbieri, G. Marandella and M. Papucci, hep-ph 0205280, hep-ph 0305044, and refs therein.

54. see for example, C. Csaki *et al.*, hep-ph 0305237, hep-ph 0308038, hep-ph 0310355; S. Gabriel, S. Nandi and G. Seidl, hep-ph 0406020 and refs therein; R. Chivukula *et al.*, hep-ph 0607124.

55. see for example, C.A. Scrucca, M. Serone and L. Silvestrini, hep-ph 0304220 and refs therein.

56. R. Barbieri, A. Pomarol and R. Rattazzi, hep-ph 0310285.

57. Y. Kawamura, Progr. Theor. Phys. **105**, 999 (2001).

58. J.L. Hewett, F.J. Petriello, T.G. Rizzo, hep-ph 0211218; C. Csaki *et al.*, hep-ph 0211124, hep-ph 0303236.

59. H-C. Cheng and I. Low, hep-ph 0405243; J. Hubisz *et al.*, hep-ph 0506042.

60. K Agashe, R. Contino, A. Pomarol, Nucl. Phys. **B719**, 165 (2005), hep-ph 0412089]; K. Agashe, R. Contino, Nucl. Phys. **B742**, 59 (2006), hep-ph 0510164]; K. Agashe, R. Contino, L. Da Rold, A. Pomarol, hep-ph 0605341].

61. For orientation, see, for example, M. Turner, astro-ph 0207297.

62. The WMAP Collaboration, D.N. Spergel *et al.*, astro-ph 0302209.

63. S. Weinberg, Phys. Rev. Lett. **59**, 2607 (1987).

64. N. Arkani-Hamed and S. Dimopoulos, hep-th 0405159; G. Giudice and A. Romanino, hep-ph 0406088.

65. N. Arkani-Hamed, S. Dimopoulos, S. Kachru, hep-ph 0501082, G. Giudice, R. Rattazzi, hep-ph 0606105

Getting Ready for Physics at the LHC

Fabiola Gianotti

CERN, Physics Department, 1211 Genève 23, Switzerland
`fabiola.gianotti@cern`

The CERN Large Hadron Collider (LHC) will start operation in November 2007. The present phase is characterized by very hard work, in order to complete the machine and experiments installation and commissioning, but also by great expectations.

Indeed, the LHC will provide pp collisions at the unprecedented centre-of-mass energy of $\sqrt{s} = 14\,\mathrm{TeV}$ and design luminosity of $L = 10^{34}\,\mathrm{cm}^{-2}\,\mathrm{s}^{-1}$. It will also deliver heavy ion collisions, for instance lead–lead collisions, at the colossal centre-of-mass energy of about $1000\,\mathrm{TeV}$. The machine is being installed in the $27\,\mathrm{km}$ ring previously used for the LEP e^+e^- collider.

Four main experiments will take data at the LHC: two general-purpose detectors, ATLAS and CMS, which have a very broad physics programme; one experiment, LHCb, dedicated to the study of B-hadrons and CP violation; one experiment, ALICE, which will study ion–ion and p–ion physics. Here only the ATLAS and CMS experiments and their physics programmes are discussed in some detail.

This paper is organized as follows. The status of the machine and detectors construction is summarized in Sect. 1. Section 2 describes how the two general-purpose experiments, ATLAS and CMS, prepare for data taking, with two explicit examples: test-beam activities and runs with cosmics. The strategy to understand the detectors and undertake the first physics measurements in the initial phases of the LHC operation is outlined in Sect. 3, whereas the possibilities for early discoveries are discussed in Sect. 4. Finally Sect. 5 is devoted to the conclusions.

1 Machine and Experiments Construction

This section summarizes the construction status of the machine and of the ATLAS and CMS experiments, as it was at the time of the IFAE workshop (April 2006). ALICE and LHCb are also on track to collect first data in 2007.

1.1 Machine Status

A high-technology machine is needed to achieve a beam energy of 7 TeV in a 27 km ring, in particular the bending power of 1232 superconducting dipole magnets providing the unprecedented field of 8.3 T. The construction is progressing well. As of April 2006, about 1050 dipoles had been delivered at CERN, and about 400 had been installed in the underground tunnel (Fig. 1). A magnet installation rate of about 25 units per week has been achieved, which would allow the completion of the machine installation by March 2007. A lot of effort is now being devoted to the complex dipole interconnection work.

The first 600 m of the cryogenic line (QRL) which cools down the LHC magnets have been successfully tested at operation temperature in September 2005, and installation is more than half way through. Hence the QRL, which in 2005 created delays and concerns due to some faulty components, is now out of the critical path.

Another important milestone was achieved in October 2004, with the first successful beam injection test from the SPS into the LHC through one of the two transfer lines. These lines consist of 5.6 km of tunnel equipped with about 700 magnets.

The machine team has also started to plan for the LHC commissioning and first operations. The present schedule, which has been revised in June 2006, foresees first pp collisions in November 2007 at a centre-of-mass energy of 900 GeV (corresponding to the beam injection energy from the SPS). This short (few weeks) pilot run will be used to debug the machine and the experiments, and will be followed by a three-month shutdown at the beginning of 2008, during which the machine commissioning for 14 TeV operation will be completed. First collisions at $\sqrt{s} = 14$ TeV are expected in June 2008 and will mark the beginning of the first physics run. The goal of this first run is to deliver an integrated luminosity of about 1 fb^{-1} by the end of 2008.

Fig. 1. A sequence of dipole magnets in the LHC underground tunnel

1.2 The ATLAS and CMS Experiments

Since it is not known how new physics will manifest, the LHC experiments must be able to detect as many particles and signatures as possible. Therefore ATLAS (A Toroidal Lhc ApparatuS [1], left panel in Fig. 2) and CMS (Compact Muon Solenoid [2], right panel in Fig. 2) are multi-purpose detectors which will provide efficient and precise measurements of e. g. electrons, muons, taus, neutrinos, photons, jets, b-jets.

The main features of the two experiments, which are complementary in several aspects, are presented in Table 1.

CMS has only one magnet, a big solenoid which contains the inner detector and the calorimeters and provides a magnetic field of 4 T in the inner detector volume. ATLAS has four magnets: a solenoid sitting in front of the electromagnetic calorimeter and producing a field of 2 T in the inner cavity, and external barrel and end-cap air-core toroids. The magnet layout determines the size, the weight and even the name of the two experiments.

The CMS inner detector consists of layers of Pixels and Silicon strips. Thanks mainly to the large magnetic field, excellent momentum resolution is expected (see Table 1). The ATLAS inner detector also contains Pixel and Silicon strip layers close to the interaction region and, in addition, a Transition Radiation Detector (TRT) at larger radii. Due to the lower magnetic field

Table 1. Main features of the ATLAS and CMS detectors

	ATLAS	CMS
Magnet(s)	Air-core toroids + solenoid in inner cavity	Solenoid
	Calorimeters in field-free region	Calorimeters inside field
	4 magnets	1 magnet
Inner detector	Si pixels and strips	Si pixels and strips
	TRT $\rightarrow$ particle identification	No particle identification
	B = 2 T	B = 4 T
	$\sigma/p_{\mathrm{T}} \sim 5 \times 10^{-4} p_{\mathrm{T}}(\mathrm{GeV}) \oplus 0.01$	$\sigma/p_{\mathrm{T}} \sim 1.5 \times 10^{-4} p_{\mathrm{T}}(\mathrm{GeV}) \oplus 0.005$
EM calorimeter	Lead-liquid argon	PbWO$_4$ crystals
	$\sigma/E \sim 10\%/\sqrt{E(\mathrm{GeV})}$	$\sigma/E \sim 2\text{–}5\%/\sqrt{E(\mathrm{GeV})}$
	Longitudinal segmentation	No longitudinal segmentation
HAD calorimeter	Fe-scintillator + Cu-liquid argon	Brass-scintillator
	$\geq 10\lambda$	$\geq 5.8\lambda$ + tail catcher
	$\sigma/E \sim 50\%/\sqrt{E(\mathrm{GeV})} \oplus 0.03$	$\sigma/E \sim 100\%/\sqrt{E(\mathrm{GeV})} \oplus 0.05$
Muon spectrometer	Chambers in air	Chambers in solenoid return yoke (Fe)
	$\sigma/p_{\mathrm{T}} \sim 7\%$ at 1 TeV	$\sigma/p_{\mathrm{T}} \sim 5\%$ at 1 TeV
	spectrometer alone	combining spectrometer and inner detector

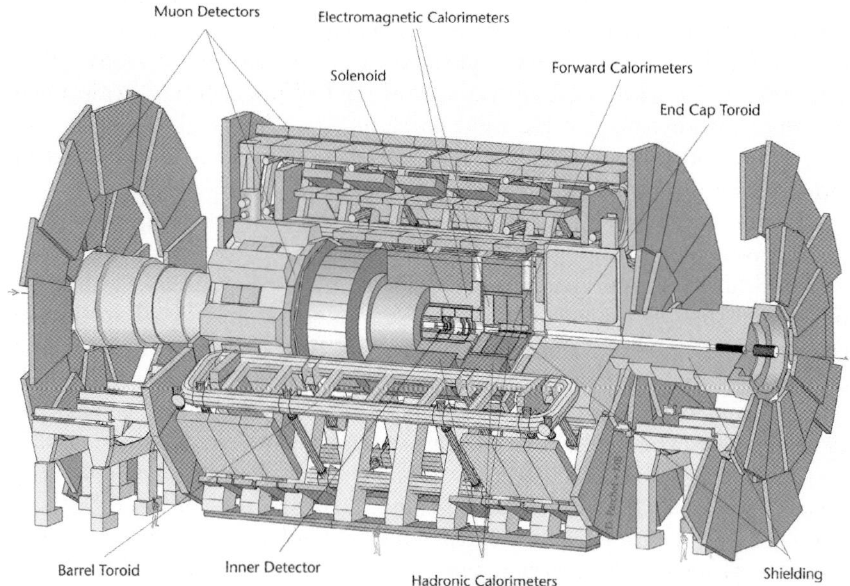

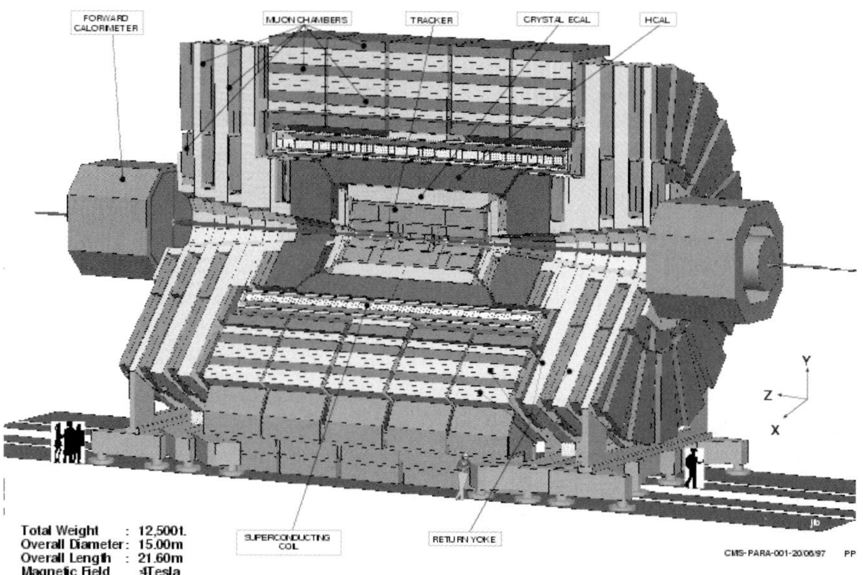

Fig. 2. Layout of the ATLAS (*top*) and CMS (*bottom*) detector

and somewhat smaller cavity, the expected momentum resolution is a factor of about three worse than that of CMS. However, the Transition Radiation Detector provides electron/pion separation capabilities.

The CMS electromagnetic calorimeter is a high-resolution crystal detector. The ATLAS calorimeter is a lead-liquid argon sampling calorimeter, therefore with a worse intrinsic energy resolution. However, thanks to a very fine lateral and good longitudinal segmentation, the ATLAS calorimeter provides more robust particle identification capabilities than the CMS calorimeter.

In both experiments the hadronic calorimeters are sampling detectors with scintillator or liquid-argon as active medium. The ATLAS calorimeter offers a better energy resolution because it is thicker (the CMS hadronic calorimeter suffers from space constraints dictated by the external solenoid) and has a finer sampling frequency.

Finally, the external Muon spectrometer of CMS consists of chamber stations embedded into the iron of the solenoid return yoke, where multiple scattering is not negligible. ATLAS has a spectrometer in air, where multiple scattering is minimised, and therefore offers the possibility of good standalone (i. e. without the inner detector contribution) measurements. The expected momentum resolution is better than 10% for muons of $p_\mathrm{T} = 1\,\mathrm{TeV}$ in both experiments. This performance is achieved by the Muon spectrometer alone in ATLAS, and by combining the information from the Muon spectrometer and the inner detector in CMS.

As of April 2006, the ATLAS barrel toroid system had been installed in the underground cavern (see Fig. 3) and cool down toward the operation temperature of 4.5 K had started. Magnetic field tests at full current (20 kA) are planned in October 2006. The barrel Silicon detector SCT, consisting of four cylindric layers of Si strips, had been inserted inside the barrel Transition Radiation Detector (Fig. 4), and this full system was ready for installation in the pit in Summer 2006.

Fig. 3. The ATLAS barrel calorimeter installed in its final position inside the barrel toroid system in the underground cavern

Fig. 4. The ATLAS barrel Silicon strip detector during insertion into the barrel Transition Radiation Detector in the surface clean room

The barrel calorimeter system (Fig. 3), consisting of the liquid-argon electromagnetic calorimeter inside its cryostat surrounded by the Tile Fe-scintillator hadron calorimeter, had been placed in its final position at $Z = 0$ (corresponding to the nominal beam-beam interaction centre) inside the barrel toroid. Both end-cap calorimeters were also in the pit, although not yet in their final positions. Finally, the installation of the barrel muon chambers (Monitored Drifted Tubes MDT for measurement and Resistive Plate Chambers RPC for trigger purposes) in the underground cavern was progressing well, with about 30% of the stations in place and with the goal of completing installation by the end of 2006.

The CMS detector is very compact and modular, therefore the integration and installation strategy is quite different from the ATLAS one. The detector is being pre-assembled at the surface, and will then be lowered in the underground cavern module by module (the module segmentation is visible in Fig. 2).

As of April 2006, the CMS solenoid, a 12.5 m long – 6 m diameter super-conducting magnet, had been inserted inside the iron structure of the return yoke (see Fig. 5) and had been cooled down to the operation temperature of 4.5 K. Full-current (20 kA) field tests in the surface assembly hall are foreseen in July–August 2006. The muon stations (Drift Tube chambers DT and RPC) are placed in the alcoves inside the iron structure visible in Fig. 5. In April 2006 the end-cap part was essentially completed and the barrel part was more than 50% done. The most critical component in the CMS path toward data taking is the electromagnetic calorimeter, a detector consisting of about 76 000 $PbWO_4$ crystals. Because of delays in the crystal delivery, only the barrel part will be installed in time for the 2007 run, whereas the end-caps will be added during the Winter 2008 shut-down. In contrast, the construction of the hadron and forward calorimeters has been completed since a long time. The CMS tracker consists of about 200 m^2 of Silicon sensors, for a total of almost 100 million channels. The assembly of the 16 000 modules for the full detector was finished in April 2006, and the emphasis was shifted to in-

Fig. 5. The CMS solenoid during insertion inside the iron structure of the return yoke

Fig. 6. One complete end-cap detector for the CMS tracker in the clean room

tegration and commissioning. Figure 6 shows one complete end-cap detector. The full tracker will be installed in the underground cavern in time for the 2007 run.

In both experiments, a lot of progress has been made in the integration and test of the detector readout electronics, trigger and data acquisition systems. Tests of the complete chain have started, on a scale of up to 10% of the full system. Final or close-to-final components are also being deployed to read out and record first cosmics data (see Sect. 2.2).

2 Preparation to First Physics

The experiments are preparing to first physics measurements in several ways, the most prominent ones being test-beam activities and cosmics runs. The ATLAS combined test beam (Sect. 2.1) and the ATLAS and CMS cosmics runs (Sect. 2.2) are discussed here, because they offer opportunities to the Collaborations to work in a coherent way as an integrated experiment (rather than a collection of individual sub-detectors), using common infrastructure and tools from on-line data taking, to software analysis, to the extraction of the "physics" results.

It should be noted that understanding unprecedented and complex detectors like ATLAS and CMS in the complex LHC environment will require a lot of time and a lot of data. The experience gained in the pre-collision phase with the above-mentioned activities should allow the experiments to save time at the beginning of the LHC operation.

2.1 The ATLAS Combined Test Beam

The combined test beam performed by ATLAS in 2004 represents a significant step toward the understanding of the complete detector and hence first physics. Indeed, a full vertical slice of ATLAS (see Fig. 7), including the Pixel detector, the Silicon strip detector, the Transition Radiation Detector, the liquid-argon electromagnetic calorimeter, the Tile hadron calorimeter, muon chambers (barrel MDT and RPC, and forward Cathode Strip Chambers CSC and Thin Gap Chambers TGC) and part of the trigger system, corresponding to a few percent of the experiment total acceptance, has been tested during six months on the CERN SPS H8 beam line. Electron, pion, proton, muon, photon data have been collected over a broad energy range (from 1 GeV up to 350 GeV in some cases), with and without magnetic field, for a total of about 90 million events (~ 4.5 TeraByte of data). A lot of global operation experience has been gained with this test, since all sub-detectors have been integrated, synchronized and run together with a common data acquisition system, and the data are being analyzed using the common ATLAS software framework. Examples of preliminary results are presented below, with emphasis on measurements where the information from several sub-detectors is combined together.

The left panel in Fig. 8 shows the momentum spectrum of 9 GeV pions reconstructed by the three tracking devices: six Pixel modules (arranged in

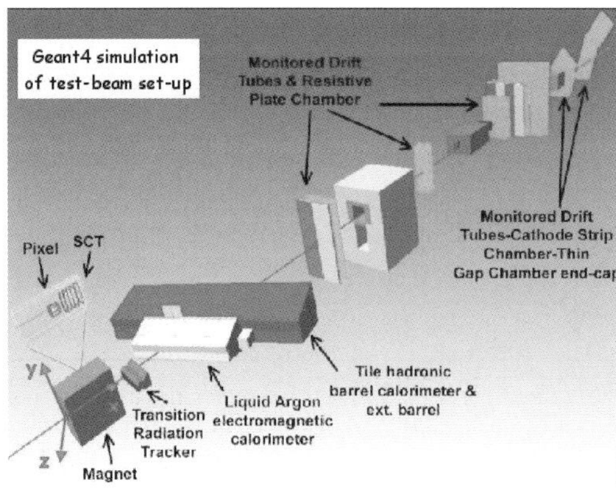

Fig. 7. GEANT4 [3] simulation of the 2004 ATLAS combined test-beam set-up, showing the beam line and the various sub-detectors

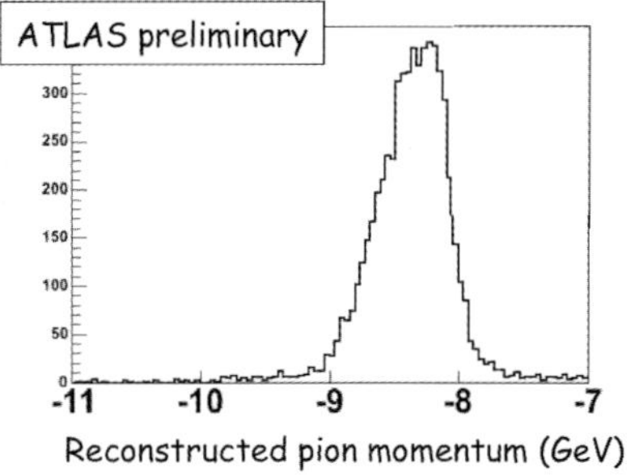

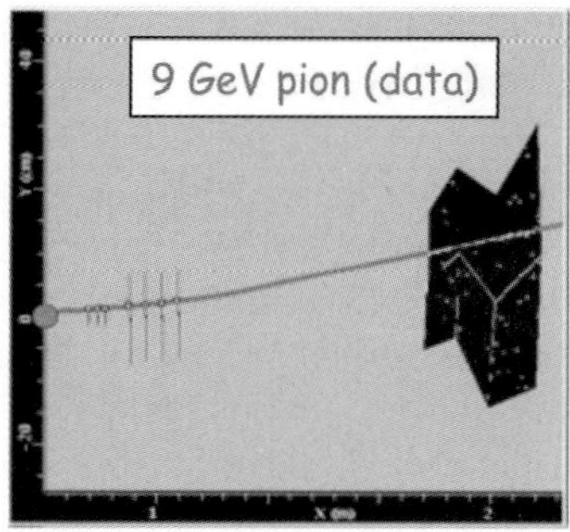

Fig. 8. *Left*: Distribution of the momentum of 9 GeV π^- reconstructed by combining the information of the Pixel, SCT and TRT detectors in the ATLAS combined test beam. *Right*: Graphic display of one of the events

three consecutive layers) and eight SCT modules (arranged in four consecutive layers) inside a 1.4 T magnetic field, followed by six TRT modules outside the field. The momentum resolution is close to that expected for the tested layout ($\sim 3.5\%$). The mean value of the distribution is shifted by half a GeV from the nominal beam energy, indicating that some systematic effects related to the detectors relative alignment and to the knowledge of the magnetic field still need to be understood. The right panel of the figure shows an event display from these data, with a pion track traversing the three detectors, bent by the magnetic field.

The nice correlation between the Muon spectrometer and the inner detector information is presented in the left panel of Fig. 9. Tracks from 180 GeV muons reconstructed in the MDT chambers have been extrapolated back-

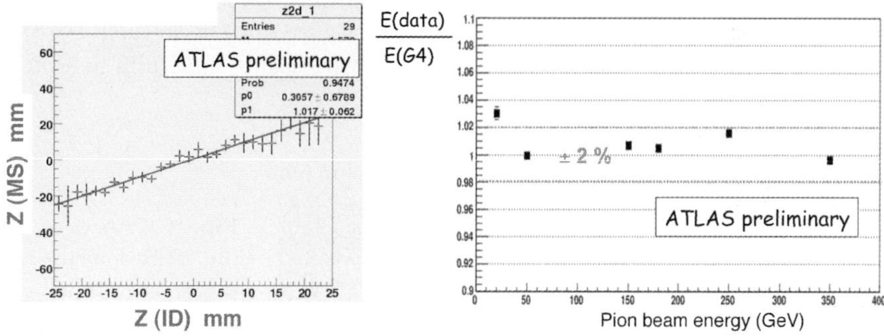

Fig. 9. *Left*: Comparison between the extrapolated position of muon tracks measured by the Muon spectrometer (MDT chambers) and that measured by the inner detector in the ATLAS combined test beam. *Right*: Ratio between data and simulation (GEANT4) for the pion energy reconstructed using the liquid-argon electromagnetic calorimeter and the hadron Tile calorimeter in the ATLAS combined test beam, as a function of the beam energy

ward (over $\sim 40\,$m) to the entrance of the inner detector (see Fig. 7), and the extrapolated position in Z (which corresponds to the LHC beam line) has been compared to that determined by the tracking devices.

A complex issue at hadron colliders, and a crucial one for a huge number of physics studies, is the determination of the jet absolute energy scale and the associated systematic uncertainty. In contrast to the electron scale, which can be established by using almost exclusively data samples (e. g. $Z \to ee$ events), the jet energy scale requires significant input from the simulation. It is therefore important that the latter reproduces the detector response to pions and jets at the percent level. Measurements performed in the combined test beam using pions reconstructed in the liquid-argon electromagnetic calorimeter and Tile hadron calorimeter show that the GEANT4 simulation describes the data to better than 2% over an energy range between 20 GeV and 350 GeV (right panel in Fig. 9). This is an important first step toward understanding the jet energy scale to the required 1% level.

2.2 Cosmics Runs

Another important phase in the preparation to physics is the commissioning with cosmics runs, allowing increasingly more complete and integrated detectors to be tested in the surface assembly halls or in the underground caverns. An example is shown in Fig. 10.

In the case of ATLAS, which is being installed in the underground cavern, cosmics runs offer the possibility to study the detector in situ in its final position. Full-simulation studies indicate that the expected rate of muons at $\sim 100\,$m below ground is of the order of 1 Hz [4]. Hence, a few million events could be recorded with the full detector in place in Summer 2007, assuming two months of cosmics data taking at 30% efficiency. These data samples will be very useful to e. g. catalog and fix problems, gain operation experience, synchronize the various sub-detectors and check their relative position, and perform alignment and calibration studies, hopefully in a more

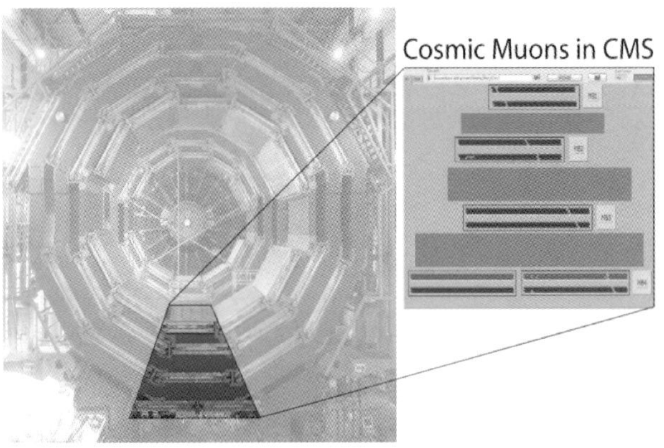

Fig. 10. View of the CMS detector in the surface assembly hall, showing the iron return yoke with several muon stations installed and a cosmic muon traversing four chambers located at the bottom of the detector

relaxed environment than later on during the initial pp phase. First cosmics data have been collected with the detector components already installed in the underground cavern, i.e. the liquid-argon electromagnetic calorimeter, the Tile hadron calorimeter and barrel muon chambers.

CMS is being assembled at the surface, where the rate of cosmics is almost three orders of magnitude larger than in the underground pits. As of April 2006 they had already recorded enough data to perform a detailed commissioning of about 80 Drift Tube stations installed at that time, by studying e. g. the detection efficiency of the individual planes as a function of the muon impact point, as shown in Fig. 11. The results demonstrate that the detector performance is in agreement with expectation.

Tests with cosmics are crucial also for the CMS electromagnetic calorimeter. Since only a handful of the 36 supermodules for the barrel calorimeter can be exposed to test beams, a dedicated cosmics stand has been put in place where all supermodules will be calibrated with cosmic muons. The goal is to achieve a crystal response uniformity at the level of 3%.

Another pre-collision activity of CMS is the Magnet Test and Cosmic Challenge (MTCC), scheduled in Summer 2006. A full vertical slice of the detector, consisting of part of the tracker, two supermodules of the electromagnetic calorimeter, and several modules of the hadron calorimeter, is being installed inside the solenoid and together with some of the muon stations already in place will collect cosmics data. Part of these data will be recorded

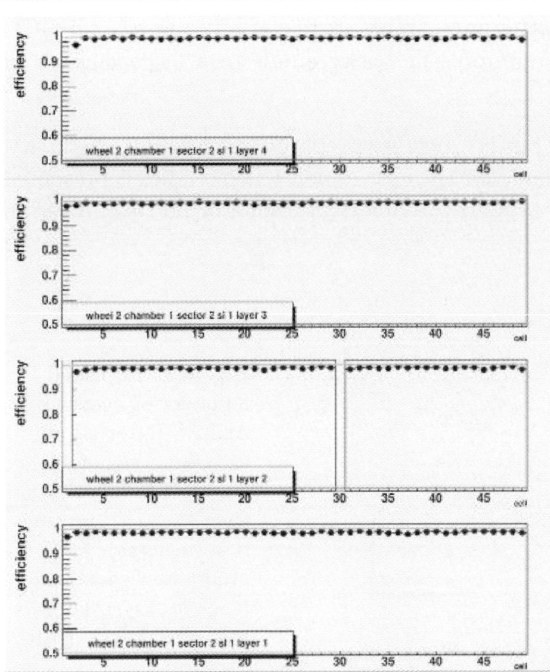

Fig. 11. Muon detection efficiency of several muon chamber planes, as a function of impact point, obtained with cosmics data collected by CMS in the surface assembly hall

with the full 4 T field on. With this test CMS will commission the magnet, exercise the combined operation of all sub-detectors with a common trigger, slow control and data acquisition system, and deploy the calibration and alignment procedures with real data.

3 First Data and First Measurements

With the advent of the first pp collisions, the most urgent goals to address will be:

- Commission and calibrate the detectors and triggers in situ using well-known physics channels. $Z \to \ell\ell$ is a gold-plated process for a large number of studies, e.g. to set the absolute electron and muon scales of the electromagnetic calorimeters and tracking detectors respectively, whereas $t\bar{t}$ events can be used for instance to establish the absolute jet scale and to understand the b-tagging performance.
- Perform extensive measurements of the main Standard Model (SM) physics processes at $\sqrt{s} = 14\,\mathrm{TeV}$, e.g. cross-sections and event features for minimum-bias, QCD di-jet, $W, Z, t\bar{t}$ production, etc., to be compared to the predictions of Monte Carlo simulations. Typical initial precisions may be 10–20% for cross-section measurements, perhaps $\sim 7\,\mathrm{GeV}$ on the top-quark mass, and will likely be limited by systematic uncertainties with integrated luminosities as low as $100\,\mathrm{pb}^{-1}$. These measurements are important "per se", but also because processes like W/Z+jets, $t\bar{t}$ and QCD multijet production are omnipresent backgrounds to a large number of new physics channels.

This phase will take time but is crucial to prepare a solid road to discovery.

The amount of data available for the above-mentioned studies is presented in Fig. 12, which shows the expected numbers of events in ATLAS, after all

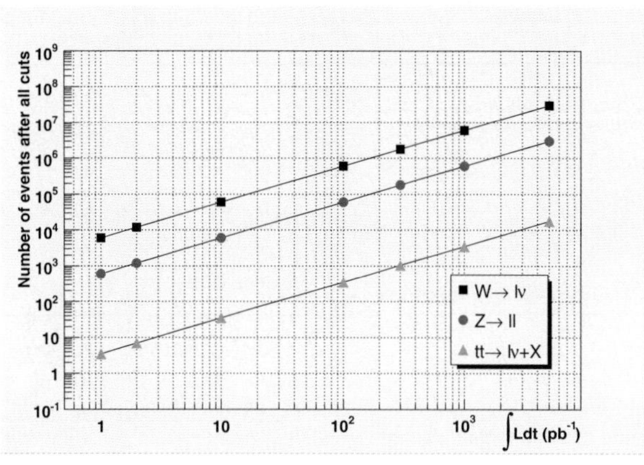

Fig. 12. Expected numbers of events in ATLAS (after all cuts), as a function of integrated luminosity, from $W \to \ell\nu$ production (*squares*), $Z \to \ell\ell$ production (*dots*) and $t\bar{t} \to b\ell\nu bjj$ production (*triangles*), with $\ell = e$ or μ

analysis cuts and as a function of integrated luminosity, for some basic SM processes, the so-called "candles": leptonic W and Z decays and semi-leptonic $t\bar{t}$ final states. It can be seen that, with only $100\,\mathrm{pb}^{-1}$, which can be collected in a few days of data taking at an initial luminosity of $\sim 10^{32}\,\mathrm{cm}^{-2}\,\mathrm{s}^{-1}$ (hence perhaps by Summer 2008), samples of $10^6\,W \to e\nu, \mu\nu$ are expected, as well as $\sim 10^5\,Z \to ee, \mu\mu$ and almost 1000 $t\bar{t} \to b\ell\nu bjj$. These samples are comparable in size to those recorded by the Tevatron experiments CDF and D0 until today. It is therefore obvious that very interesting detector and physics studies shall be performed with so much (or even less) integrated luminosity. Some examples are discussed below.

3.1 Understanding the Detector Performance

An illustration of the detector performance to be expected on "day 1", i.e. at the moment when data taking will start, is presented in Table 2. These predictions are based on construction quality checks, on the known precision of the hardware calibration and alignment systems, on test-beam measurements and on simulation studies. The initial uniformity of the electromagnetic calorimeters should be at the level of 1% for the ATLAS liquid-argon calorimeter and 3% for the CMS crystals, where the difference comes from the different techniques and from the limited time available to CMS for test-beam measurements. Prior to data taking, the jet energy scale may be established to about 10% from a combination of test-beam measurements and simulation studies. The tracker alignment in the transverse plane is expected to be known at the level of $20\,\mu\mathrm{m}$ in the best case from surveys, from the hardware alignment systems, and possibly from some studies with cosmic muons.

This performance should be significantly improved as soon as the first data will be available (see last column in Table 2) and, thanks to the huge event rates expected at the LHC, the ultimate statistical precision should be achieved in principle after a few days of data taking. Then the painful battle

Table 2. Examples of expected ATLAS and CMS detector performance at the time of the LHC start-up, and of physics samples which will be used to improve this performance

	expected performance on "day 1"	data samples (examples) to improve the performance
ECAL uniformity	$\sim 1\%$ ($\sim$3%) in ATLAS (CMS)	minimum-bias, $Z \to ee$
electron energy scale	$\sim$2%	$Z \to ee$
HCAL uniformity	3%	single pions, QCD jets
jet energy scale	$\leq 10\%$	$Z(\to \ell\ell)$+jet, $W \to jj$ in $t\bar{t}$ events
tracker alignment	20–200 μm in $R\phi$	generic tracks, isolated μ, $Z \to \mu\mu$

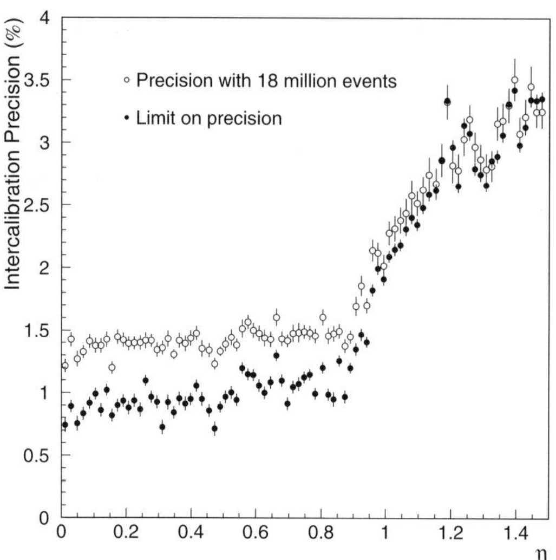

Fig. 13. Inter-calibration precision of the CMS electromagnetic calorimeter achievable with 18 million minimum-bias events [5], as a function of rapidity (*open circles*). The dots show the limit coming from the non-uniformity of the upstream material

with the systematic uncertainties will start. This is illustrated in Fig. 13 which shows that, by measuring the energy flow in about 18 million minimum-bias events (which can be collected in principle in a few hours of data taking), the non-uniformity of the CMS electromagnetic calorimeter should be reduced from the initial 3% to about 1.5% in the central part of the barrel detector. Therefore the systematic limit coming from the non-uniformity of the upstream tracker material will be hit very quickly.

Similarly, 10^5 $Z \to ee$ events, which should become available with an integrated luminosity of $100\,\mathrm{pb}^{-1}$ (see Fig. 12), would allow the non-uniformity of the ATLAS electromagnetic calorimeter to be reduced from $\sim 1\%$ to a few permil, hence to satisfy one of the performance requirements to observe a possible $H \to \gamma\gamma$ signal.

3.2 First Physics Measurements

An example of early measurements of SM physics is $t\bar{t}$ production. Indeed, a top-quark signal can be observed with few pb^{-1} of data, with a very simple analysis and a detector still in the commissioning phase [6]. In turn such a signal can be used to improve the knowledge of the detector performance and physics. The feasibility of this early measurement is due to the large ($\sim 250\,\mathrm{pb}$) cross-section for the gold-plated semileptonic $t\bar{t} \to b\ell\nu bjj$ channel (where $\ell = e, \mu$) and the clear signature of these events. A simple analysis has been performed by ATLAS [7], using a $t\bar{t}$ sample fully simulated with GEANT4, requiring an isolated electron or muon with $p_\mathrm{T} > 20\,\mathrm{GeV}$, large missing transverse energy, and four and only four jets with $p_\mathrm{T} > 40\,\mathrm{GeV}$. The additional constraint that two of the jets have an invariant mass compatible with the W mass was imposed. The resulting mass spectrum of the three

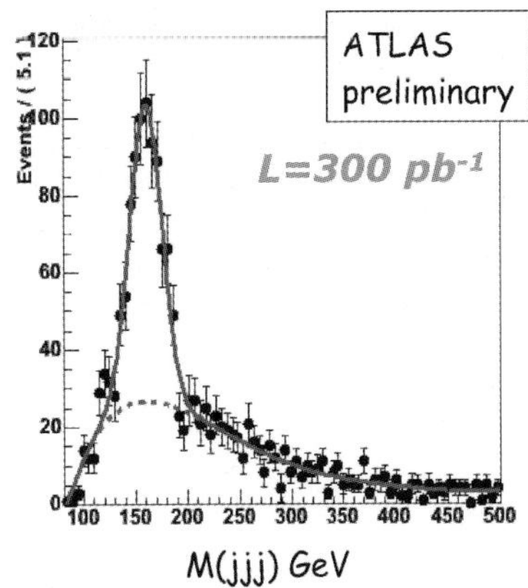

Fig. 14. Three-jet invariant mass distribution for events selected as described in the text, as obtained from a full simulation of the ATLAS detector [7]. The *dots with error bars* show the expected signal from $t\bar{t}$ events plus the background, the *dashed curve* indicates the background alone. The number of events corresponds to an integrated luminosity of 300 pb^{-1}

jets giving the highest p_{T} of the top quark is presented in Fig. 14. It should be noted that no b-tagging of two of the jets was required, assuming (conservatively) that the performance of the vertex detector would not be well understood at this early stage. Figure 14 shows that, even in these pessimistic conditions, a clear top signal should be observed above the background. An integrated luminosity of less than 30 pb^{-1}, which should be collected by mid 2008, would be sufficient.

Such a top sample will be very useful to understand several aspects of the detector performance. For example, the two b-jets in the final state can be used to study the efficiency of the b-tagging procedure, and the jet energy scale of the experiment can be established in a preliminary way from the reconstructed $W \rightarrow jj$ mass peak. Furthermore, the (reconstructed) p_{T} spectrum of the top-quark is very sensitive to higher-order QCD corrections, and this feature can be exploited to test the theory and tune the Monte Carlo generators.

4 Early Discoveries

Only after the steps outlined in Sect. 3 will have been fully addressed, can the LHC experiments hope to extract convincing discovery signals from their data. Three examples are discussed briefly below, ranked by increasing difficulty: an easy case, namely a possible $Z' \rightarrow e^+e^-$ signal, an intermediate case, Supersymmetry, and a difficult case, a light Standard Model Higgs boson.

4.1 $Z' \rightarrow e^+e^-$

A particle of mass 1–2 TeV decaying into e^+e^- pairs, such as a possible new gauge boson Z', is probably the easiest object to discover at the LHC, for three main reasons. First, if the branching ratio into leptons is at least at the percent level as for the Z boson, the expected number of events after all experimental cuts is relatively large, e. g. about ten for an integrated luminosity as low as $300\,\mathrm{pb}^{-1}$ and a particle mass of 1.5 TeV. Second, the dominant background, di-lepton Drell-Yan production, is small in the TeV region, and even if it were to be a factor of two-three larger than expected today (which is unlikely for such a theoretically well-known process), it would still be negligible compared to the signal. Finally, the signal will be indisputable, since it will appear as a resonant peak on top of a smooth background, and not just as an overall excess in the total number of events. These expectations are not based on ultimate detector performance, since they hold also if the calorimeter response is understood to a conservative level of a few percent.

4.2 Supersymmetry

Extracting a convincing Supersymmetry (SUSY) signal in the early phases of the LHC operation is not as straightforward as in the previous case, since good calibration of the detectors and detailed understanding of the numerous backgrounds are required. As soon as these two pre-requisites will be satisfied, however, observation of a possible TeV-scale SUSY signal should be relatively easy and fast. This is because of the huge cross-section for squark and gluino pair production, with about ten events per day expected in each experiment at instantaneous luminosities of only $10^{32}\,\mathrm{cm}^{-2}\,\mathrm{s}^{-1}$ and for squark and gluino masses as large as $\sim 1\,\mathrm{TeV}$. In addition, cascade decays of (heavy) squarks and gluinos should give rise to clear-signature final states, containing several high-p_T jets, leptons and, in dark-matter motivated scenarios, large missing transverse energy coming from the escaping stable neutralinos (χ_1^0). Figure 15 shows that with only $100\,\mathrm{pb}^{-1}$ of data, and provided the detectors and the backgrounds are well understood, ATLAS and CMS should be able to discover gluinos up to masses of about 1.2 TeV, whereas the ultimate LHC reach extends up to masses of 2.8 TeV.

Particle physics and the planning for future facilities would greatly benefit from a quick determination of the scale of new physics. If squarks and gluinos are discovered at the LHC with only $100\,\mathrm{pb}^{-1}$, Supersymmetry is relatively light, therefore a good part of the spectrum (e. g. charginos, neutralinos, sleptons) should be accessible for detailed measurements at a 1 TeV International Linear Collider (ILC). On the other hand if nothing is found at the LHC with $100\,\mathrm{pb}^{-1}$ of well-understood data, it is likely that Supersymmetry, if it exists at all, is too heavy to be detected at an ILC, because the lightest supersymmetric particle (the lightest neutralino χ_1^0) would be heavier than 300 GeV, as shown in Fig. 15.

It should be noted that understanding the detectors and the backgrounds at the level needed to discover Supersymmetry will take time, and will likely require a larger amount of data than $100\,\mathrm{pb}^{-1}$ (see Refs. [1, 2] for more details).

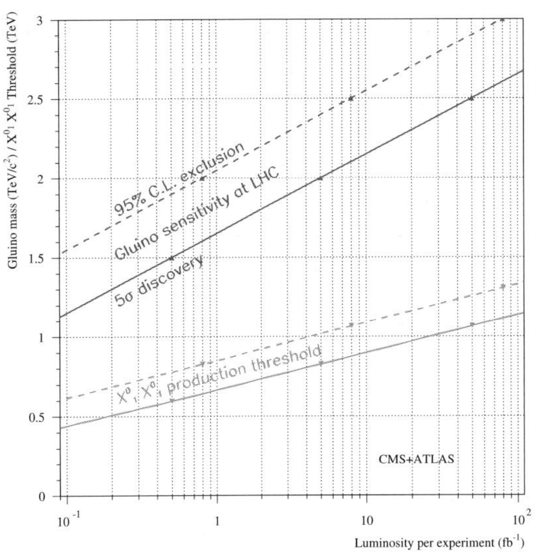

Fig. 15. The reach for gluino detection at the LHC, and the corresponding threshold for the production of pairs of lightest neutralinos (χ_1^0) at linear colliders, as a function of the LHC integrated luminosity per experiment. The *full lines* indicate the 5σ discovery reach, the *dashed lines* the 95% C.L. exclusion. From [8]

4.3 Standard Model Higgs Boson

Figure 16 shows the needed integrated luminosity per experiment, as a function of the Higgs boson mass, to discover a possible Higgs signal (5σ excess required) or to exclude it at the 95% C. L. Two conclusions can be drawn from these projections. First, with a few fb^{-1} of well-understood data the LHC can say the final word about the SM Higgs mechanism, i. e. discover the Higgs boson or exclude it over the full allowed mass range. Second, ignoring

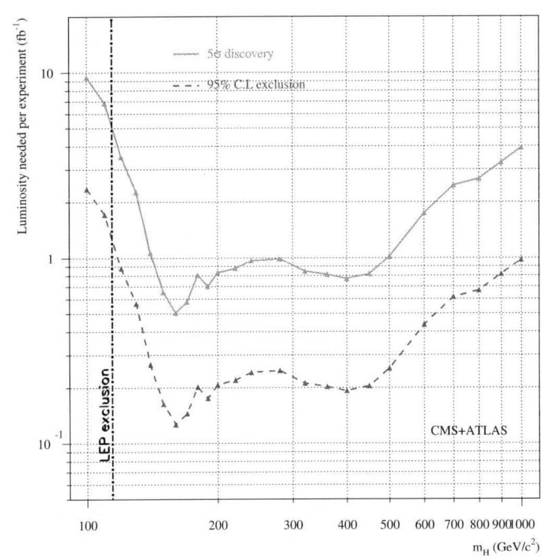

Fig. 16. Integrated luminosity per experiment, as a function of the Higgs boson mass, needed for 5σ discovery (*upper curve*) and for 95% C.L. exclusion (*lower curve*) of a SM Higgs boson signal at the LHC. From [8]

masses above 300 GeV, which are disfavoured by the electroweak data [9], two regions can be identified. If the Higgs mass is larger than 180 GeV, discovery should be easier thanks to the gold-plated $H \to 4l$ channel which is essentially background free. If, on the oher hand, the Higgs mass is around 115–120 GeV, i. e. just above the experimental lower limit coming from LEP, more luminosity is needed and observation of a possible signal is less straightforward. This is because in this mass region the experimental sensitivity is equally shared by three different channels ($H \to \gamma\gamma$, $t\bar{t}H$ production with $H \to b\bar{b}$, and Higgs production in vector-boson fusion followed by $H \to \tau\tau$) which all require close-to-ultimate detector performance and a control of the huge backgrounds at the few percent level. More details can be found in Refs. [1, 2].

In conclusion, discovery of a light SM Higgs boson at the LHC is not likely to happen before the end of 2009.

5 Conclusions

LHC operation will start in about one year, and the machine and the experiments are progressing at full speed toward this goal.

There have been impressive achievements in the machine construction and installation over the last months. About 30% of the dipoles have been installed in the underground cavern, the problems with the cryogenic line have been solved, and a first planning for the machine commissioning and operation has been developed. The present schedule foresees first pp collisions at $\sqrt{s} = 900$ GeV in November 2007, followed by collisions at $\sqrt{s} = 14$ TeV in Summer 2008. The experiments are on track to meet this calendar.

With the first pp data, the most urgent tasks will be to understand the detectors in detail and to perform first measurements of SM physics. The latter include minimum-bias events, accessible in a few hours of data taking; QCD jets and their underlying event; W and Z cross-sections, perhaps with a precision of 10% for 100 pb^{-1}, and constraints of parton distribution functions (in particular gluons at low x) using angular distributions of leptons from W/Z decays; observation of a top-quark signal with (less than) 30 pb^{-1} and measurements of the $t\bar{t}$ cross-section to $\sim 20\%$ and of the top mass to ~ 7 GeV with 100 pb^{-1}; etc.

With more time and more data, the LHC will be able to explore the highly-motivated TeV scale in detail, with a direct discovery potential up to particle masses of ~ 5–6 TeV. Hence, if new physics is there the LHC will find it. It will also provide definitive answers about the SM Higgs mechanism, Supersymmetry, and several other TeV-scale predictions that have resisted experimental verification for decades. Finally, and perhaps more importantly, the LHC will tell us which are the right questions to ask and how to continue.

References

1. ATLAS Collaboration: *Detector and Physics Performance Technical Design Report*, CERN/LHCC/99-15
2. CMS Collaboration: *Physics Technical Design Report, Vol. I – Detector Performance and Software*, CERN/LHCC/06-01
3. S. Agostinelli et al.: Nucl. Inst. Meth. A **506** 250 (2003)
4. M. Boonekamp et al.: ATLAS Note ATL-GEN-2004-001
5. D. Futyan: CMS Note CMS CR 2003/005
6. M. Cobal and S. Bentvelsen: ATLAS Note ATL-PHYS-PUB-2005-024
7. I. Van Vulpen, W. Verkerke and S. Bentvelsen: *Top physics during ATLAS commissioning*, http://agenda.cern.ch/askArchive.php?base=agenda&categ=a044738&id=a044738s11t3/moreinfo, ATLAS Note in preparation
8. J.-J. Blaising et al.: *Potential LHC contributions to Europe's future strategy at the high-energy frontier*, input n. 54 to the CERN Council Strategy group, http://council-strategygroup.web.cern.ch/council-strategygroup/
9. G. Altarelli, these Proceedings

Lattice QCD and Numerical Simulations

Raffaele Tripiccione

Dipartimento di Fisica, Università di Ferrara and INFN, Sezione di Ferrara
`tripiccione@fe.infn.it`

Summary. We briefly review the theoretical background, the computational techniques and the phenomenological predictions made possible by Lattice Gauge Theories (LGT) studies of the strong interactions.

1 Overview

Quantum field theory (QFT) is the theoretical framework in which elementary particles are described and studied. Most quantitative results in QFT are derived in perturbation theory, in which a systematic expansion in a small parameter (typically the coupling constant(s), $\alpha_{em}(q), \alpha_s(q), \ldots$) is made. Perturbation theory is however not applicable in physics rich dynamic regimes, where the coupling constant is no longer a small parameter. In QCD, the theory of strong interactions in which we are mainly interested here, this occurs in the phenomenologically rich low energy regime. Key physical quantities - the physical hadron spectrum to name just one - cannot be treated in this way. Lattice Gauge Theory (LGT) comes to the rescue, providing several handles, relevant at both the theoretical and phenomenological level. The theory defined on a discrete lattice provides a clean, intrinsically non perturbative regularization of the divergencies of the theory. Divergencies of course reappear as the lattice spacing a goes to zero, but ratios of physical observables remain finite. At the phenomenological level, this theoretical framework becomes a source of quantitative predictions when combined with the power of fire of numerical techniques enabled by the use of powerful computer systems. Lattice simulations provide numbers which are relevant for two somehow different purposes. First they corroborate our belief on the correctness of the theory (e.g., by computing accurate figures for the experimentally well known spectrum of the low lying hadrons), second, they contribute in the phenomenological arena and in the analysis of experimental data, by providing often crude but critical predictions of key physical parameters. As an example of the latter perspective, consider the determination of

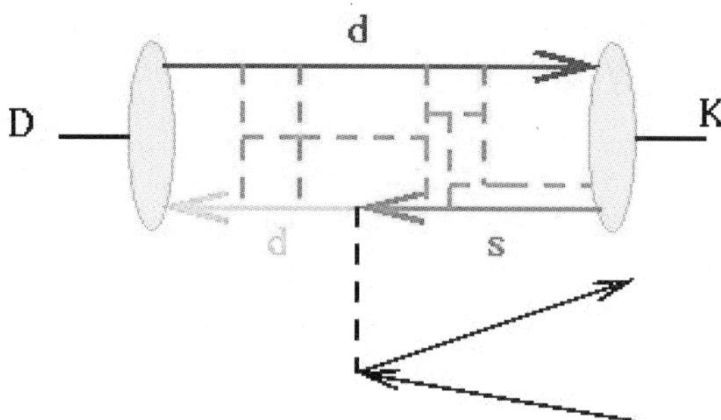

Fig. 1. Typical diagram associated to the decay $D \to Kl\nu$

all entries of the CKM matrix. A value for the entry associated to the mixing of the s-quark and d-quark can be extracted in principle from experimental data on processes like $D \to Kl\nu$ or $D^* \to Kl\nu$ only if the matrix element $\langle D|J_W^\mu|K\rangle$ of the weak current can be controlled (see Fig. 1): something that we can do only on the lattice. This is why lattice results are becoming more and more relevant in experimental contexts.

This paper is a very concise introduction to the main ideas and to the key computational tools of LGT. The aim of the paper is much more to provide a general introduction to this field for an experimental physicists, so (s)he can grasp the origin and limitations of lattice data (s)he may use for analysis, rather than a theoretically and technically sound description of the theory. Recent results are also fully neglected. The interested reader may refer to specific contributions [1]. The following of this paper is structured as follows: in Sect. 2 I describe the basic features of the theory, in Sect. 3 I look at the tecniques used to measure a typical hadron mass on the lattice. Section 4 sketches the numerical machinery that makes LGT a predictive theory, and 5 contains my conclusions, including some comments on the role of high-end computers in the field.

2 LGT Basics

A field theory is defined on the lattice using the approach of the functional integral, with the key added ingredient that all elements of the theory are defined only at the sites of a discrete lattice. The lattice spacing (usually equal in the space and time directions) is a. Derivatives become finite differences:

$$\partial_\mu \phi \to \Delta_\mu \phi = \frac{\phi(x + a\mu) - \phi(x)}{a} \,, \tag{1}$$

integrals become sums ($\int \mathrm{d}^4 x \to \sum_x a^4$) and the integration measure in the functional space is reduced to a discrete set of integration variables

$$\Delta\phi \to \prod_x \mathrm{d}\phi(x) \,. \tag{2}$$

Expectation values of arbitrary correlation functions for a theory involving a generic field ϕ can be formally defined as

$$\langle Q_1(x_1)\cdots Q_k(x_k)\rangle = \frac{1}{Z}\int \prod_x \mathrm{d}\phi(x)\langle Q_1(\phi(x))\cdots Q_1(\phi(x))\rangle\,\mathrm{e}^{-S[\phi]} \qquad (3)$$

where Z is the partition function, S is the action and a Wick rotation to imaginary time is implied. On a finite lattice the quantities defined above can be numerically computed, so in principle all abservables of the theory can be worked out, to arbitrary precision. We will see later that pursuing this program is not an easy task.

A gauge invariant theory on the lattice requires a more careful definition. Finite differences couple the fields at different sites, so the corresponding terms are no longer gauge invariant, since $\bar{\psi}(x) \to \bar{\psi}(x)g^{-1}(x)$ and $\psi(y) \to g(y)\psi(x)$. We try to keep gauge invariance on the lattice (we are not happy to just recover it at the continuum limit, as we do for Lorentz invariance) because the former introduces severe constraints and allow to keep the theory under control. For instance, if we drop gauge invariance we are no longer guaranteed that all the vertices of the theory have the same coupling constant. We need to introduce objects like $U(x,y)$ such that $\bar{\psi}(x)U(x,y)\psi(x)$ is gauge invariant (that is, we require that, for a gauge transformation $g(x)$, $U(x,y) \to g(x)U(x,y)g^{-1}(y)$. The most popular choice for $U(x,y)$, originally made by Wilson [2], uses elements of the gauge group in the adjoint representation (e. g., 3×3 SU(3) matrices for QCD) that we associate with the elementary path (called link in the trade) connecting nearby elements of the lattice $U(x, x + \hat{\mu}) = U_\mu(x)$. Closed loops written in terms of U's are obviously gauge-invariant. The smallest such object is called the plaquette, and its trace $P_{\mu\nu}$ is the building block for a pure-gauge action:

$$P_{\mu\nu}(x) = \mathrm{Re}\,\mathrm{Tr}[1 - U_\nu(x + \hat{\mu})U_\mu(x + \hat{\nu})U_\mu^\dagger(x)U_\nu^\dagger(x)] \qquad (4)$$

In fact, we define (for an SU(N) gauge group) the action as

$$S_g = \beta/(2N)\sum_{x,\mu,\nu} P_{\mu\nu}(x) \qquad (5)$$

(β is the inverse of the coupling constant g_0, $1/g_0^2 = \beta/2N$). It is easy to show that, in the continuum limit of $a \to 0$,

$$S \to \int \mathrm{d}^4x(F_{\mu\nu})^2 + O(a^2) \qquad (6)$$

the well-known action of a pure gauge theory.

Introducing fermionic degrees of freedom (that is, quarks) implies some additional headaches. First, fermions are defined in terms of anticommuting field variables. Luckily enough, as long as fermions appear quadratically in the action (as they do in the Dirac term), they can be integrated analytically, so they disappear from the functional integral, leaving behind the determinant of the fermion operator. So, for QCD on the lattice, we have:

$$S = \int DU_\mu \det M[U]\, \mathrm{e}^{-S_{\mathrm{g}}[U]} = \int DU_\mu\, \mathrm{e}^{-S_{\mathrm{g}}[U]+\log(\det M)} \,. \qquad (7)$$

Details on the fermion operator M have not been specified yet. It turns out that the naive discretization of the Dirac term is not a viable choice. Indeed, the corresponding free field propagator has 16 poles, corresponding to additional unwanted degeneracies in the fermion states. This desease is cured in several ways. A popular solution are the so-called Wilson fermions associated to additional terms in the action (vanishing as $a \to 0$), that solve the problem but explicitly break chiral invariance. The latter is recovered in the continuum limit. However, in principle independent limits, vanishing quark mass and lattice spacing are inextricably connected together in this scheme.

3 A typical LGT measurements

As discussed above, expectation values of all correlation functions can be meaured on the lattice. In this section, I discuss in a pedagogically detailed way an example of such a measurement. Assume that we want to measure on the lattice the mass of the pion. We choose $\langle \sum_x Q(x,t)Q(0,0)\rangle$ as our correlation function ($Q = A_4 = \bar{\psi}\gamma_4\gamma_5\psi$) is the axial current). We make this choice because, after expanding on a complete set of energy eigenstates,

$$\left\langle \sum_x Q(x,t)Q(0,0) \right\rangle = \sum_n \frac{\langle 0|Q|n\rangle\langle n|Q|0\rangle}{2E_n}\, \mathrm{e}^{-E_n t} \,, \qquad (8)$$

we can take the limit of large t that projects onto the lowest state, the physical pion:

$$\left\langle \sum_x Q(x,t)Q(0,0) \right\rangle \to \frac{\langle 0|Q|\pi\rangle\langle \pi|Q|0\rangle}{2M_\pi}\, \mathrm{e}^{-M_\pi t} \,. \qquad (9)$$

So the physical pion mass is measured on the lattice by measuring the above defined $\langle QQ\rangle$ correlation and fitting with an exponential function at large t. The mass of other particles is extracted in principle in the same way, by choosing appropriate operators.

Note that the procedure described above in principle requires to measure fermionic operators. This is done in practice by writing the correlation functions in term of quark propagators, so what we really need to measure is $\langle \sum_x S_{\mathrm{F}}[U](0,0;x,t)\gamma_4\gamma_5 S_{\mathrm{F}}[U](x,t;0,0)\gamma_4\gamma_5\rangle$, where $S_{\mathrm{F}}[U](x,t;0,0)$ is the quark propagator (in which we have explicitly shown the dependence on the gauge field configuration U). The propagator, in turn, is the inverse of the Dirac operator M appearing in the original action, so our measurement program can be actually put in practice once the following linear problem is solved.

$$M_{ab}^{ij}[U](x_\alpha; y_\beta) S_{bc}^{jk}[U](y_\beta \rangle 0) = \delta_4(x_\alpha - 0)\delta^{ik}\delta_{ac} \tag{10}$$

I have explicitly written down all indices, associated to the lattice sites (x_α) to the spinor (ij) and color (ab) degrees of freedom, in order to stress the size of the linear problem that we have to solve, a matrix whose dimension is of the same order of magnitude as the sites of our lattice. Note also that, thanks to translation invariance, we only need to compute the propagators from point $(0, 0)$.

Note that the determinant associated to the operator of (10) is in principle needed to correctly evaluate the functional integral of (7). This is a major computational effort that, in many cases, is alleviated by the so-called quenched approximation, in which it is assumed that the value of the determinant equals unity. This in principle unsatisfactory approximation has worked surprisingly well sofar, and only recently have simulations of the full theory been carried out in details.

Summing up, a typical "lattice experiment" involves the following steps:

- Pick up an arbitrary gauge configuration U (that contributes with weight proportional to $e^{-S_g(U)}$).
- Compute the quark propagator, by (numerically) inverting the Dirac operator (for the given field U).
- Measure the value of the correlation function(s) for the given configuration.

We perform these steps on a large number of field configurations, take the average of all measurements and finally fit the "lattice-experimental" result to an exponential behaviour, computing the pion mass (or any other quantity we are interested in). The next section will discuss some of the problems that we encounter when performing this steps.

4 Numerical methods for LGT

The averaging procedure associated to any lattice measurement is carried out in practice using Monte Carlo techniques, using powerful supercomputers. Sampling the configuration space randomly is not viable in practice, and inefficient in principle, since we would sample dynamically irrelevant configurations (associated to a very small value of $e^{-S_g[U]}$). We solve the problem by resorting to importance sampling, that is by choosing Monte Carlo configurations U_c with probability $P([U_c(x)]) \simeq e^{-S_g[U_c]}$ (this is possible, since S in imaginary time is bounded from below). Measurements become simply

$$\langle Q_1 \cdots Q_k \rangle = \frac{1}{C} \sum_c^C Q_1[U_c] \cdots Q_k[U_c]\,. \tag{11}$$

In other words, we have reduced quantum field theory to the realization of a (multi-variable) random number generator with probability distribution $\mathrm{e}^{-S_\mathrm{g}[U]}$.

In principle, this programs gives us the results of the theory, with a statistical error that we can bring to zero as $1/\sqrt{N}$, by increasing the number of measurements. In practice, we are still far away from this port of heaven, for several different reasons, some of which we discuss here.

Lattice meausurements give us numbers that are adimesionalized by appropriate powers of the lattice spacing (the only dimensioned quantity in the system). So we have for instance $m_\pi a$. The lattice spacing is connected to the coupling constant by the beta function of the renormalization group (that we parametrize in term of its scale in momentum space Λ). As the coupling constant changes, so does the lattice spacing. The point is that our theory has physically many scales associated to light (m_q) and heavy (m_Q) quarks($m_\mathrm{q} \ll \Lambda \ll m_\mathrm{Q}$). A lattice of size $L = Na$ has even more scales. Ideally, we would like to have

$$L_{-1} \ll m_\mathrm{q} \ll \Lambda \ll m_\mathrm{Q} \ll \pi/a \tag{12}$$

while, in practice we can only hope for:

$$L_{-1} < m_\mathrm{q} \ll \Lambda \ll m_\mathrm{Q} \simeq \pi/a \tag{13}$$

A brute force approach would require small enough a and large enough Na. Computationally we are not yet able to satisfy these conditions. Memory reqirements for a lattice with N^4 lattice points grow as $(L/a)^4$, while the computating time needed to explore the configuration space grows as $1/a^{4+z}$. The term z (typical values are $z \simeq 1 \cdots 3$) is associated to critical slowing down, the fact that information moves on the lattice on scale a, while the dynamics has a natural scale $\simeq \Lambda^{-1}$. Inverting the Dirac operator is the key numerical point, especially in the physically important regime of small quark masses. Efficient iterative solvers are used, with a computational complexity of the order $\min[1/(m_\mathrm{q}a)^p, (L/a)^p]$, with $p \simeq 1 \cdots 3$.

Coping with these requirements is a complicated tasks. On the one hand, one tries to modify the theory on the lattice to improve its convergence to the continuum limit. For instance, since $\partial = \Delta + a^2\Delta^3 + O(a^4)$, one might introduce additional operators that cancel the a^2 term in the expansion (or, even more important, the a term encountered in the fermion sector [4]. In this way, a larger lattice spacing can be used for real simulations. On the other hand, contact with ideas coming from elsewhere is often necessary to produce reasonable results, epsecially in the regime of low-mass quarks. Since, it is today hopeless to sit at $L \gg 1/m_\mathrm{q}$, one may resort to chiral perturbation theory. Functional relations, like $m_N = c_0 + c_1 m_\pi + c_2 m_\pi^2 + c_3 m_\pi^3$ can be used to exptrapolate from measurements made at unphysically large values of the π mass.

Progress in the area is therefore a combined effort, of better algorithmic tools and better match with different physics models on the one hand

and increased computer power on the other hand. The overall computational complexity C of the problem has been estimated using very crude modelling. A popular paramtrization reads

$$C \sim \left(\frac{m_\rho}{m_\pi}\right)^6 \times (L/L_0)^5 \times (a/a_0)^7 \qquad (14)$$

and stipulates that doubling the size of the lattice and extracting all the physical information available in the system needs a hundred-fold increase in computing costs. Even if this may seem hopeless, computers used for lattice QCD have increased their performances of nearly 4 orders of magnitude in the last 20 years [5], making lattice QCD able to provide quantitative predictions to be compared with experimental results.

5 Conclusions

In this paper I have briefly covered the main ideas behind the lattice formulation of a field theory. The lattice approach is theoretically sound and phenomenologically relevant to evaluate theoretical predictions to be confronted with experimental results. This is particularly relevant in areas such as the weak interaction of hadronic states, where epxerimental numbers have relatively large uncertainties and a theoretical bridge is needed to use experimentally observable quantities in order to constrain parameters of the standard model. This approach to the theory is strongly tied to progress in the computer power available for simulations. This unbreakable connection has lead to the development of several special computer systems [6] accurately tailored to this computational problem.

Acknowledgments

I would like to thank the organizers of IFAE2006 for the opportunity to attend a very interesting and lively conference. This paper is dedicated to the memory of Giuseppe (Beppe) Curci, who gave several important contributions to lattice physics in the last 20 years and unexpectedly passed away recently.

References

1. For an in-depth review, see for instance, R. Gupta, Lecture Notes (Les Houches Summer School 1997) hep-lat/9807028
2. K.G. Wilson, Phys. Rev D **10** 2445 (1974)
3. K.G. Wilson, Rev. Mod. Phys. **55** 583 (1983)
4. B. Sheikholeslami and R. Wohlert, Nucl. Phys. B **259** 572 (1985)
5. R. Tripiccione, Comp. Phys. Comm. **169** 442 (2005)
6. F. Belletti *et al.*, Computing in Scienze and Engineering **8** 18 (2006)

The double life of the X meson

A.D. Polosa

INFN Sez. di Roma 1, P.le A. Moro 2, I-00185 Roma, Italy
antonio.polosa@roma1.infn.it

Three years have passed since the BELLE discovery of the $X(3872)$, and there are still (at least) two competing interpretations of this particle, which resembles a charmonium but behaves in a dramatic different way from it. Is X a molecule of two D mesons or a compact four-quark state? Are these two pictures really resolvable?

The quantum mechanical intuition can also lead to more refined pictures: the X could be a sort of breathing mode of a charmonium oscillating into a molecule and back.

Other particles have been discovered since then: the $X(3940)$, $Y(3940)$, $Z(3930)$ (amazingly the first two have the same mass and both decay to charmonium but with a different decay pattern to open charm) and $Y(4260)$. The latter also decays into J/ψ and could be an 'hybrid' particle (two quarks and a constituent gluon), likely the most experiment-proof interpretation so far.

In this talk I will not try to describe all the experimental facts and theoretical ideas, thoroughly reported and commented elsewhere in the literature. I will rather comment on the first question here raised, namely, how far are molecules, four-quarks or charmonium-molecule oscillating states distinguishable in principle and in the experimental practice? The suspect that the competing scenarios fade into one another could dangerously leave this field in a confused and controversial situation similar to that existing for sub-GeV scalar mesons (and for their super-GeV partners).

1 Sewing Quarks

The prominent decay mode of $X(3872)$ is $X \to J/\psi\rho$. Several studies conclude that the X cannot be an ordinary $c\bar{c}$ state, even though the J/ψ invokes a charmonium assignation. The next-to-easy interpretations can be: (1) X is a $D\bar{D}^*$ bound object, with the correct 1^{++} quantum numbers. Such a molecule could decay at the same rate to $J/\psi\rho$ and $J/\psi\omega$, which is actually what surprisingly happens in nature (this was not a prediction though). This molecule lives for a while, until the two heavy quarks get close enough to form a J/ψ, leaving the light quarks free to hadronize to a ρ^0. (2) X is a four-quark $c\bar{c}q\bar{q}$ meson. The four quarks could be diving in some potential bag but, if so, we should get $\mathbf{3} \otimes \mathbf{\bar{3}} \otimes \mathbf{3} \otimes \mathbf{\bar{3}}$, i.e., 81 particles. This is the obvious problem of exotic structures: a copious number of states is predicted. Moreover such multiquark structures could fall apart and decay at an immense rate (resulting in very broad and experimentally elusive states): at the lowest $1/N_c$ order a propagating state of four quarks in a color-neutral, gauge-invariant combination $q^i\bar{q}_i q^j\bar{q}_j$, is indistinguishable from two freely propagating $q\bar{q}$ mesons. On the other hand, it turns out that quarks (and antiquarks) can be bound in qq ($\bar{q}\bar{q}$) diquarks (anti-). As for the color, a diquark is equivalent to an anti-quark and the anti-diquark is equivalent to a quark. A diquark-antidiquark meson is therefore pretty much the same as an ordinary meson, as for the strong force.

A $\mathbf{\bar{3}_c}$ spin zero diquark is antisymmetric in flavor, $\mathbf{\bar{3}_f}$, because of the Fermi statistics, as long as $q = u, d, s$. Therefore a four-quark system made up of two diquarks involves $\mathbf{3_f} \otimes \mathbf{\bar{3}_f}$ states, 9 states versus the 81 given before (a crypto-exotic) which is much better, although X is a 1^{++} state and two spin zero diquark cannot do the job.

The X should however contain two c quarks. The heavy quark Q is not indistinguishable from q and spin-spin interactions between an heavy quark and a light one are $O(1/M_c)$, so that, even if non-perturbative dynamics tends to favor the formation of a spin zero diquark (as it has been proved by lattice studies focused on light-light diquarks), an heavy-light diquark can be equally well spin zero or one and its flavor group structure is determined by the light quark only. Again 9 states, but with spin one. The other quantum numbers follow easily.

On the other hand, the number of ways of sewing quarks into a four-quark structure is not exhausted by the possibilities just described. Two $\mathbf{3_c}$ quarks can either attract or repel each other in the $\mathbf{\bar{3}_c}$ or $\mathbf{6_c}$ color channels according to the one-gluon exchange model (which qualitatively reproduces the lattice indications). According to the same model, a $\mathbf{6_c}$ diquark and a $\mathbf{\bar{6}_c}$ anti-diquark could form a color neutral object with the *same* binding energy of the $\mathbf{3_c}$–$\mathbf{\bar{3}_c}$ diquark–anti-diquark. This object looks like the non-abelian analog of a system of two electrons and two protons in some closed region: an H_2 molecule is formed as a result of the binding of two individual hydrogen atoms.

The one-gluon exchange model of the strong interactions in a hadron is just a qualitative oversimplification, yet it gives the feeling of how the molecule and four-quark languages could dangerously be interchangeable[1].

---

[1] The oscillating $c\bar{c}$-molecule picture is a smart refinement of the basic molecule description with a stronger adaptability to the sometimes adverse climatic conditions of the experimental situation.

2 Tracing differences

Building four-quark mesons made up of two diquarks requires 9 states: charged X's should be visible, as well as strange X_s states, according to $SU(3)$. An entire spectrum of these states has been calculated. Only one neutral non-strange X has been observed so far; similarly no charged partners of the higher mass X, Y, Z are observed. This is usually addressed as the weakest point in the tetraquark picture. But, (1) even if an attempt to calculate the X mass spectrum in the four-quark picture has been made, it is not at all easy to predict the widths of these nonet states, most of which could turn out to be very broad. (2) $D\bar{D}^*$ molecules could as well occur in 9 states, though it seems that binding potentials can be tuned to account for the 'reduced' observed spectrum.

At any rate, molecules are very loose bound states: consider for example that $m_D + m_{\bar{D}^*} = 3871.3 \pm 0.7$ MeV. Then we can expect that the typical size of such a molecule is $r \sim 1/\sqrt{2 M_X E_\mathrm{bind.}} \sim 3 - 4$ fm. Charm quarks have to recombine into a J/ψ (kind of ~ 0.25 fm object) starting from a configuration in which they are up to 4 times the typical range of strong interactions apart. In the tetraquark picture, instead, the c quarks are as close to each other as two quarks in a standard meson.

A $D\bar{D}^*$ molecule should have a decay width $X \rightarrow D^0 \bar{D}^0 \pi^0$ comparable to the $\Gamma(D^* \rightarrow D\pi) \sim 70$ KeV width. This decay mode has been very recently observed to occur at a rate about nine times larger than the $J/\psi\rho$ mode, in bold contradiction with the basic molecular picture where $J/\psi\rho$ was predicted to be by far the dominant one. The tetraquark X is allowed to decay $X \rightarrow D^0 \bar{D}^0 \pi^0$ with a rate almost two times larger than the $J/\psi\rho$. This experimental fact, if confirmed, seriously challenges both models.

All these semi-quantitative considerations are not definitive in deciding neatly between the two options: molecule or tetraquark? In many respects one could so far object that the two scenarios seem quite contiguous. But, in the tetraquark picture the $X(3872)$ has a 'double life', two different X's are required (call them X_l and X_h) to account for the observed isospin violation: $\mathcal{B}(X \rightarrow J/\psi\rho)/\mathcal{B}(X \rightarrow J/\psi\omega) = 1.0$. In what follows I will sketch the latter point.

Consider the states $X_\mathrm{u} = [cu][\bar{c}\bar{u}]$ and $X_\mathrm{d} = [cd][\bar{c}\bar{d}]$, where the square parentheses indicate a diquark binding. The B^+ could decay as $B^+ \rightarrow K^+ X_\mathrm{u}$ and $B^+ \rightarrow K^+ X_\mathrm{d}$ with some undetermined relative frequency. Let us call A_1 and A_2 the two decay amplitudes. Data on the production of $X(3872)$ in $B^+ \rightarrow K^+ X$ reasonably show that only one single state is produced in this channel. Therefore either $A_1 \gg A_2$ or $A_1 \ll A_2$. Whatever the actual situation is, a naive quark analysis shows that in $B^0 \rightarrow KX$, A_1 and A_2 would be interchanged: if X_u is produced in B^+ decay, then X_d is produced in B^0 decay and vice-versa.

Actually, the real X's can be superpositions of X_u and X_d. In a standard mixing scheme we can introduce two orthogonal superpositions, X_l and X_h, mixed by an angle θ. The annihilation diagrams describing $u\bar{u} - d\bar{d}$ transitions are reasonably quite small at the m_c scale so that we expect θ to be small. X_l and X_h are therefore unbalanced superpositions of $I = 1/2$ and $I = -1/2$ states (at $\theta = \pi/4$, X_l and X_h are $I = 0$ and $I = 1$ respectively; X_l could, e. g., only decay to $\omega J/\psi$) opening the way to isospin violations in the decays of X_l and X_h. On the other hand the $D\bar{D}^*$ molecule is per-se a single isospin impure state.

X_l and X_h are expected to have a difference in mass, ΔM, proportional to $(m_d - m_u)$ and inversely proportional to $\cos\theta$ (which can be fixed by decay data). Such a mass difference is under experimental study (the mass of the X produced in B^+ is confronted to the mass of X produced in B^0) but the error on data still does not allow to draw a definitive conclusion. A $\Delta M \sim O(1)$ MeV would clearly unveil the double life of X excluding the molecule (and all the way around).

Resolving this molecule-tetraquark dichotomy is not only a matter of taxonomy. Diquarks have an interesting role in QCD. An entire region of the QCD phase diagram in the (μ, T) plane has been found to exist in a phase of color superconductor where the analogous of the ordinary Cooper pairs are diquarks. Diquarks also enter in diverse QCD considerations. Just to mention one, recall for example the argument to explain the limit $F_2^n/F_2^p \longrightarrow 1/4$ as $x \to 1$ of the DIS structure functions of neutron and proton. Diquarks could also help to explain the fact that the Regge trajectories of mesons and baryons happen to have the same slope.

3 Counting Quarks

Obtaining direct experimental evidence that the X is a multiquark object would certainly be rather useful. A new method to investigate the quark nature of the X and of all those states missing a clear quark-identikit, like $f_0(500)$, $a_0(980)$, $f_0(980)\ldots$, could be obtained by the analysis of certain heavy-ion collision observables.

A stunning fact emerged at RHIC is that the number of protons divided by the number of pions counted in a $p_\perp$ region $1.5\,\mathrm{GeV} \leq p_\perp \leq 4\,\mathrm{GeV}$ is ≥ 1, against any expectation based on fragmentation functions which would predict an opposite pattern. In such experimental situation, fragmentation is insufficient at producing high $p_\perp$ hadrons. In the standard fragmentation picture, an off-shell parton loses energy via a perturbative 'shower' of gluon emissions until the energy scale of Λ_{QCD} is approached, where the non-perturbative domain opens. At this stage all the partons formed will get together in separated clusters tending to neutralize the color and generating, via some non-perturbative mechanism, a collection of real hadrons. The energy of the initial parton is repartitioned among all the final state hadrons produced. High $p_\perp$ hadrons in the final state must be originated by very high $p_\perp$ initial partons which, in usual conditions, are not abundantly produced: pQCD spectra are steeply falling with $p_\perp$. Moreover, the standard fragmentation function approach predicts that, for a generic parton a, $D_{a\to p}/D_{a\to\pi} \lesssim 0.2$ in the above $p_\perp$ range.

But, suppose that a rapidly expanding state, overflowed in phase space with partons, is created in a heavy-ion collision. Neighboring partons in phase space could be able to recombine into hadrons whose momenta are just the algebraic sums of the parton momenta involved. In this case we could state that $[p \text{ spectrum}] \sim \exp[-p_\perp^{(a)}/3]^3 \approx [\pi \text{ spectrum}] \sim \exp[-p_\perp^{(a)}/2]^2$, $p_\perp^{(a)}$ being a parton momentum; this is the essential point about the so called 'coalescence' picture.

Attempts have been made to device models of fragmentation/coalescence (f/c) and to calculate the $p_\perp$ dependence of certain experimental observables. One of these observables, the so called 'nuclear modification ratio', is:

$$R_{AA} = 1/N_{\text{coll}}(b=0)\left[\mathrm{d}N_{\mathrm{H}}(b=0)/\mathrm{d}^2p_\perp)|_{AA}/(\mathrm{d}N_{\mathrm{H}}/\mathrm{d}^2p_\perp)|_{pp}\right] ,$$

where N_{H} is the number of hadrons counted, b is the impact parameter of the heavy-ion collision ($b = 0$ means maximum centrality), AA labels nucleus-nucleus collision (pp for proton-proton) and N_{coll} is the number of nucleon-nucleon collisions occurred in AA. Such a quantity can be measured experimentally and calculated in a f/c model. The results are given in Fig. 1.

As shown, R_{AA} has the ability to discriminate between mesons and baryons, as baryons tend to be higher in R_{AA} than mesons. The curves are instead the result of a theoretical calculation in a f/c model.

Let us consider here the case of the $f_0(980)$ scalar meson which also evades any standard $q\bar{q}$ interpretation. Two possibilities are examined: the f_0 is (1) a $q\bar{q}$, (2) a diquark-antidiquark meson (a molecular picture in which the f_0 is a kind of $K\bar{K}$ molecule is as old as the discovery of the f_0 itself). The $R_{AA}(f_0)$ at RHIC has not yet been analyzed. We provide a couple of theoretical curves to eventually compare to data.The X will be produced at the LHC where an $R_{AA}(X)$ analysis might be performed.

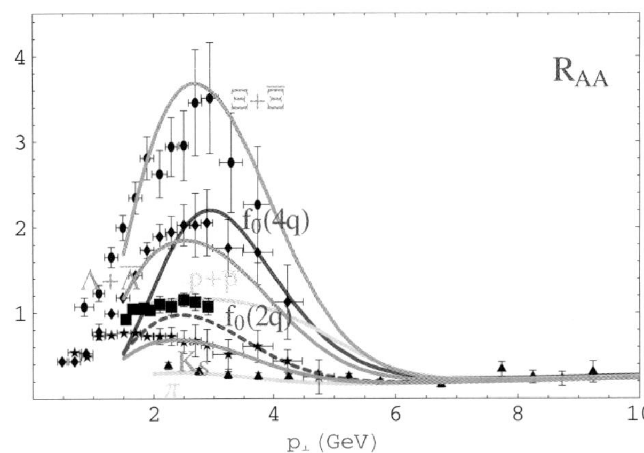

Fig. 1. The R_{AA} value as a function of $p_\perp$ for various hadrons. The *solid lines* are the theoretical results obtained in the f/c model

4 Conclusions

It would be an error if collaborations like BELLE and BaBar gave up the investigation of a possible $\Delta M \neq 0$ between the X produced in B^0 and B^+, or the search for charged X's. Clarifying the nature of the X and its 'similes' gives an opportunity to learn some new fundamental aspects of quantum chromodynamics.

Acknowledgement. I whish to thank L. Maiani for fruitful and enjoyable collaboration and R.L. Jaffe and R. Faccini for many stimulating discussions. I conclude by thanking the organizers O. Nicrosini, G. Montagna and C. Vercesi for their kind invitation and their valuable work.

References

1. *The X discovery*: S. K. Choi *et al.* [Belle], Phys. Rev. Lett. **91**, 262001 (2003) [arXiv:hep-ex/0309032]; D. Acosta *et al.* [CDF II], Phys. Rev. Lett. **93**, 072001 (2004); V. M. Abazov *et al.* [D0], Phys. Rev. Lett. **93**, 162002 (2004); B. Aubert *et al.* [BABAR], Phys. Rev. D **71**, 071103 (2005).
2. *Molecules*: M. B. Voloshin and L. B. Okun, JETP Lett. **23**, 333 (1976) [Pisma Zh. Eksp. Teor. Fiz. **23**, 369 (1976)]; N. A. Tornqvist, Phys. Rev. Lett. **67**, 556 (1991); F. E. Close and P. R. Page, Phys. Lett. B **578**, 119 (2004) [arXiv:hep-ph/0309253]; E. S. Swanson, Phys. Lett. B **588**, 189 (2004) [arXiv:hep-ph/0311229]; E. Braaten and M. Kusunoki, Phys. Rev. D **72**, 054022 (2005) [arXiv:hep-ph/0507163]; M. Suzuki, Phys. Rev. D **72**, 114013 (2005) [arXiv:hep-ph/0508258]; E. S. Swanson, Phys. Rept. **429**, 243 (2006) [arXiv:hep-ph/0601110].
3. *Diquarks & Tetraquarks*: R. L. Jaffe and F. Wilczek, Phys. Rev. Lett. **91**, 232003 (2003) [arXiv:hep-ph/0307341]; L. Maiani, F. Piccinini, A. D. Polosa and V. Riquer, Phys. Rev. Lett. **93**, 212002 (2004) [arXiv:hep-ph/0407017]; Phys. Rev. D **70**, 054009 (2004) [arXiv:hep-ph/0407025]; Phys. Rev. D **71**, 014028 (2005) [arXiv:hep-ph/0412098]; Phys. Rev. D **72**, 031502 (2005) [arXiv:hep-ph/0507062]; AIP Conf. Proc. **814**, 508 (2006) [arXiv:hep-ph/0512082]; arXiv:hep-ph/0604018; H. Hogaasen, J. M. Richard and P. Sorba, Phys. Rev. D **73**, 054013 (2006) [arXiv:hep-ph/0511039]; R. Jaffe, Phys. Rev. D **72**, 074508 (2005) [arXiv:hep-ph/0507149]; M. Karliner and H. J. Lipkin, R. D. Matheus, S. Narison, M. Nielsen and J. M. Richard,
4. *Lattice and Diquarks*: C. Alexandrou, Ph. de Forcrand and B. Lucini, arXiv:hep-lat/0609004.
5. *Hybrid*: F. E. Close and P. R. Page, Nucl. Phys. B **443**, 233 (1995) [arXiv:hep-ph/9411301]; E. Kou and O. Pene, Phys. Lett. B **631**, 164 (2005) [arXiv:hep-ph/0507119]; F. E. Close and P. R. Page, Phys. Lett. B **628**, 215 (2005) [arXiv:hep-ph/0507199].
6. *Counting Quarks*: L. Maiani, A. D. Polosa, V. Riquer and C. A. Salgado, arXiv:hep-ph/0606217.

Physics with Neutrino Beams

Mauro Mezzetto

INFN – Sezione di Padova

1 Introduction

Artificial neutrino beams had been first introduced in high energy physics at Brookhaven in the 60's with the classical experiment that led to the discovery of the two neutrino families [1]. The first neutrino beam setup as we know today was realized at CERN in the 70's, and led to another milestone in h.e.p.: the discovery of the weak neutral currents [2]. Since then neutrino beams had been widely used to measure the electroweak parameters, structure functions, neutrino cross sections etc.

It is from the 90's that neutrino beams focus on neutrino oscillations.

Muon neutrino beams are produced through the decay of π and K mesons generated by a high energy proton beam hitting needle-shaped light targets. Positive (negative) mesons are sign-selected and focused (defocused) by large acceptance magnetic lenses into a long evacuated decay tunnel where ν_μ's ($\overline{\nu}_\mu$'s) are generated. In case of positive charge selection, the ν_μ beam has typically a contamination of $\overline{\nu}_\mu$ at few percent level (from the decay of the residual π^-, K^- and K^0) and $\sim 1\%$ of ν_e and $\overline{\nu}_e$ coming from three-body $K^\pm$, K_0 decays and μ decays.

2 Neutrino Oscillations

The neutrino oscillation probability depends on 3 mixing angles, $\theta_{12}, \theta_{23}, \theta_{13}$, 2 mass differences, $\Delta m_{12}^2 = m_2^2 - m_1^2$, $\Delta m_{23}^2 = m_3^2 - m_2^2$, and a CP phase δ_{CP}. Furthermore, the neutrino mass hierarchy, the ordering with which mass eigenstates are coupled to flavor eigenstates, can be fixed by measuring the sign of Δm_{23}^2.

Two experimental parameters are relevant for neutrino oscillations: the neutrino energy E_ν and the baseline L (distance of the neutrino source from the detector); in the oscillation formulas they are combined into the L/E_ν ratio.

Table 1. Neutrino oscillation parameters [3]

'solar parameters'	$\Delta m_{12}^2 = (7.92 \pm 0.72) \cdot 10^{-5}$ eV^2	$\sin^2 \theta_{12} = 0.314^{+0.030}_{-0.025}$
'atmospheric parameters'	$\Delta m_{23}^2 = \pm(2.4^{+0.5}_{-0.6}) \cdot 10^{-3}$ eV^2	$\sin^2 \theta_{23} = 0.44^{+0.18}_{-0.10}$
still unknown	$\theta_{13}, \delta_{CP}, \mathrm{sign}(\Delta m_{23}^2)$	

The present values of oscillation parameters are summarized in Table 1.

The phenomenon of CP (or T) violation in neutrino oscillations manifests itself by a difference in the oscillation probabilities of say, $P(\nu_\mu \to \nu_e)$ vs $P(\overline{\nu}_\mu \to \overline{\nu}_e)$(CP violation), or $P(\nu_\mu \to \nu_e)$ vs $P(\nu_e \to \nu_\mu)$ (T violation).

The leading parameter of $P(\nu_\mu \to \nu_e)$ is θ_{13}, whose value is still unknown. It is of full evidence that the first priority of future neutrino experiments will be to measure the value of θ_{13} by looking for experimental evidence of ν_e appearance in $\nu_\mu$beams. The present limit $\sin^2 2\theta_{13} \leq 0.09$, coming from a fit of all the neutrino oscillation experiments [4], translates into a $\nu_\mu \to \nu_e$ appearance probability less than 10% at the appearance maximum in a high energy muon neutrino beam.

3 The recent past

The recent past of neutrino beam experiments has been characterized by the search of $\nu_\mu \to \nu_\tau$ oscillations at CERN and by the LSND experiment at LAMPF.

The search for $\nu_\mu \to \nu_\tau$ oscillations at CERN had been the last act of a a-posteriori loosing strategy to search for neutrino oscillations. While neutrino beams experiment could explore a wide range of oscillation amplitudes in a restricted range of Δm^2 values, atmospheric neutrino experiments were capable to explore a wide range of Δm^2 for very large amplitude values only. At the time a very strong prejudice against large oscillation amplitudes was radicated due to the small mixing values in the quark sector, motivating the Chorus and Nomad experiments looking for $\nu_\mu \to \nu_\tau$ oscillation [5, 6] and $\nu_\mu \to \nu_e$ oscillations [7] for $\Delta m^2 > 10$ eV^2.

The LSND experiment reported evidence of $\overline{\nu}_\mu \to \overline{\nu}_e$ oscillations with a Δm^2 of 0.3–20 eV^2 detecting a $\sim 4\sigma$ excess of $\overline{\nu}_e$interactions in a neutrino beam produced by π^+ decays at rest where the $\overline{\nu}_e$component is highly suppressed ($\sim 7.8 \times 10^{-4}$) [8]. The KARMEN experiment [9], with a very similar technique but with a lower sensitivity (a factor 10 less for the lower Δm^2), did not confirm the result: a combined fit of the two experiments still exhibits a sizable signal region [10]. The LSND result doesn't fit the overall picture of neutrino oscillations and several non-standard explanations, as for instance sterile neutrinos, have been put forward to solve this experimental conflict.

4 Present generation of long-baseline experiments

The MiniBooNE experiment at the FNAL Booster, presently taking data, is designed to settle the LSND puzzle with a $3-5\sigma$ sensitivity [11] by looking for $\nu_\mu \to \nu_e$ transitions in a ν_μ beam of $0.7\,\text{GeV}$ average energy.

K2K at KEK [12] and MINOS [13] at the NuMI beam from FNAL are designed to confirm the atmospheric evidence of oscillations and measure $\sin^2 2\theta_{23}$ and $|\Delta m^2_{23}|$ within 10–15% of accuracy by measuring the ν_μ survival probability as a function of the neutrino energy.

OPERA [14] at the CNGS beam from CERN to LNGS will search for evidence of ν_τ interactions in a pure ν_μ beam, the final proof of $\nu_\mu \to \nu_\tau$ oscillations.

K2K has already published the final data analysis [12], while MINOS has started data taking beginning 2005 and recently shown preliminary results [15], Fig. 1-left. CNGS started operations July 2006.

Present generation experiments can look for $\nu_\mu \to \nu_e$ even if they are not optimized for θ_{13} studies. MINOS is expected to reach a sensitivity of $\sin^2 2\theta_{13} = 0.08$ (90% C.L.), OPERA $\sin^2 2\theta_{13} = 0.06$ [16]. A sketch of θ_{13} sensitivities as a function of the time, following the schedule reported in the experimental proposals, computed for the approved experiments, is reported in Fig. 1-right.

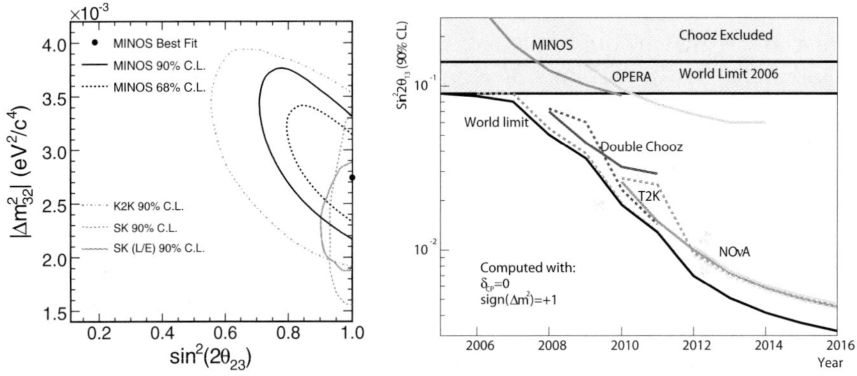

Fig. 1. *Left*: Fits to the atmospheric parameters from Super-Kamiokande, K2K and Minos (from [15]). *Right*: evolution of sensitivities on $\sin^2 2\theta_{13}$ as function of time. For each experiment are displayed the sensitivity as function of time (*solid line*) and the world sensitivity computed without the experiment (*dashed line*). The world overall sensitivity along the time is also displayed

5 Next future

The next future generation of long baseline experiments will focus on ν_e appearance searches optimized for the measure of θ_{13}.

They will use neutrino beams with the axis tilted by a few degrees with respect to the position of the far detector (off-axis beams) [17, 18]. These neutrino beams, optimized for the two body π decays, have several advantages with respect to the corresponding on-axis ones: they are narrower, lower energy and with a smaller ν_e contamination (since ν_e mainly come from three body decays) although the neutrino flux can be significantly smaller.

The T2K experiment [17] will aim neutrinos at an angle of 2.5° from the direction of the Super-Kamiokande detector (295 km away), assuring a ν_μ peak energy of 0.6 GeV. The beam line is equipped with a set of dedicated on-axis and off-axis detectors at the distance of 280 meters from the target. The main goal of the experiment is the search for ν_e appearance to detect $\nu_\mu \to \nu_e$ oscillations with a sensitivity of $\sin^2 2\theta_{13} = 0.006$ (90% C.L.). Furthermore the disappearance measurements of ν_μ will improve measurement of Δm_{23}^2 down to a precision of about $10^{-4}\,\mathrm{eV}^2$. Neutral current disappearance (in events tagged by π° production) will allow for a sensitive search of sterile neutrino production. T2K is planned to start in 2009 with a beam intensity reaching 1 MW beam power on target after a couple years.

The phase II of the experiment, often called T2HK, foresees an increase of beam power up to 4 MW, antineutrino runs, and a very large water Čerenkov detector, Hyper-Kamiokande, with a rich physics programme in its own like proton decay, atmospheric and supernova neutrinos etc.

The NOνA experiment [19] with an upgraded NuMI off-axis neutrino beam ($E_\nu \sim 2$ GeV and a ν_e contamination lower than 0.5%) and with a baseline of 810 km (12 km Off-Axis), has been recently proposed at FNAL. The expected proton intensity is 6.5×10^{20} pot/year. In a 5 years ν_μ run, with 30 kton liquid scintillator far detector, it could reach a sensitivity on $\sin^2 2\theta_{13}$ very similar than T2K, as well as a precise measurement of $|\Delta m_{23}^2|$ and $\sin^2 2\theta_{23}$. NOνA can also allow to solve the mass hierarchy problem in a limited range of the θ_{13} and δ_{CP} parameters [19].

6 Long term future

T2K and NOνA will certainly improve a lot the explored range of θ_{13}, but it's clear from now they will have very reduced sensitivity in measuring both δ_{CP} and $\mathrm{sign}(\Delta m_{23}^2)$ [20]. The role of thumb for experiments capable to perform a sensitive search for leptonic CP violations will be to accumulate $\mathcal{O}(100)$ more neutrinos than T2K. This can be done with upgrades of existing facilities, as in the case of T2HK, of with new very intense conventional neutrino beams, as in the case of the CERN SPL neutrino beam [21, 22]. These experiments could reach a sensitivity of $\sin^2 2\theta_{13} \simeq 10^{-3}$ (3σ) and be sensitive to δ_{CP} if $\theta_{13} \geq 3 \cdot 10^{-3}$, limited by their intrinsic ν_e contamination with its associated systematic errors [23].

These limitations are overcome if the neutrino parents can be fully selected, collimated and accelerated to a given energy. This can be attempted
within the muon or a beta decaying ion lifetimes. The neutrino beams from
their decays would then be pure and perfectly predictable. The first approach
brings to the Neutrino Factories, the second to the BetaBeams.

6.1 Beta Beams

BetaBeams have been introduced by P. Zucchelli in 2001 [24]. The idea is
to generate pure, well collimated and intense ν_e ($\overline{\nu}_e$) beams by producing,
collecting, accelerating radioactive ions and storing them in a decay ring. The
resulting BetaBeam fluxes could be easily computed by the properties of the
beta decay of the parent ion and by its Lorentz boost factor γ and would be
virtually background free. The best ion candidates so far are ^{18}Ne and ^{6}He
for ν_e and $\overline{\nu}_e$ respectively. The schematic layout of a Beta Beam is shown in
Fig. 2-left.

A baseline study for a Beta Beam complex has been carried out at
CERN [25]. The neutrino beam could be fired to a 0.5 Mt water Čerenkov de-
tector, Memphys [26], that could be located under Frejus 130 km away from
CERN.

The most updated sensitivities for the baseline Beta Beam are computed
in a scheme where both ions are accelerated at $\gamma = 100$, the optimal setup
for the CERN-Frejus baseline, [23, 27]. BetaBeams could reach a sensitivity
of $\sin^2 2\theta_{13} \simeq 4 \cdot 10^{-4}$ (3σ) and be sensitive to δ_{CP} if $\theta_{13} \geq 10^{-3}$.

Exciting new concepts of Beta Beams have been proposed in literature:
high energy Beta Beams based on accelerators more powerful than SPS [28],
new innovative schemes of radioactive ion productions [29], experiments based
on different ions than ^{6}He and ^{18}Ne [30] and monochromatic neutrino beams
based on the electron capture process [31].

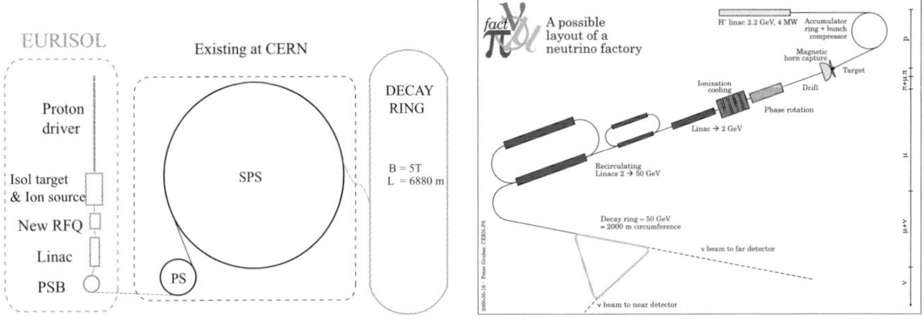

Fig. 2. *Left*: A schematic layout of the BetaBeam complex. At *left*, the low energy
part is largely similar to the EURISOL project. The central part (PS and SPS) uses
existing facilities. At *right*, the decay ring has to be built. *Right*: Expected layout
for a neutrino factory at CERN

6.2 Neutrino Factories

The neutrino production by muon decay from a pure muon beam has been considered since 1998 [32]: this is indeed a perfectly well known weak process and the μ beam can be well measured in momentum and intensity.

The CERN present layout for a Neutrino Factory (νF) [33] is sketched in Fig. 2-right. The decay $\mu^+ \to e^+ \nu_e \overline{\nu}_\mu^- \to e^- \overline{\nu}_e \nu_\mu)$ produces a pure well collimated neutrino beam with equal numbers of $\overline{\nu}_\mu$, $\nu_e(\nu_\mu, \overline{\nu}_e)$. Storing $50\,\text{GeV}/c$ muons in the decay ring it will be possible to extend the baseline to several thousand kilometers of distance.

The search for $\nu_e \to \nu_\mu$ transitions ("golden channel") appears to be very attractive at νF, because this transition can be studied in appearance mode looking for μ^- (appearance of wrong-sign μ) in neutrino beams where the neutrino type that is searched for is totally absent (μ^+ beam in νF). With a $40\,\text{Kt}$ magnetic detector (MINOS like) exposed to both polarity beams and 10^{21} muon decays, it will be possible to explore $\sin^2 2\theta_{13}$ down to 10^{-5} (3σ) and δ_{CP} if $\theta_{13} \geq 10^{-4}$ [34].

References

1. G. Danby *et al.*, Phys. Rev. Lett. **9**, 36 (1962).
2. F. J. Hasert *et al.* [Gargamelle Neutrino Collaboration], Phys. Lett. B **46**, 138 (1973).
3. G. L. Fogli, E. Lisi, A. Marrone and A. Palazzo, Prog. Part. Nucl. Phys. **57**, 742 (2006).
4. T. Schwetz, Acta Phys. Polon. B **36** (2005) 3203
5. E. Eskut *et al.* [CHORUS Collaboration], Phys. Lett. B **424** (1998) 202 and Phys. Lett. B **434** (1998) 205.
6. P. Astier *et al.* [NOMAD Collaboration], Nucl. Phys. B **611** (2001) 3
7. P. Astier *et al.* [NOMAD Collaboration], Phys. Lett. B **570** (2003) 19
8. A. Aguilar *et al.* [LSND Collaboration], Phys. Rev. D **64** (2001) 112007
9. B. Armbruster *et al.* [KARMEN Collaboration], Phys. Rev. D **65** (2002) 112001
10. K. Eitel, New J. Phys. **2** (2000) 1
11. E. Church *et al.* [BooNe Collaboration], nucl-ex/9706011.
12. M. H. Ahn *et al.* [K2K Collaboration], hep-ex/0606032.
13. E. Ables *et al.* [MINOS Collaboration], Fermilab-proposal-0875
14. [OPERA Collaboration], CERN-SPSC-P-318, LNGS-P25-00; CERN-SPSC-2000-028.
15. [MINOS Collaboration], hep-ex/0607088.
16. M. Komatsu, P. Migliozzi and F. Terranova, J. Phys. G **29** (2003) 443 P. Migliozzi and F. Terranova, Phys. Lett. B **563** (2003) 73
17. Y. Itow *et al.*, hep-ex/0106019.
18. The E889 Collaboration, BNL Report No. 52459. A. Para and M. Szleper, hep-ex/0110032.
19. D. S. Ayres *et al.* [NOvA Collaboration], hep-ex/0503053.
20. P. Huber *et al.*, Nucl. Phys. Proc. Suppl. **145** (2005) 190

21. B. Autin *et al.*, CERN-2000-012.

22. J. J. Gomez-Cadenas *et al.*, Proceedings of "Venice 2001, Neutrino telescopes", vol. 2*, 463-481, hep-ph/0105297. A. Blondel *et al.*, Nucl. Instrum. Meth. A **503** (2001) 173. M. Mezzetto, J. Phys. G **29** (2003) 1771. J. E. Campagne and A. Cazes, Eur. Phys. J. C **45**, 643 (2006)

23. J. E. Campagne, M. Maltoni, M. Mezzetto and T. Schwetz, hep-ph/0603172.

24. P. Zucchelli, Phys. Lett. B **532** (2002) 166.

25. B. Autin *et al.*, physics/0306106. M. Benedikt, S. Hancock and M. Lindroos, Proceedings of EPAC 2004, http://accelconf.web.cern.ch/AccelConf/e04.

26. A. de Bellefon *et al.* hep-ex/0607026.

27. M. Mezzetto, J. Phys. G **29** (2003) 1771. J. Bouchez, M. Lindroos, M. Mezzetto, AIP Conf. Proc. **721** (2004) 37. M. Mezzetto, Nucl. Phys. Proc. Suppl. **155** (2006) 214.

28. J. Burguet-Castell *et al.*, Nucl. Phys. B **695** (2004) 217. J. Burguet-Castell *et al.*, Nucl. Phys. B **725**, 306 (2005)

29. C. Rubbia, A. Ferrari, Y. Kadi and V. Vlachoudis, hep-ph/0602032.

30. A. Donini and E. Fernandez-Martinez, hep-ph/0603261. C. Rubbia, hep-ph/0609235.

31. J. Bernabeu *et al.*, hep-ph/0505054; J. Sato, hep-ph/0503144.

32. S. Geer, Phys. Rev. D **57** (1998) 6989 [Erratum-ibid. D **59** (1999) 039903],

33. M. Apollonio *et al.*, hep-ph/0210192. A. Baldini *et al.* [BENE Steering Group], CERN-2006-005.

34. J. Burguet-Castell *et al.*, Nucl. Phys. B **608** (2001) 301; P. Huber, M. Lindner, M. Rolinec and W. Winter, hep-ph/0606119.

Status and perspectives of Dark Matter and Astroparticle searches

Oliviero Cremonesi

Sez. di Milano and Università degli studi di Milano Bicocca,
20126 Milano – Italy
oliviero.cremonesi@mib.infn.it

Present status and future perspectives of astroparticle experiments are reviewed. Possible strategies for the next decade are also briefly outlined in particular for Dark Matter and Neutrino Physics searches.

1 Introduction

Astroparticle Physics can be considered as a perfect symbiosis of methods and interests between particle physics, astronomy and cosmology. In this interdisciplinary environment in which the borders between a given discipline and the other are somewhat faint, the assignment of certain types of experiments to either astroparticle physics, or particle physics or cosmology are often debatable. The experiments belonging to specific disciplines (cosmic rays, dark matter, neutrino physics, etc.) has been historically considered part of this field. Now the attribution criteria has been enlarged, to include all the searches seeking answers for a number of basic questions such as: i) what is the Universe origin and composition; ii) what are the properties of neutrinos and what is their role in cosmic evolution; iii) what can neutrinos tell us about the internal structure and evolution of the Earth, stars and cosmological structures; iii) what are the stability properties of ordinary matter; iv) what is the explanation for the observed asymmetry of the Universe; v) what is the origin of cosmic rays; vi) what is the view of the sky at different wavelengths and with different probes (multi-messenger analysis); vii) what is the nature of gravity; viii) are gravitational waves detectable; etc.

While allowing a better understanding of the Universe origin, structure and evolution, the answers to these questions could mark a major breakthrough in our understanding of the infinetely small universe of particle properties.

2 Dark Matter

One of the most relevat questions concerning Astroparticle Physics is the Universe composition. Only 4% of the Universe seems actually made of ordinary matter, while 73% of it seems to consist of "dark energy" and 23% of Dark Matter (DM) clustered around cosmic objects (galaxies and clusters/superclusters of galaxies) and influencing or sharing their evolution. Several particle candidates can be considered for this sizeable dark component. The simplest solution is based on the assumption of Weakly Interacting Massive Particles (WIMPs), produced in the Early Universe, whose most natural candidate is the neutralino, the lightest super-symmetric particle. As many other exotic particles suggested as possible DM candidates, WIMPS can be also searched for at LHC experiments [1], even if in this case the evidence for super-symmetric particles would not imply their existence as Dark Matter.

The particle structure of the dark halo embedding our galaxy can be detected with both direct and indirect methods [2]. Direct methods are mainly based on the observation (recoil products) of WIMPs scattering off target nuclei in deep underground detectors. The annual modulation (due to the movement of the Earth) of the rate of these signals together with the A-dependence of signal rate and shape, and a possible directional signature (due to the movement of the Sun through the galactic halo) are the only available signatures for the WIMP character of the observed signals. An annual modulation of the event rate has been actually reported by the DAMA group in Gran Sasso [3]. The DAMA detector (100 kg of NaI scintillating crystals) has collected an exposure larger by orders of magnitude with respect to competing experiments and is continuing data acquisition with DAMA/LIBRA, an improved version totalling 250 kg. The DAMA observed signal modulation is by itself a model independent signature even if, interpreting the signal as a standard neutralino, would lead to contradictions with limits obtained by other experiments. The verification of the DAMA signal should one of the primary goals of the future program of Dark Matter searches. Besides DAMA/LIBRA upcoming results, the presently ongoing experiment ANAIS (at a different site in Spain) might provide a valuable cross check.

A number of experiments aiming at the direct observation of WIMPs candidates have been developed in recen years. By using powerful experimental techniques to isolate recoil events, they have reached unprecedented sensitivities. It should be stressed however that they lack (so far) any correlation with the dark halo components and are unfortunately completely blind to other possible candidates characterized by interactions other than elastic scattering.

Cryogenic detectors are characterized by thresholds of few keV and an excellent background suppression. With a mass of order one ton they could cover an important fraction of the parameter space of existing models and eventually be sensitive to WIMPs with an interaction cross section as low as 10^{-10} pb. Present best limits are of the order of few 10^{-7} pb (CDMS in the USA) and are expected to improve by an order of magnitude within the next

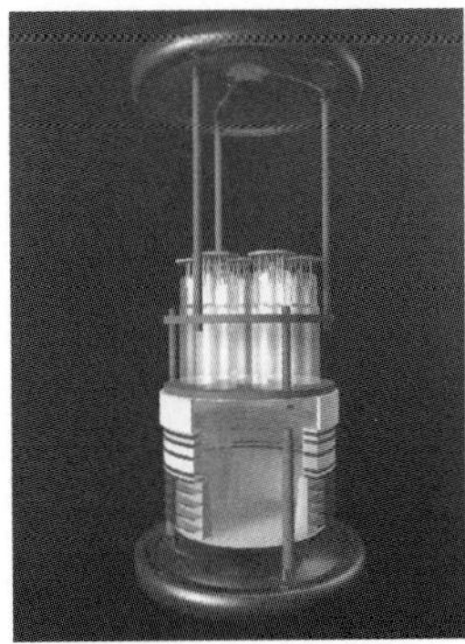
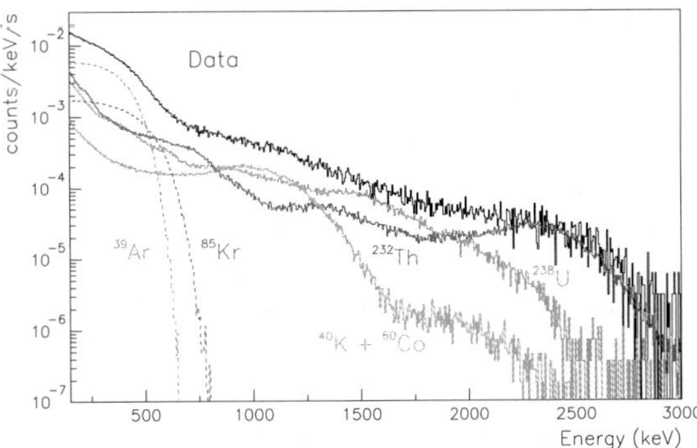

Fig. 1. Scheme of the WArP 2.3 l prototype with the observed background energy spectrum interpretation

two years. Two other advanced projects also using bolometric techniques are presently taking data at Gran Sasso Laboratory (CRESST) and in the Fréjus Tunnel (EDELWEISS). They could eventually converge in the next future to a single project on the scale of a few hundred kg to one ton (EURECA).

Another rather promising approach is based on noble liquid techniques to identify low iolnizing events mainly by a comparison of the primary and secondary scintillation signals: ZEPLIN (Boulby mine, UK) and XENON (to be installed in Gran Sasso) use Xenon while WArP (Gran Sasso) and ArDM would explore the feasibility of Argon. Relevant results on background level and composition, quenching factor and efficiency of the suppression method have been in particular obtained by the WARP collaboration in a series of measurements carried out at LNGS with a 2.3 liters prototype [4] (Fig. 1). A 100 liter WArP is under construction and future larger realizations can be foreseen to improve the experimental sensitivity to DM candidates. The WArP detector will be provided with a highly efficient active shield allowing to tag and measure the neutron induced background and is expected to come into operation during 2007 with a sensitivity of the order of 10^{-8} pb.

Noble liquid techniques could provide a complementary path to reach detectors with a ton-scale and should also converge towards a single proposal for a large-scale facility with 10^{-10} pb sensitivity.

A 1-ton DAMA type detector (10 times larger than the original DAMA but much cheaper than 1-ton cryogenic detectors), is an additional option when first conclusions from DAMA/LIBRA will be drawn.

Although not yet explored, the possibility to use the WIMPs directional dependence has been recently taken into account (DRIFT project in the Boulby mine). One possibility would be provided by the detection of a clear signal in a number of non-directional, large mass detectors. Further development of this technique are therefore important.

The progress made over the last few years is impressive, suggesting that there is a significant chance to detect WIMPs in the next decade.

Indirect evidence for WIMPs can be obtained by observing the effects of their interactions in astrophysical objects with the help of gamma telescopes, space based cosmic ray detectors and neutrino detectors. This attempt is complementary to the direct methods and, for some DM models, even superior.

A number of DM candidates other than WIMPs have been theoretically proposed. In some cases (notably CAST for the axion) dedicated experiments have been constructed. The search for thesse exotic components should be a relevant part of the future program of DM esarches.

One of the most important present problems in Cosmology is the nature of dark energy, which can be explored through its influence on cosmic evolution. Such kind of observations are traditionally based on astronomical techniques without contributions from the particle physics area. because of its relevance and strict connection with the Astroparticle questions, particle physicists have however recently joined this new field in which they wish to play a relevant role. Early projects for the detection of such an evanescent cosmic component are now flourishing.

3 Neutrino properties

By demonstrating the finiteness of neutrino mass and mixing, neutrino oscillations have provided us with the first clear evidence of phenomena beyond the reach of Standard Model (SM). Other questions concerning neutrino properties have remained however unsolved and become thereafter subject of increasing interest as a unique tool to see what new Physics lies beyond SM predictions. Although we know in fact that neutrinos are massive we still ignore the absolute scale and ordering of their masses as well as their Dirac/Majorana nature.

Only kinematic measurements of the β spectrum end-point, neutrinoless double beta decay and cosmological measurements can give direct informations on the neutrino mass absolute scale (actually matter effects in neutrino oscillations could give access to the mass difference sign). They all measure in fact different combinations of the neutrino mass eigenvalues: $\langle m_\beta \rangle = \sqrt{\sum_{i=1}^{3} |U_{ei}|^2 m_i^2}$, $\langle m_{\beta\beta} \rangle = \sum_{i=1}^{3} |U_{ei}|^2 m_i^2 \epsilon_i^{CP}$ and $\langle m_{\mathrm{Cosm}} \rangle = \sum_{i=1}^{3} m_i$.

$\langle m_{\mathrm{Cosm}} \rangle$ is constrained to values from 0.17 to 2.3 eV from recent cosmological observations [5]. Although cosmological values are more constraining than upper limits of 2.2 eV for $\langle m_\beta \rangle$ obtained so far in experiments on single-beta decay, they are strongly model dependent and therefore less robust with respect to laboratory measurements. On the other hand the best sensitivity expected on $\langle m_\beta \rangle$ for next generation experiments is of the order of ~ 0.2 eV (KATRIN [6], Karlsruhe) and can just aim at studying the degenerate mass hierarchy. A positive effect measured by KATRIN would mildly violate limits obtained from present precision cosmology but would certainly challenge more rigid upper limits like those expected from the PLANCK satellite. On the other hand, bolometric techniques to measure the electron spectrum (Re μ–bolometers [7]) do not suffer from the principle limitations of the KATRIN technique but seem to have not yet reached their technological limit. After a proper technological R&D they could eventually go beyond the projected sensitivity of KATRIN.

Neutrinoless Double Beta Decay searches play a unique role giving the possibility to probe the Majorana character of neutrinos while obtaining informations on the neutrino mass hierarchy and Majorana phases. If neutrinos are Majorana particles more stringent constraints, or a positive value for the effective neutrino mass, can in fact be obtained [8,9]. No evidence has been obtained for $\beta\beta(0\nu)$ so far, with the only exception of a claim of a subset of the Heidelberg-Moscow Collaboration headed by Klapdor-Kleingrothaus [10,11] (KDKC, corresponding to a $\langle m_{\beta\beta}\rangle$ value of the order of $400\,\mathrm{meV}$) which could need however a deeper and more statistically significant verification, possibly through the study of candidate isotopes other than ^{76}Ge. Existing experiments like CUORICINO and NEMO-3 are exploring $\langle m_{\beta\beta}\rangle$ values of the order of $500\,\mathrm{meV}$, belonging to the range of the degenerate hierarchy. They could address (but not fully disprove) the KDKC positive claim. Next generation experiments are GERDA/MAJORANA (^{76}Ge), CUORE (^{130}Te), EXO (^{130}Xe) and Super-NEMO (mainly ^{100}Mo) [8]. GERDA-I and CUORE are presently under construction at Gran Sasso. GERDA-I aims at validating the KDKC claim within 2010 while CUORE could probe the inverted hierarchy starting from 2011. CUORE will consist of 988 natural TeO_2 bolometers arranged in a cylindrical configuration of 19 towers (each made of a stack of thirteen, 4-detector modules) and exploiting the bolometric technique for a calorimetric approach. Each bolometer will consist of a temperature sensor (NTD thermistor) glued to a cubic $5 \times 5 \times 5\,\mathrm{cm}^3$ TeO_2 crystal with a mass of about $750\,\mathrm{g}$. Due to its high transition energy ($252\,8.8 \pm 1.3\,\mathrm{keV}$), the favourable nuclear matrix elements and the large natural isotopic abundance (33.8%) ^{130}Te is in fact one of the best candidates for DBD searches and allows to perform a sensitive experiment even using natural tellurium. CUORE expected sensitivities (extrapolated from CUORICINO results) are in the range of 20–$100\,\mathrm{meV}$ for $\langle m_{\beta\beta}\rangle$.

ν's are also a very important tool to probe Universe structures. Future measurements with neutrinos from the Sun, supernovae or other astrophysical objects, coupled with those generated in the Earth's interior and atmosphere besides providing a deeper understanding of their sources, will give improved information on the neutrino mixing and properties. In particular, precise measurements of the low-energy part of the solar neutrino spectrum could improve our understanding of neutrino oscillations, allow a fine-tuning of our picture of nuclear fusion processes inside the Sun and help understanding its long-term variations. On the other hand, supernovae neutrinos would give detailed insight in the catastrophic processes governing the explosion of supernovae and provide the best sensitivity to many intrinsic properties of neutrinos.

Precision data on neutrino mixing, in particular the mixing angle θ_{13} and the CP-violating phase δ, are expected from dedicated experiments with neutrinos generated in reactors (e. g. *Double CHOOZ* and *Daya Bay*) and in accelerators (T2K and Noνa).

4 Cosmic rays origin and properties

After their discovery nearly a century ago, cosmic rays have soon shown energies a hundred million times larger than those available at terrestrial accelerators. Since then, their observation has raisen several questions concerning their possible origin and nature, and their propagation properties. In particular, physicists are still puzzled by the cosmic accelerating processes able to boost particles to the observed extremely large energies. Great interest is moreover risen by the possibility that the cosmic ray spectrum can extend beyond the maximum allowed energy for a proton travelling large cosmic distances in a sea of microwave background radiation (CZK cutoff). A finite flux above this limit should in fact definitely sign new cosmic phenomena.

Cosmic rays unsolved questions [12] are going to be answered by an interplay of detectors for high energy gamma rays [13], neutrinos [14] and charged cosmic rays [15].

High-energy cosmic rays in the "knee region" between a few 10^{14} and a few 10^{16} GeV continue being studied by a number of air-shower experiments. On the other hand, cosmic rays below this region are studied by balloon (e. g. TRACER and CREAM) and satellite detectors (e. g. PAMELA and AMS2) while the upper region, extending up to 10^{18} GeV, presently domain of the Auger detector, will be partially covered by square kilometre air shower detectors like KASCADE-Grande, TUNKA-133 (Siberia) and IceTop (South Pole).

High energy neutrinos can provide an uncontroversial proof of the hadronic character of the source and reach us from cosmic regions which cannot be escaped by other types of radiation. The reason for their interest is therefore obvious. Pioneering neutrino telescopes (NT200 in Lake Baikal and AMANDA at the South Pole) have shown the viability of the technique without reaching unfortunately the sensitivities required for a direct observation of a distinctive signal. Larger size projects (km^3) are now being proposed. Successor of AMANDA, IceCube is presently being constructed under the South Pole. Prototype underwater detectors (ANTARES, NEMO and NESTOR) are on the other hand producing the first results under the Mediterranean sea. They all aim at a common KM3 project.

Finally, high-energy gamma-ray astronomy is showing (H.E.S.S. and MAGIC) that high-energy phenomena are ubiquitous in the sky. New Cherenkov Telescope Array (CTA) are being developed to boost the sensitivity by about an order of magnitude and overlap the GLAST satellite operation around 2010.

5 Gravitational waves

While negligible at microscopic scales, gravitation governs the large scale behaviour of the Universe. The emission of gravitational waves (GW) from accelerated masses is one of the central predictions of the theory of General Relativity. Their observation would represent a fundamental test of the theory and would provide fundamental information on strong field gravity around black holes and astrophysical systems. GW's could moreover represent a cosmological probe, in particular to test the evolution of dark energy.

GWs detectors include interferometers with broad-band sensitivity as well as resonant detectors [16]. At present, the world s most sensitive interferometer is LIGO (USA), the other interferometers being VIRGO in Italy, GEO600 in Germany and TAMA in Japan. Future efforts should include a quasi-continuous observations program together with a constant upgrade of the existing detectors as well as the design and construction of new ones.

References

1. M.Drees: hep-ph/0210142
2. L. Baudis: astro-ph/0511805
3. R. Bernabei et al.: La Rivista del Nuovo Cimento **26** 1 (2003) (astro-ph/0307403)
4. P. Benetti et al.: astro-ph/0603131
5. Fogli et al., hep-ph/0608060
6. V.M. Lobashev: Nucl. Phys. A **719** 153 (2003)
7. C. Arnaboldi et al., MARE: Microcalorimeter Arrays for a Rhenium Experiment. Milano Internal Note, Spring 2005
8. O. Cremonesi: Proc. of *Lepton and Photon 2005*, ed. by R.Brenner et al. (World Scientific 2006) pp n310–323
9. K. Zuber, Acta Phys. Pol. B **37** 1905 (2006)
10. H.V. Klapdor-Kleingrothaus, A. Dietz, I.V. Krivosheina, O. Chkvorets: Nucl. Instr. and Meth. A **522** 371 (2004)
11. H.V. Klapdor-Kleingrothaus, A. Dietz, I.V. Krivosheina, Ch. Doerr, C. Tomei: Lett. B **578** 54 (2004)
12. T.K. Gaisser astro-ph/0501195; E. Zas astro-ph/0103370
13. T. Weekes, astro-ph/0508253; H. Voelk, astro-ph/0401122 and astro-ph/0603501
14. C. Spiering, Rev. Sci. Inst. 75, 293 (2004) [astro-ph/0311343]
15. A. Watson, astro-ph/0511800
16. J. Hong, S. Rowan, B.S. Sathyaprkash: gr-qc/0501007

Future Perspectives of High Energy Experimental Physics and the Role of INFN

Umberto Dosselli

INFN Sezione di Padova, Via F. Marzolo, 8 – I35131 Padova, Italy
umberto.dosselli@pd.infn.it

The assessment of the future of HEP starts from what the field has achieved so far; the last fifty years have seen an exceptional number of discoveries whose focal point can be well represented by the Standard Model. In fact, instead of the many dozens of different elements that characterize the macroscopic world of the chemical elements, the HEP field has reached a remarkable degree of synthesis with the Standard Model, capable to describe the entire Universe with an handful of elementary particles and forces. Unfortunately we now know that this wonderful scheme describes only less than 5% of the known Universe, the remaining 95% being attributed to something that today generically we call "Dark Energy" and "Dark Matter" . And is the understanding of the existence of this large area where the map still reports *hic sunt leones* and that we have to describe in terms of particles and interactions that justifies the strong belief that HEP has a very interesting future.

This introduction is sufficient to explain why the exact description of the future perspectives of the field is a difficult task but the hope is that the landscape will be quite different after few years of data taking at the Large Hadron Collider (LHC).

In the meantime, during the long period of the preparation for that event, the international community tried to assess the future of the field preparing roadmaps in order to prioritize the different initiatives. One of the first high-level initiatives was organized few years ago by the US DoE with the title "Facilities for the future of science: a twenty years outlook" and it was one of the first of such reviews that tackled a broad spectrum of initiatives in the field and tried to put them in a prioritized list. A more focused approach was taken by the Consultative Group in High Energy Physics of the OECD Organization in 2002 that, by concentrating the study in the domain of what is needed in order to make significant steps forward in the field of Accelerator Based Particle Physics ended up in a strong support for an electron-positron linear collider with center of mass energies up to 1 TeV. This recommendation was subsequently endorsed by the OECD Research Ministers and has been at the beginning of the ILC (International Linear Collider) project. During the work

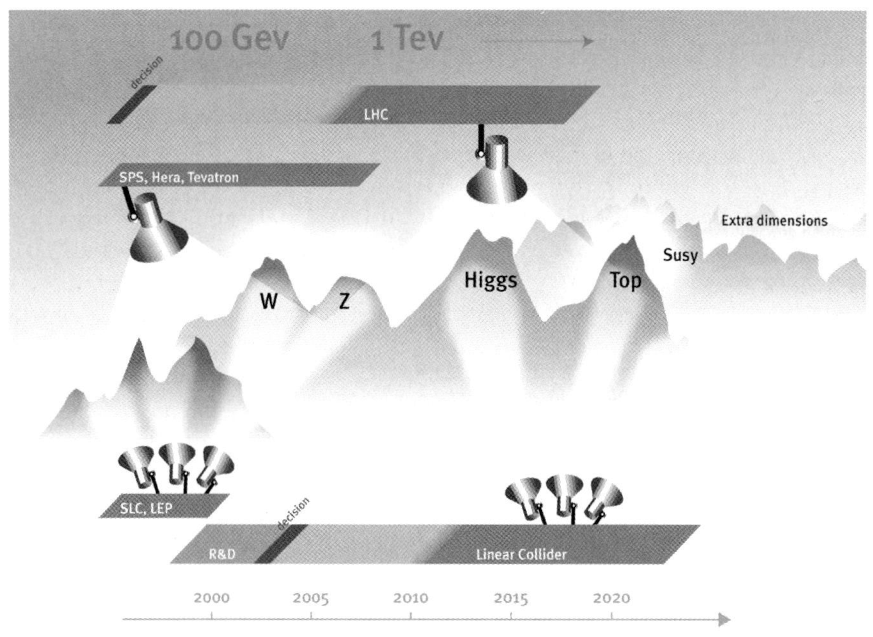

Fig. 1.

in preparation of the document the group work in the direction of explaining the different roles that various colliders have in advancing the frontiers of science in order to have arguments for convincing policy makers about the need of such facilities, and in Fig. 1 one can see an example of such pictorial description. Recently also the Cern Council has launched a Strategy Group about the future needs in the field of HEP aiming to act as the European coordinating body for such a research field.

In a tentative of summarizing some of the main points still open in particle physics one can say that the next experiments should try to answer the following questions:

1. are there new symmetries and/or physical laws?
2. can we give a description of the dark energy and dark matter in terms of particles and fields?
3. do extra-dimensions exists?
4. does a Grand Unification for the forces exist?
5. what can neutrino tell us about the general picture?
6. where is all antimatter gone?

As we can see this is a very rich and attracting program and how do we tackle this challenge? First of all we have to exploit to their limit the existing facilities, like the B-factories of KEK and SLAC and the Tevatron at Fermilab, since they are now at the peak of their potentialities and can still yield very interesting results. The next step is the startup of the experimentation at the LHC in Geneva; since quite a few years there is an unprecedented experimental effort being carried out by international collaborations in order to prepare the gigantic detectors that will study the events at the energy frontier for about a decade. This phase will start about in 2007 and it is the first priority of the field since a huge scientific return is expected. It is anticipated that, after a first phase of data taking at the nominal luminosity (i. e. about $10^{34}\,\mathrm{cm}^{-2}\,\mathrm{s}^{-1}$), about 5 to 7 years long, the error halving time will be so large that a luminosity upgrade will be necessary with a consequence upgrade of the detectors. Also this part of the experimentation is a priority for the field since it allows the full exploitation of the LHC investment; to allow such a detector upgrade the necessary R&D should start relatively soon, since the technical problems to be solved are severe.

But what else is needed by HEP to run at the same time of the LHC and why? The main avenues that will lead to the exploration of (possible) new physics are the high energy frontier, represented nowadays by the LHC, but also the high intensity frontier; in fact the former is indicated in order to study the gauge sector of the theory whereas the latter explores the flavor part, i. e. neutrino mixings, CP violation, lepton flavor violating phenomena and others. A link amongst these two faces is provided by a lepton collider. As we have seen the international community has already indicated the ILC, i. e. an e^+e- linear collider with center of mass energy up to 1 TeV, as the next facility necessary to shed light on the fundamental questions of Nature; the physics case for such a machine has been studied in detailed and found compelling and it is very important that the ILC can run for a good fraction of its time concurrently with the LHC. In fact the synergy between the two machines is such that if the LHC finds new physics at the TeV frontier than its flavor structure must be studied carefully, and the ILC will allow such studies. But, and maybe even more important, if the LHC does NOT finds new physics at the TeV scale the ILC can find deviations from the Standard model in the flavor sector from which one can derive the next scale for new physics. It is my opinion that the ILC should be approved in the near future in such a way to be ready for its construction by the end of the current decade, after a vigorous R&D phase on all its technological details.

Together with the LHC and the ILC other dedicated machines are necessary in order to explore the high intensity frontier to fully understand the discoveries done at the energy frontier; these machines can also serve as centres of excellence needed in order to maintain active and scientifically productive regional centres that are very important for the field. Amongst the interesting facilities a SuperB Factory with the capability of delivering integrated luminosities in the ballpark of $50ab^{-1}$ has a physics potentiality of first choice and perfectly in agreement with the possibilities in this sector at the LHC. Also measurements of the rare K decays, especially the golden channel

$K^0_L \to \pi^0 \nu\nu$, together with EDM and LFV dedicated experiments, would surely shed light on the beyond the standard model sector. A possibility in this area could come from CERN: in fact one off the scenario that is being investigated in the quest of reaching the ultimate luminosity for LHC is the refurbishing of various machines in the injection chain of the collider; if this path could lead to a new rapid-cycling high intensity PS and to an energy upgrade SPS (up to about 1 TeV) the perfect conditions for new experiments in the line described above could be found.

Experimentation at the energy and intensity frontiers are nicely complemented by a globally coordinated effort on neutrino experiments dedicated to measure their mass hierarchy, the mixing parameters and the CP phase. The results of this experiments, that could be available in a decade, will tell us about the feasibility and physics reach of a future neutrino factory.

In all the scenarios described so far INFN must continue, as it did so far, to play an important role by supporting the most promising experiments, by forming the researchers that will be able to take leadership positions in the collaborations and be proponents for the next generation of experimentation. Important sectors that are strategic for INFN will be a good preparation to the analysis of the data coming from LHC and an effort, coordinated with the international partners, on R&D for future accelerators.

The field of HEP is rapidly evolving and accordingly must also evolve the structure of a Funding Agency like INFN, always targeted to excellence in science but also ready to identify the spin-offs that could be useful for everyday life and that can encourage the society at large to continue to support the field.

I would like to thank the organizers and expecially V. Vercesi, O. Nicrosini and L. Trentadue for their enthusiasm.

References

1. H. Ibach, H. Lüth: *Solid-State Physics*, 2nd edn (Springer, Berlin Heidelberg New York 1996) pp 45–56
2. D.M. MacKay: Visual stability and voluntary eye movements. In: *Handbook of Sensory Physiology*, vol 3, ed by R. Jung, D.M. MacKay (Springer, Berlin Heidelberg New York 1973) pp 307–331
3. S. Preuss, A. Demchuk Jr, M. Stuke et al: Appl. Phys. A **61**, 33 (1995)
4. D.W. Ross: Lysosomes and storage diseases. MA Thesis, Columbia University, New York (1977)

REVIEW TALKS OF PARALLEL SESSIONS

Status of the Standard Model

Patrizia Azzi

INFN Sezione di Padova, Via Marzolo 8, 35100 Padova
patrizia.azzi@pd.infn.it

1 Introduction

The results of the electroweak precision tests have been recently updated and
are nearly finalized. Along with the measurement of the m_t and m_W from
the Tevatron and LEP, and other data from older experiments, they provide
a set of very precise constraints to be compared with the Standard Model
(SM). The short summary is that the agreement with the SM is excellent and
there are no hints of new physics that emerge from the data. In this paper
we'll review the latest results of Standard Model parameters and test from
low energy experiment up to hadron collider searches. More details can be
found in the specific contributions to the SM session of this conference also
published here.

2 Electroweak Fits

Probably the most complete analysis of the consistency of the standard
model comes from the global fits to the electroweak variables collected with
many different experimental setups spanning a vaste range in energy and
processes investigated. The latest updated numbers introduced in the electroweak fits come from $\Gamma(W) = 2.115 \pm 0.058$ at the Tevatron, $m(W) =
80.404 \pm 0.030\,\mathrm{GeV}/c^2$ from LEP-2 and $m(\mathrm{top}) = 172.5 \pm 2.3\,\mathrm{GeV}/c^2$ from
the Tevatron experiments. From this analysis the current best fit to the Higgs
mass is $m(H) = 89^{+42}_{-30}\,\mathrm{GeV}/c^2$ with a 95% C.L. limit of $m(H) < 175\,\mathrm{GeV}$ [1].
It has to be noted that the global analysis hides some local discrepancies that
are still present. For example the variable $\sin^2 \theta_{\mathrm{eff}}^{\mathrm{lept}}$ that has only a logarithmic dependence on $m(H)$ is in perfect agreement with the result from $m(W)$
but a difference remains with its determination in leptonic and hadronic
measurements. New theoretical calculations have been performed recently [1]
(two-loop electroweak fermionic corrections and two-loop electroweak corrections with Higgs-mass dependence) to try to improve the agreement, however

the evaluation of the two-loop bosonic correction to $\sin^2\theta_{\text{eff}}^{\text{lept}}$ is still an important missing piece. The other open problem is that the Higgs mass upper bound is small compared to the lower limit from direct searches at LEP of $m(H) > 114\,\text{GeV}/c^2$.

3 Low energy physics

For what concerns the contribution to the knowledge of SM parameters from low energy physics there are interesting developments in the determination of the muon anomalous magnetic moment and its comparison with theory. The current world average is a very precise number $a_\mu = 116\,592\,080(63) \times 10^{-11}$ [2], that is a precision of 0.5 parts per million. On the theory side there has been improvement as well. In the case of the electron $g - 2$ the experiment provides a check of QED at the fourth loop level, however in the case of the muon there is only the calculation for the electroweak contribution at the one loop level, and one loop plus higher-order. The problems arise from the contributions of the hadronic sector: in particular from the latest results it appears that the measurements from the e^+e^- data show a better agreement among themselves and a clear inconsistency with the results of the analysis with τ's. Comparing to the latest theoretical calculations this translates in a discrepancy $\Delta(\text{Exp} - \text{SM})$ between 2.0 and 3.0 σ if the e^+e^- data are used, that reduces to only $1\,\sigma$ using the τ data. Clearly new data are needed to better undertand the sector of the hadronic contributions. For what concerns the determination of R the theoretical error (0.50%) is now bigger than the experimental one (0.3%): lots of effort is going into the development of new event generators for the study of the luminosity in events of Bahbha scattering at large angle, among those the BABAYAGA [3] code has a new implementation of the radiative correction with a matching between the parton shower and the next to leading order.

4 QCD at hadronic colliders: theory and experiment

In order to the test the standard model at high energies and improve the knowledge of its parameters we need to use hadronic colliders data from the Tevatron and later on from the LHC. In particular at the LHC we can expect improvements in the measurements of the W mass, top mass, electroweak mixing angle and also the vector boson self-coupling and the Higgs mass (if discovered). However, the precision of the measurement will rely also on many other concurrent factors such as: knowledge of PDF's on a vaste range in x^2 and Q^2, the study of minimum bias and underlying events that is always present in high luminosity proton-proton collisions, the theoretical expectations and their implementation in MonteCarlo programs available to the experimental community. In recent years a lot of progress has been achieved in matrix element based MonteCarlo such as ALPGEN [4] and MADEVENT [5,6] and their interface to general purpose Parton Shower event generators such as HERWIG [7] and PYTHIA [8]. Despite the importance of MonteCarlo simulations, for the sake of brevity the subject will not be further discussed in this contribution.

5 Parton Distribution Functions

Almost every event recorded at the LHC will involve collisions of partons, mostly gluons, carrying a relatively small proportion of the longitudinal momenta of the colliding beams. Even benchmark cross sections such as W and Z production are largely made up of contributions from partons carrying rather small values of x. The major source of information on low-x physics in the last decade has been the data from the HERA $e^{+/-}p$ collider, which is the ideal machine to measure the structure functions, due to the large kinematic region accessible much bigger than fixed target experiment or the Tevatron collider. At HERA the experiments H1 and Zeus can measure the structure functions: F_2, F_L and xF_3. The measurement of F_2 spans between $6.32 \times 10^{-5} < x, 0.65$ and $1 < Q^2 < 30\,000\,\mathrm{GeV}^2$ with a precision of 2–3% [9]. Moreover from the measurement of $\sigma(cc)$ and $\sigma(bb)$ it is possible to extract also F_2^{cc} and F_2^{bb}: the two experiments show agreement between themselves and pQCD. H1 has obtained the first measurement of F_2^{bb}, at HERA-II also Zeus will be equipped with a vertex detector and soon will be able to perform the same measurement.

In summary today's knowledge about quark, anti–quark PDF's comes from the lepton–hadron deep inelastic scattering (DIS) experiments, such those at HERA, and from Drell–Yan lepton pair production in hadron collisions, while most information about the gluon distribution functions is extracted by hadron–hadron interactions with photons in the final state. The theoretical interpretations of these measurements has resulted in various sets of PDF's which are the basis for cross section predictions at the LHC. Although these PDF's are widely used for LHC simulations their uncertainties are difficult to estimate and various quantitative methods are being developed now also at the Tevatron experiments. The Drell–Yan like production of W boson represents one of the cleanest processes with a large cross section at the LHC. This reaction is not only well suited for a precise determination of the W boson mass but also yelds valuable information on the parton structure of the proton. The need for a higher precision of the theoretical expectation in this case is reflected in the fact that new Monte Carlo generators have been developed that incorporate the electroweak corrections at the next to leading order [10]. Parton distribution functions inside the proton determine, for instance, the W longitudinal momentum, and therefore affect the transverse mass distribution trough the lepton acceptance effects. It has been shown that PDF can be constrained to a few percent at the LHC using mainly the pseudorapidity distribution of leptons produced in W and Z decays. For instance a CMS study shows that considering all the available PDF's sets the uncertainty on $d\sigma/dy$ is of about 8% for central rapidity [11]. This means that if the systematic uncertainty in the data is kept below 5% it'd be possible to use this tecnique to distinguish the best set of PDF's using lepton rapidity distribution and the asymmetry. The impact of LHC on the PDF uncertainty could be as big as a reduction of 50% on $\delta\lambda_{\mathrm{g}}$ (for gluon at low x) with only a statistic of $0.1\,\mathrm{fb}^{-1}$.

Finally another fundamental input to the evaluation of PDF comes from the measurement of the inclusive high-p_{T} jet cross section at the Tevatron. The latest results are based on a large statistic of about one femtobarn of data. The large statistic allows a precise test of QCD NLO up to jet with transverse energies of $600\,\mathrm{GeV}$ in the central region, and, for the first time, different rapidity regions have been also explored up to the forward region of $|\eta| < 2.1$. This last check is particularly important since measurement on forward jets constrain the gluon distribution in a kinematic region where no effect from new physics is expected [12].

6 QCD tests at hadron colliders

The W and Z bosons are well understood objects that can used as standard candles to study QCD and jet production with high precision at hadron colliders. Leptonic W and Z decay modes provide clean signals with low background and high statistics samples are available in the Run II Tevatron data. Production rates and properties of the W and Z are precisely predicted within the SM and this is one of the few areas in which NNLO theory can be tested to high precision. Studies of the production of associated jets with the W and Z allow a test of perturbative QCD and are inmportant for tuning the simulation at the Tevatron and LHC. W and Z plus jets are also the most important background for top physics, Higgs searches and many searches for new phenomena. Recent Tevatron results from CDF and D0 use a data set of about $300\,\mathrm{pb}^{-1}$ and compare the transverse energy of the jet produced in association with a boson, the differential cross section as a function of the di-jet mass,and as a function of the event jet multiplicity. The comparison with the latest ME MonteCarlo predictions, such as those of Alpgen [4], shows overall a good agreement but the increasing precision of the data measurements suggests regions where improvements to the theory prediction can be made. Clearly an important limiting factor to the precision of these tests is also the modelling of the minimum bias processes and the underlying event activity in hadronic collisions. These are related to the soft-QCD aspects of hadronic collisions and are the first fundamental pieces that need to be tuned in the MonteCarlo simulation since they affect the occupancy of the detector and the background to all the other processes. To clarify, the underlying event that is comprised by the activity due to the interaction of the spectator partons, initial state radiation, final state radiation, and the multiple partonic interactions (MPI). Instead, with the name minimum bias, we refer to a generic particle interaction that includes all those events that could be recorded with a fully inclusive trigger (elastic, inelastic and diffractive processes). They are characterized by a a soft transverse momentum spectrum and a low particle multiplicity. These are the kind of events that appear as extra interaction in a single beam crossing: their activity and their production vertex is independent of those of the signal event of interest. A very detailed study and tuning of the CDF data has been now applied to the simulated data of the CMS experiment [13], the main strategies employed are based on the study of the charged particle and energy density in the trasverse regions defined by the muons direction in Drell–Yan processes or the jets direction in purely hadronic events.

7 Top and Higgs physics

After 10 years from its discovery obtained with only $100\,\text{pb}^{-1}$ of data, now the good performances of the Tevatron collider have allowed the experiments to enter the realm of a precise study of top properties, that will be complete once the LHC collider will beome operational. The top is produced at the Tevatron mainly in $t\bar{t}$ pairs: the current uncertainty on the combined cross section is of about 11% using a dataset of about $750\,\text{pb}^{-1}$ [14], this is an important test of QCD, especially since now we are able to check the consistency of the production cross section in all the different decay modes, including the most challenging ones. The top can also be produced singly through an electroweak process, while this process has not been observed yet, the discovery is not very far: extrapolation curves based on current analyses by CDF and D0 show that an observation can be made with about $1.5\,\text{fb}^{-1}$ of data per experiment [15]. By far the most important information we can obtain from the top quark is a precise measurement of its mass that is one of the fundamental parameter of the Standard Model. The most recent world average value, published in July 2006 [16], is $m_{\text{top}} = 171.4 \pm 2.1\,\text{GeV}/c^2$, and the most impressive feature of this result is the extrapolation of the uncertainty on the top mass measurement as a function of integrated luminosity. With the latest analysis' improvements that include a "in situ" determination of the light quark jet energy scale (JES) using the hadronic W in a lepton plus jet event, the systematic uncertainty due to the JES now scales as $1/\sqrt{N}$ with increasing data statistics. The expected projection for $4\,\text{fb}^{-1}$ is an uncertainty of $1.5\,\text{GeV}/c^2$.

The discovery of the Higgs boson, in its SM interpretation or not, is one of the major goals for LHC physics program. The recent excellent performances of the Tevatron collider are giving the CDF and D0 experiment a chance to reoptimize their trigger and analysis configuration to push farther the sensitivity for the Higg,s but it is still a very tough call. Given the current published results the LHC seems still the best place for a discovery. The production process with the highest cross section is the one of $gg \to H$ for which is available also the MonteCarlo generator at NLO. For the other channels, $qq \to qqH$(VBF) and $qq \to W(Z)H$, and for most of the background the simulation is possible only with LO generators and the corresponding correction factors, $K = \sigma_{\mathrm{NLO}}/\sigma_{\mathrm{LO}}$, need to be applied in the calculation of the signal significance and of the analysis sensitivity [17]: for instance in the case of the gluon fusion the significance increases by 50% from LO to NLO. The other major problem is that even in the case of processes with calculated K-factors they are useless for what concern the differential distribution: right now there are a few options, the NLO parton level MonteCarlo or the application of reweighting techniques withing LO MonteCarlo generators in order to provide differential effective K-factors. The latest results from Atlas and CMS predict the possibility of a $5\,\sigma$ observation with an integrated luminosity of $15\,\mathrm{fb}^{-1}$ but combining the result from various channels. In order to study other interesting parameters such as the Higgs mass, CP, spin and partial width it will be necessary to have a much larger dataset, corresponding to more than $100\,\mathrm{fb}^{-1}$ [17].

References

1. S. Uccirati, *these proceedings.*
2. A. Passera, *these proceedings*
3. G. Balossini et al., hep-ph/0607181, *these proceedings*
4. M. L. Mangano, M. Moretti, F. Piccinini, R. Pittau and A. D. Polosa, *JHEP* **0307**, 001 (2003) [hep-ph/0206293].
5. T. Stelzer and W. F. Long, *Comput. Phys. Commun.* **81**, 357 (1994) [hep-ph/9401258].
6. F. Maltoni and T. Stelzer, *JHEP* **0302**, 027 (2003) [hep-ph/0208156].
7. G. Marchesini, B. R. Webber, G. Abbiendi, I. G. Knowles, M. H. Seymour and L. Stanco, *Comput. Phys. Commun.* **67**, 465 (1992).
8. T. Sjostrand, *Comput. Phys. Commun.* **82**, 74 (1994).
9. A. Parenti, *these proceedings*
10. C. M. Carloni Calame et al., *these proceedings*
11. S. Bolognesi, *these proceedings*
12. S. Vallecorsa, *these proceedings*
13. F. Ambroglini, *these proceedings*
14. A. Gresele, *these proceedings.*
15. S. Rolli, *these proceedings.*
16. CDF Collaboration, D0 Collaboration and TEVEWWG, hep-ex/0608032
17. S. Rosati, *these proceedings.*

New physics

Andrea Perrotta[1] and Alessandro Strumia[2]

[1] INFN Bologna `perrotta@bo.infn.it`
[2] Dipartimento di fisica di Pisa and INFN `astrumia@mail.df.unipi.it`

1 Introduction

A lively parallel session on physics beyond the Standard Model was organized during IFAE 06, with interesting presentations on both the theoretical and experimental side. Still in 2006, an appropriate title for the session continues to be: "new physics without new physics". In this edition we tried to focus the attention on the preparation for LHC: within a few years its results will hopefully allow removing the second part of the title, or maybe the whole session. We outline in this contribution the status of the field.

2 Theory

The Standard Model (SM), proposed around 1970, successfully anticipated experimental results: the key theoretical guideline was *gauge invariance*. Since then, theorists tried proceeding beyond the SM in an autistic way: models of new physics are proposed without having experiments that demand new physics. These attempts rely on a different guideline: *naturalness*: the SM should be replaced by some more natural theory where quantum corrections to the Higgs mass are not much larger than the Higgs mass itself. The main thing we hope to learn from LHC is whether naturalness is a correct guideline[1].

In the past, naturalness gave correct hints: electromagnetism gives ultraviolet (UV) divergent corrections to i) the electron mass and to ii) the mass difference between charged and neutral π. In the first case, imposing naturalness (i. e. $\delta m_{\rm e} \sim \alpha_{\rm e.m.} \Lambda \lesssim m_{\rm e}$ where Λ is an UV cut-off introduced to mimic the unknown true cut-off) suggested new physics at $\Lambda \lesssim 10\, m_{\rm e}$: today

---

[1] Officially, LHC is built to discover the Higgs. However, discovering just a light Higgs would mean no new physics, and discovering no light Higgs would be crazy new physics.

we know that the cut-off is provided by a new symmetry (chiral symmetry broken by the electron mass, so that $\delta m_{\rm e} \sim \alpha_{\rm e.m.} m_{\rm e}$) that manifests as a new particle, the positron. In the second case, naturalness (i. e. $\delta m^2_{\pi^\pm} \sim \alpha_{\rm e.m.} \Lambda^2$ not much larger than the measured value of $m^2_{\pi^\pm} - m^2_{\pi^0}$) suggested $\Lambda \lesssim 1\,{\rm GeV}$: today we know that the cut-off arises because pions are composite particles with size $\Lambda \sim \Lambda_{\rm QCD}$.

Present attempts of addressing the Higgs mass hierarchy problem fall in two similar classes.

2.1 Class 1: there is a new symmetry

Similarly to the case of the electron, some new symmetry could keep the Higgs massless; the Higgs mass is controlled by its breaking. The main concrete possibilities are:

- *The symmetry directly acts on the Higgs, that might be a pseudo-Goldstone boson.* However, a Goldstone boson has zero potential (zero mass term and zero quartic coupling); we want instead to suppress only the Higgs mass term while allowing a sizeable quartic Higgs coupling and top Yukawa coupling. One can proceed at any cost by inventing complicated models (known as "little Higgs"), where a specific set of particles is added to implement the desired selection. Unfortunately, these particles ruin the agreement with precision electroweak data, unless one considers specific ranges of the parameter space or adds an extra ad-hoc symmetry (known as "T-parity").
- *Vectors can be kept massless by gauge symmetry: one can invent a new symmetry that links the Higgs to vectors.* Typically one considers a SU(3) electroweak gauge symmetry in 5 dimensions, and identifies the Higgs as the $5^{\rm th}$ component of SU(3)/SU(2) vectors. These models are problematic because the Higgs Yukawa couplings are different (most smaller, one larger) than gauge couplings, and disfavored by precision data, as in the previous case.
- *Fermions can be kept massless by chiral symmetries: one can invent a new symmetry (known as supersymmetry) that links the Higgs to fermions.* This is considered the most plausible scenario: we skip its well known long list of successes and problems.

2.2 Class 2: the Higgs is an extended object

Similarly to the case of the pions, the Higgs might be an extended object with size not much larger than its mass, solving the Higgs mass hierarchy problem. Concrete possibilities are

- *The Higgs is the bound state of some QCD-like "technicolor" gauge interaction.* This possibility is disfavored by precision measurements of the S and T parameters. Technicolor models exist with very few technicolors and techniquarks [1], such that S is not too big. To compute T one needs to know how to introduce the top Yukawa coupling, but adding flavor to technicolor models seems to be a problem.

- *The ultimate quantum gravity scale is around the TeV scale.* This can be achieved in models with branes in large extra-dimensions: the Higgs and all SM particles might e. g. be strings with TeV-scale length. If something like this were true, one naively expects that we should have already seen form factors, while precision measurements agree with the SM where all particles are pointlike. We have no theory of quantum gravity able of telling if the above qualitative guess is true.

- *A warped extra-dimension.* The AdS/CFT conjecture means that this is equivalent to walking technicolor, e. g. the Kaluza–Klein gravitons predicted by a warped extra-dimension can be reinterpreted as spin 2 mesons, analogous the ones present also in QCD. It is possible to build a model [2] where arbitrary choices (where to localize particles in the extra-dimension, etc) suppress calculable unwanted effects.

2.3 Another possibility: we misunderstood something

Waiting for LHC, negative results from previous experiments cast doubts on the previous possibilities, and in recent years theorist explored three different possibilities, from mild to drastic.

- *The Higgs is heavy: $m_h \sim$ few hundred GeV.* This makes the naturalness condition $\delta m_{\rm h}^2 \lesssim m_{\rm h}^2$ less restrictive, relaxing the tension of the above scenarios with data. Within this scenario the fact that precision data agree with the SM and a light Higgs is considered as more-or-less accidental.

- *The Higgs does not exist.* Within the SM this gives breakdown of unitarity below ~ 1 TeV (although we do not know what LHC would see in practice). Extra-dimensional models allow to maintain unitarity up to a few TeV, but cause a lot of problems with precision data.

- *Naturalness is a wrong assumption. We spent 30 years on a wrong track.*

Most theorists ignored this last possibility until a few years ago, but the following three developments made it more plausible.

The first comes from cosmology: *dark energy* is consistent with being just a small cosmological constant $V_0 \sim 10^{-123} M_{\rm Pl}^4$. Nobody has ever found a possible solution to this other hierarchy problem, and proposed tentative solutions fail for generic reasons. Furthermore, V_0 is comparable to the energy density ρ of the universe today. Well established cosmology tells that matter can cluster forming galaxies only when ρ is dominated by non-relativistic matter, and suggests that if V_0 were 10 (100) times bigger than what it is, our (any) galaxy would not have been formed (we skip caveats and controversies). This not fully successful anthropic interpretation is the best/only idea we have today.

Second, collider data suggest *a problem with the Higgs mass hierarchy problem*: as already discussed, most tentative natural models of EWSB can survive only by reintroducing some uncomfortable amount of the same fine-tuning that they should eliminate. At the moment this is only a hint; we hope that LHC will clarify this point.

Finally, *the string demographic explosion*. Theorists attached quantum gravity (despite being possibly experimentally irrelevant) hoping that it leads to a unique "theory of everything" that predicts something at low energy. String theory was considered a promising attempt and gained a strong influence on theorists. At the moment, the outcome seems to be that there is one "M-theory" in 11 dimensions. After eliminating one extra-dimension, it reduces to 5 string theories in 10d. After reaching 4d, one has $10^{\mathcal{O}(500)}$ string models. Predictivity seems lost in the following way: SM particles are different vibrations of a string; to compute their spectrum one needs to know both string dynamics and the background on which strings move: the complicated SM physics mostly comes by assuming a complicated enough higher dimensional geography, rather than from theory. This makes string theory the best candidate for an anthropic theory: there are many vacua with different values of v and Λ, and we live in some vacuum that accidentally has small v and Λ, because their smallness is necessary for life. This is a plausible scenario, but we do not know how to get physics from it; e. g. recent attempts along this line lead to the following qualitative predictions for sparticle masses: they are around M_Z, or around $4\pi M_Z$, or at the string scale, or anywhere, or split, or super-split.

3 Experiments

From the point of view of the experiments, the search for new physics follows two main directives: the direct search for new particles in the frontier high-energy colliders, and the search for tiny but nevertheless significant deviations from the SM predictions in precision measurements, mostly carried out at lower energies. Other constraints or evidences may also arise from the cosmological observations, but it goes beyond the scope of this review.

3.1 Searches for anomalous effects in precision measurements

Careful measurements of observables precisely computable in the SM framework become stringent tests of the SM when the experimental precision is better than the size of the possible deviations due to non-SM physics. New physics effects are usually hidden in loop diagrams, and can be calculated as function of the parameters of the model. Nowadays, there are a few places/experiments in the world where to look for new physics effects as deviations from the SM predictions. Amongst the most interesting precision measurements where to look for hints of (either constrain) new physics there are:

- Precision electroweak measurements from LEP/SLC/TeVatron [3]: rich harvest of results collected in past years, and no significant deviations from the SM expectations observed.
- Rare FCNC B decays [4]: while some of the results collected in the B-factories (CLEO, BABAR, BELLE experiments) or at the TeVatron already severely challenge possible new physics (e. g. the radiative $b \to s\gamma$ decay, which measurements of the branching ratio and asymmetries, even at the present level of precision, pose strong constraints on many supersymmetry and other new physics models), several other measurements are still limited by statistics and systematics. Further data from the current devices will help, as well as the high statistics and complementar methods of the dedicated LHCb experiment at LHC.
- $g_\mu - 2$, precisely measured at BNL [5]
- $\mathrm{BR}(\mu \to e\gamma)$: this FCNC muon decay is forbidden in the SM, not in several new physics scenarios, where it can happen with a tiny BR. The MEG experiment [6], now in progress at PSI, plans to explore it down to 10^{-13}, thus covering a region motivated by some supersymmetric scenarios.
- The electric dipole moment of the neutron is also another quantity which is strongly suppressed in the SM but with possible relevant contributions from new physics. The present limit [7] already ruled out several proposed extentions of the SM; planned improvements in the experiment will increase its sensitivity.

3.2 Searches in the data collected by high-energy experiments

While (negative) LEP results still dominates the direct searches for new particles in many domains [8], are the searches at the TeVatron [9] that do focus the attention of the specialists in these days. Not only because as more data are collected and analyzed the possibility of observing new particles increases, but also because, being the TeVatron a hadron collider, it can also be viewed as a playground where to test and specialize tools and methods that will be used in the forthcoming searches at the LHC.

An interesting issue coming out from the analysis of the data at CDF and D0 (and also from the H1 detector at HERA [10]) is: since we have no idea of what the phenomenology of the physics beyond SM can be, should we specialize searches at colliders to any specific new physics scenario one can imagine, or should we plan "model independent" (signature based) searches instead? In principle, signature based searches are the most obvious tools to answer the "inverse problem" of the searches: if we notice a signature, or a deviation from what we expect, which type of new physics can likely supposed to be responsible of it? Such "inverse problem" is what we expect to encounter at LHC, when first evidences of something different from the SM will start to appear in the data and people will start wondering whether it comes from SUSY (which SUSY?), or from technicolor, or extra-dimensions, or else. However, experience at present colliders shows that the more traditional method of studying carefully a specific phenomenology and look for it in the data can be a much more efficient way of spotting new effects, since a dedicated search can always reach the best discriminating power for a given scenario. This is supposed to work for every possible new physics scenario provided:

- every possible "anomalous effect" is searched for, regardless if it derives from a theory which seems to be ruled out by some different consideration[2];
- when evidence of an effect results from an analysis, say the search for technicolor, we don't run at claiming that we have discovered technicolor, but only something which is compatible with it, and then start exploring all possible scenarios that predict this particular phenomenology;
- we combine with all other results at our disposal until a solid scenario beyond SM can be assessed.

3.3 Searches at LHC and beyond

LHC is expected to start colliding at the end of 2007. While first data will be used for the commissionig of detectors and computing, maybe already in the 2008 run at 14 TeV we will have the first glimpses of what may happen in such an yet unexplored energy domain. Figures on the initial luminosity the machine will provide in the high-energy phase are not yet fully decided, but it is likely that during the first year of operations at 14 TeV a few fb^{-1} will be collected by each experiment. Careful studies were performed in the ATLAS and CMS collaborations to evaluate the performance of the detectors in discovering SUSY [11] or other new physics scenarios [12]. ATLAS and CMS potentials are substantially similar, and they both would allow the discovery, for example, of squark and gluino masses below 1.5 TeV with 1 fb^{-1} of integrated luminosity. Those studies assume however perfectly known SM physics backgrounds and ideal detectors, with nominal asymptotic performance. It is clear, therefore, that SUSY discovery capabilities will depend not solely on statistics, but mostly on the understanding of the SM physics backgrounds and detector systematics with the early data. The use of Matrix Elements to complement the usual Parton Shower MC's is a step towards a better understanding of SM processes at the LHC energy, while careful confrontations with "candle" processes in the data itself will be the best way to certify them. Detectors calibrations and alignements will also improve with the first data.

In the spirit of looking for new physics everywhere in the data, the experimental collaborations at LHC are preparing tools able to discriminate un-

---

[2] Past experience at LEP tells us that whatever strange signal seems to show up in the experimental data, a few theoreticians capable to justify it can always be found.

conventional effects. One example is the search for long-lived heavy charged particles [13], where the subdetectors are planned to be used in an "improper" way: timing information will be extracted from calorimeters and muon detectors; electromagnetic showers will be searched for also in the muon stations; spatial asymmetries in the shower shapes will be evaluated. Moreover, as far as larger statistics will be collected, LHC will also be used as a *top*-factory, where to look for rare FCNC *top* decays, or for deviations from the expected differential cross-sections [14].

Finally, it is clear to everybody that LHC alone will not allow a detailed spectroscopy of a possible new physics scenario, unless nature has put us in a very fortunate condition. Thus, to fully solve the "inverse problem" of the searches we will have to complement the LHC results with the precision measurements achievable at a possible future e^+e^- Linear Collider.

4 Conclusions

LHC will surely be a turning point for both experimental and theoretical high-energy physics. We hope to learn soon in which directions data will force us to go.

References

1. F. Sannino, these proceedings
2. R. Contino, these proceedings.
3. W.M. Yao, *et al.*, Journal of Physics **G33**, 1 (2006).
4. S. Vecchi, these proceedings.
5. G.W. Bennet, *et al.*, Phys. Rev. Lett. **89**:101804 (2002), erratum ibid. **89**:129903 (2002); G.W. Bennet, *et al.*, Phys. Rev. Lett. **92**:161802 (2004).
6. http://meg.web.psi.ch
7. C.A. Baker, *et al.*, hep-ex/0602020, submitted to Phys. Rev. Lett.
8. See, for example, F.L. Navarria, these proceedings.
9. S. Rolli, these proceedings.
10. A. Aktas, *et al.*, Phys. Lett. **B602**, 14 (2004).
11. T. Lari, these proceedings.
12. L. Menici, these proceedings.
13. S. Viganò, these proceedings.
14. M. Cobal, these proceedings.

Flavour Physics

Stefano Giagu and Luca Silvestrini

Dipartimento di Fisica, Università di Roma "La Sapienza" and
INFN Sezione di Roma, Piazzale Aldo Moro 2, I-00185 Roma, Italy
stefano.giagu@roma1.infn.it, luca.silvestrini@roma1.infn.it

In the last decade, flavour physics has witnessed unprecedented experimental and theoretical progress, opening the era of precision flavour tests of the Standard Model (SM). The advent of B factories, with the measurements of the angles of the Unitarity Triangle (UT), has opened up the possibility of the simultaneous determination of SM and New Physics (NP) parameters in the flavour sector. Detailed reviews of recent theoretical and experimental results can be found in the proceedings of this conference [1], and so will not be repeated here. On the experimental side, we will focus on the very recent results on B_s^0-$\bar{B}_s^0$ oscillations by the CDF collaboration. We will then briefly present the impact of the present experimental data on the SM and on several NP scenarios.

1 B_s^0-$\bar{B}_s^0$ Mixing in CDF

The precise determination of the B_s^0-$\bar{B}_s^0$ oscillation frequency Δm_s from a time-dependent analysis of the B_s^0-$\bar{B}_s^0$ system has been one of the most important goals of heavy flavor physics in the last 20 years. As shown in Sects. 2–4, this frequency can be used to strongly improve the knowledge of the Cabibbo–Kobayashi–Maskawa (CKM) matrix [2], and to constrain contributions from new physics. Very recently, the CDF collaboration reported [3] the direct observation of B_s^0-$\bar{B}_s^0$ oscillations with more than 5σ significance, yielding the definitive observation of time-dependent B_s^0-$\bar{B}_s^0$ oscillations.

CDF has access to B_s decays in hadronic ($\bar{B}_s^0 \to D_s^+\pi^-$, $D_s^+\pi^-\pi^+\pi^-$) and semileptonic ($\bar{B}_s^0 \to D_s^{+(*)}\ell^-\bar{\nu}_\ell$) modes. Moreover signal statistics is also improved by adding partially reconstructed hadronic decays in which a photon or π^0 is missing: $\bar{B}_s^0 \to D_s^{*+}\pi^-$, $D_s^{*+} \to D_s^+\gamma/\pi^0$ and $\bar{B}_s^0 \to D_s^+\rho^-$, $\rho^- \to \pi^-\pi^0$, with $D_s^+ \to \phi\pi^+$. Signal yields are optimized by using particle identification techniques and by employing an artificial neural network (ANN) to improve candidate selection. The signal yelds obtained by CDF using a sample corresponding to 1 fb^{-1} of data are of about 6000 fully hadronic B_s decays, about 62 000 semileptonic decays, and about 3000 partially reconstructed decays.

The proper decay time in the B_s rest frame is measured in CDF as $t = m_{B_s} L_{\mathrm{T}}/p_{\mathrm{T}}^{\mathrm{recon}}$, where L_{T} is the transverse decay lenght of the B_s, and $p_{\mathrm{T}}^{\mathrm{recon}}$ is the transverse momentum of the reconstructed decay products. The distribution of σ_t for fully reconstructed decays has an average value of 87 fs, which corresponds to one fourth of an oscillation period at $\Delta m_s = 17.8\,\mathrm{ps}^{-1}$, and an rms width of 31 fs. For the partially reconstructed hadronic decays the average σ_t is 97 fs, while for semileptonic decays, σ_t is worse due to decay topology and the much larger missing momentum of decay products that were not reconstructed.

The flavor of the $\bar{B}_s^0$ at production is determined in CDF using both opposite-side and same-side flavor tagging techniques. Lepton charge, jet charge and the charge of identified opposite side kaons are used as tags, combining the available information using an ANN. CDF implemented also a kaon based same-side flavor tag, using an ANN to combine kaon particle-identification likelihood with kinematic quantities of the kaon candidate into a single tagging variable. The combined opposite-side tag effectiveness is $\epsilon D^2 = 1.8 \pm 0.1\%$, while the effectiveness of the same side kaon tag is $\epsilon D^2 = 3.7\%$ (4.8%) in the hadronic (semileptonic) decay sample.

An unbinned maximum likelihood fit has been used by CDF to search for B_s^0-$\bar{B}_s^0$ oscillations. Following the method described in [4], the oscillation amplitude $\mathcal{A}$ has been fitted while fixing Δm_s to a probe value. The oscillation amplitude is expected to be consistent with $\mathcal{A} = 1$ when the probe value is the true oscillation frequency, and consistent with $\mathcal{A} = 0$ when the probe value is far from the true oscillation frequency. Figure 1 shows the fitted value of the amplitude as a function of the oscillation frequency for the semileptonic candidates alone, the hadronic candidates alone, and the combination. The sensitivity is $31.3\,\mathrm{ps}^{-1}$ for all decays combined. At $\Delta m_s = 17.75\,\mathrm{ps}^{-1}$, the observed amplitude $\mathcal{A} = 1.21 \pm 0.20$ (stat.) is consistent with unity, indicating that the data are compatible with B_s^0-$\bar{B}_s^0$ oscillations with that frequency, while the amplitude is inconsistent with zero: $\mathcal{A}/\sigma_{\mathcal{A}} = 6.05$, where $\sigma_{\mathcal{A}}$ is the statistical uncertainty on $\mathcal{A}$ (the ratio has negligible systematic uncertainties). The small uncertainty on $\mathcal{A}$ at $\Delta m_s = 17.75\,\mathrm{ps}^{-1}$ is due to the superior decay-time resolution of the hadronic decay modes.

The significance of the signal has been evaluated by using the logarithm of the ratio of likelihoods for the hypothesis of oscillations ($\mathcal{A} = 1$) at the probe value and the hypothesis that $\mathcal{A} = 0$, which is equivalent to random production flavor tags. Figure 1 shows Λ as a function of Δm_s. The significance of the signal has been estimated to be 8×10^{-8} corresponding to $5.4\,\sigma$. To measure Δm_s, CDF fix $\mathcal{A} = 1$ and fit for the oscillation frequency. The result of this procedure is $\Delta m_s = 17.77 \pm 0.10$ (stat.) ± 0.07 (syst.)$\,\mathrm{ps}^{-1}$. The only non-negligible systematic uncertainty on Δm_s is from the uncertainty on the absolute scale of the decay-time measurement.

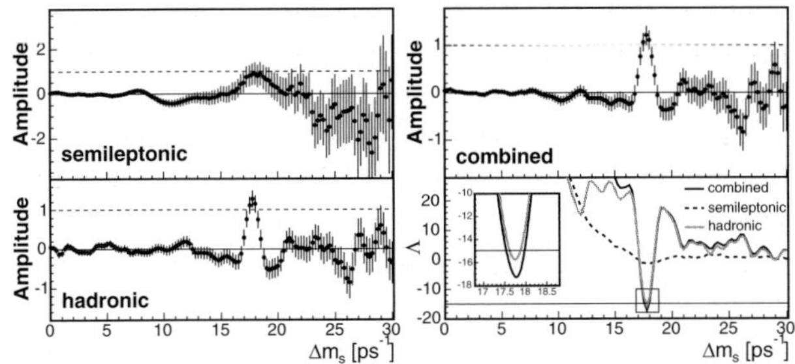

Fig. 1. The measured amplitude values and uncertainties versus the B_s^0-$\bar{B}_s^0$ oscillation frequency Δm_s. (*Upper Left*) Semileptonic decays only. (*Lower Left*) Hadronic decays only. (*Upper Right*) All decay modes combined. (*Lower Right*) The logarithm of the ratio of likelihoods for amplitude equal to one and amplitude equal to zero versus the oscillation frequency

2 Impact of experimental data on the SM

The most efficient way to combine all available experimental data on flavour physics within the SM is the UT analysis. Indeed, all FCNC and CP violating phenomena in the SM are driven by two parameters of the CKM matrix, $\bar{\rho}$ and $\bar{\eta}$, which identify the apex of the UT. Let us now list the most relevant measurements entering the SM UT fit: i) The rates of charmed and charmless semileptonic B decays, which allow to measure the ratio $|V_{ub}| \, / \, |V_{cb}|$. ii) The mass difference between the light and heavy mass eigenstates of the $B_{(s)}^0 - \bar{B}_{(s)}^0$ systems Δm_d and Δm_s, which give access to $|V_{td}|$ and $|V_{td}|$. iii) The ε_K parameter, which measures CP violation in the neutral kaon system. iv) The angle β extracted from $b \to c\bar{c}s$ modes and from $B^0 \to D^0\pi^0$. v) The angle α, that can be obtained from the $B \to \pi\pi$ and $B \to \rho\rho$ decays, assuming isospin symmetry and neglecting the contributions of electroweak penguins. It can also be obtained using a time-dependent analysis of $B \to (\rho\pi)^0$ decays on the Dalitz plane. vi) The angle γ that can be extracted from the tree-level decays $B \to DK$, using the fact that a charged B can decay into a $D^0(\overline{D}^0)K$ final state via a $V_{cb}(V_{ub})$ mediated process. CP violation occurs if the D^0 and the $\overline{D}^0$ decay to the same final state. The same argument can be applied to $B \to D^*K$ and $B \to DK^*$ decays.

In Table 1 we summarize the output of the fit including all the constraints (a table of the input values can be found in [5]). A graphical view of the fit result in the $(\bar{\rho}, \bar{\eta})$ plane is shown in the left plot in Fig. 2. This figure clearly displays the great success of the SM in flavour physics: all measurements do agree in constraining the apex of the UT at an astonishing level. However, by looking in more detail at Fig. 2, it is interesting to note that the 95% probability regions depicted by the $\sin 2\beta$ and $|V_{ub}| \, / \, |V_{cb}|$ constraints, two of the most precise ones used in the fit, show just a bare agreement. In particular, in our analysis we find that while the experimental value of $\sin 2\beta$ is in good

Table 1. SM UT fit results

Parameter	Output	Parameter	Output	Parameter	Output
$\bar{\rho}$	0.163 ± 0.028	$\bar{\eta}$	0.344 ± 0.016	$\alpha[°]$	92.7 ± 4.2
$\beta[°]$	22.2 ± 0.9	$\gamma[°]$	64.6 ± 4.2	$\Delta m_s\,[\text{ps}^{-1}]$	17.77 ± 0.12

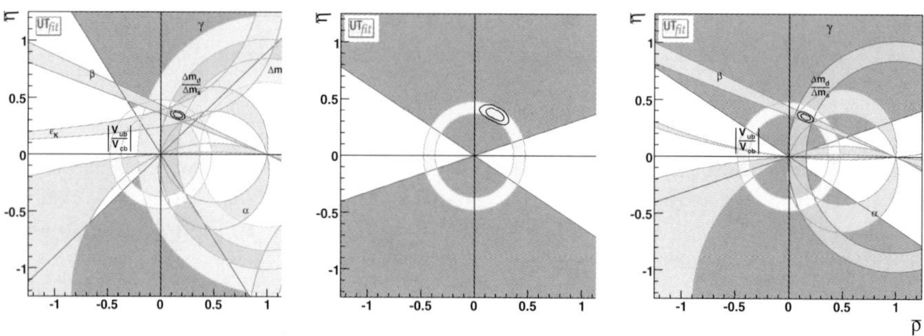

Fig. 2. Determination of $\bar{\rho}$ and $\bar{\eta}$ within the SM (*left*), in the generalized NP analysis (*middle*) and in the UUT analysis (*right*)

agreement with the rest of the fit, the same does not hold for $|V_{ub}|\,/\,|V_{cb}|$, which is rather on the high side. It can be shown that this is due to a large value of the inclusive determination of $|V_{ub}|$. Unless this discrepancy should be considered as a hint of NP, it has to be explained by the uncertainties of the theoretical approaches needed to determine $|V_{ub}|$ [6].

3 Impact of experimental data on NP

We now turn to assessing the impact of presently available experimental data on NP models. Also in this case, the UT analysis is the most efficient way of combining all constraints. Different assumptions about the flavour structure of NP lead of course to different results for the UT and to different constraints on NP. Let us first present the most general case of arbitrary loop-mediated NP. Then we will consider the much more restrictive case of Minimal Flavour Violation (MFV). Finally, as a concrete example of non-MFV NP, we will discuss the impact of the measurement of Δm_S on SUSY models with non-diagonal squark mass matrices (Sect. 4).

Following [7], we incorporate general NP loop contributions in the fit in a model independent way, parametrizing the shift induced in the $B_q - \bar{B}_q$ mixing frequency (phase) with a parameter C_{B_q} (ϕ_{B_q}) having expectation value of one (zero) in the SM:

$$C_{B_q}\,e^{2i\phi_{B_q}} = \frac{\langle B_q|H_{\text{eff}}^{\text{full}}|\bar{B}_q\rangle}{\langle B_q|H_{\text{eff}}^{\text{SM}}|\bar{B}_q\rangle}, \quad (q=d,s) \qquad C_{\epsilon_K} = \frac{\text{Im}[\langle K^0|H_{\text{eff}}^{\text{full}}|\bar{K}^0\rangle]}{\text{Im}[\langle K^0|H_{\text{eff}}^{\text{SM}}|\bar{K}^0\rangle]}.$$

Table 2. Determination of UT and NP parameters from the NP generalized fit

Parameter	Output	Parameter	Output	Parameter	Output
C_{B_d}	1.25 ± 0.43	$\phi_{B_d}[^\circ]$	-2.9 ± 2.0	C_{B_s}	1.13 ± 0.35
$\phi_{B_s}[^\circ]$	$(-3 \cup 94) \pm 19$	C_{ϵ_K}	0.92 ± 0.16		
$\bar{\rho}$	0.20 ± 0.06	$\bar{\eta}$	0.36 ± 0.04	$\alpha[^\circ]$	93 ± 9
$\beta[^\circ]$	24 ± 2	$\gamma[^\circ]$	62 ± 9		

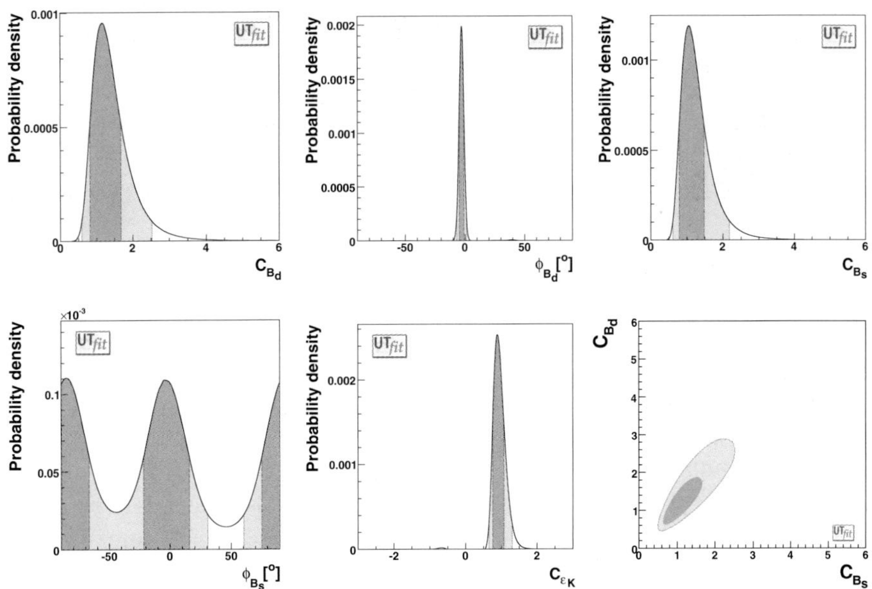

Fig. 3. Constraints on ϕ_{B_q}, C_{B_q} and C_{ϵ_K} coming from the NP generalized analysis, and correlation between C_{B_d} and C_{B_s}

We present here the results obtained in [8]. The fit is summarized in Table 2. The bound on $\bar{\rho}$ and $\bar{\eta}$ is also shown in Fig. 2 (middle). The distributions for C_{B_q}, ϕ_{B_q} and C_{ϵ_K} are shown in Fig. 3. We see that the *non-standard* solution for the UT with its vertex in the third quadrant, which was present in previous analyses [7], is now absent thanks to the improved value of $A_{\rm SL}$ by the BaBar Collaboration and to the measurement of $A_{\rm CH}$ by the D0 Collaboration. Furthermore, the measurement of Δm_s strongly constrains C_{B_s}, so that C_{B_s} is known better than C_{B_d}. Finally, A_{CH} and $\Delta\Gamma_s$ provide stringent constraints on ϕ_{B_s}. Taking these constraints into account, we obtain

$$S_{J/\Psi\phi} = 0.09 \pm 0.60 \,, \tag{1}$$

leaving open the possibility of observing large values of $S_{J/\Psi\phi}$ at LHCb. We point out an interesting correlation between the values of C_{B_d} and C_{B_s} that

Table 3. Determination of UUT parameters from the constraints on α, β, γ, $|V_{ub}/V_{cb}|$, and $\Delta m_d/\Delta m_s$ (UUT fit)

Parameter	Output	Parameter	Output	Parameter	Output
$\bar\rho$	0.154 ± 0.032	$\bar\eta$	0.347 ± 0.018		
$\alpha[^\circ]$	91 ± 5	$\beta[^\circ]$	22.2 ± 0.9	$\gamma[^\circ]$	66 ± 5

can be seen in Fig. 3. This completely general correlation is present since lattice QCD determines quite precisely the ratio ξ^2 of the matrix elements entering B_s and B_d mixing amplitudes respectively.

We conclude this discussion by noting that the fit produces a nonzero central value of ϕ_{B_d}. This is due to the difference in the SM fit between the angles measurement (in particular $\sin 2\beta$) and the sides measurement (in particular V_{ub} inclusive) that we mentioned in Sect. 2. Further improvements in experimental data and in theoretical analyses are needed to tell whether this is just a fluctuation or we are really seeing a first hint of NP in the flavour sector.

In the context of MFV extensions of the SM [9], it is possible to determine the parameters of the CKM matrix independently of the presence of NP, using the Universal Unitarity Triangle (UUT) construction [10], which is independent of NP contributions. In particular, all the constraints from tree-level processes and from the angle measurements are valid and the NP contribution cancels out in the $\Delta m_d/\Delta m_s$ ratio; the only NP dependent quantities are ϵ_K and (individually) Δm_d and Δm_s, because of the shifts δS_0^K and δS_0^B of the Inami-Lim functions in K–$\bar K$ and $B_{d,s}$–$\bar B_{d,s}$ mixing processes. With only one Higgs doublet or at small $\tan\beta$, these two contributions are dominated by the Yukawa coupling of the top quark and are forced to be equal. For large $\tan\beta$, the additional contribution from the bottom Yukawa coupling cannot be neglected and the two quantities are in general different. In both cases, one can use the output of the UUT given in Table 3 and in the right plot of Fig. 2 to obtain a constraint on $\delta S_0^{K,B}$ using ϵ_K and Δm_d. We get $\delta S_0 = \delta S_0^K = \delta S_0^B = -0.12 \pm 0.32$ for small $\tan\beta$, while for large $\tan\beta$ we obtain $\delta S_0^B = 0.26 \pm 0.72$ and $\delta S_0^K = -0.18 \pm 0.38$. Using the procedure detailed in [11], these bounds can be translated into lower bounds on the MFV scale Λ:

$$\Lambda > 5.9\,\text{TeV@95\% Prob. for small } \tan\beta$$
$$\Lambda > 5.4\,\text{TeV@95\% Prob. for large } \tan\beta \tag{2}$$

significantly stronger than previous results $\Lambda > 3.6\,\text{TeV}$ and $\Lambda > 3.2\,\text{TeV}$ for small and large $\tan\beta$ respectively [7].

4 Constraints on SUSY from $B_S - \bar{B}_S$ mixing

The last item we would like to present is the impact of the CDF measurement of Δm_s on SUSY sources of $b \to s$ transitions, following [12].

To fulfill this task in a model-independent way we use the mass-insertion approximation. Treating off-diagonal sfermion mass terms as interactions, we perform a perturbative expansion of FCNC amplitudes in terms of mass insertions. The lowest nonvanishing order of this expansion gives an excellent approximation to the full result, given the tight experimental constraints on flavour changing mass insertions. It is most convenient to work in the super-CKM basis, in which all gauge interactions carry the same flavour dependence as SM ones. In this basis, we define the mass insertions $(\delta^d_{ij})_{AB}$ as the off-diagonal mass terms connecting down-type squarks of flavour i and j and helicity A and B, divided by the average squark mass.

The constraints on $(\delta^d_{23})_{AB}$ have been studied in detail in [13], using as experimental input the branching ratios and CP asymmetries of $b \to s\gamma$ and $b \to s\ell^+\ell^-$ decays, and the lower bound on $B_s\text{–}\bar{B}_s$ mixing previously available. An update using the summer 2005 data has been presented in [14]. We perform the same analysis using the most recent CDF result for $B_s\text{–}\bar{B}_s$ mixing reported in Sect. 1, and refer the reader to [13] for the details of the procedure.

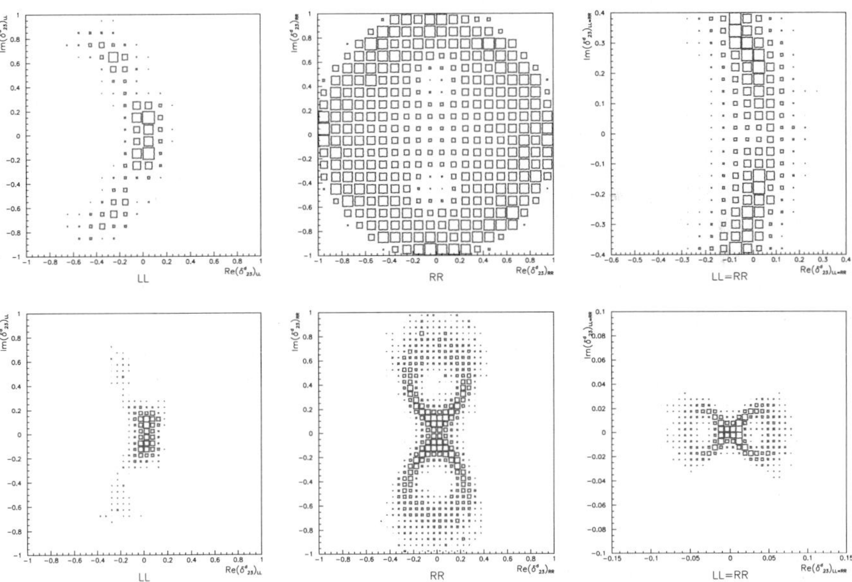

Fig. 4. Allowed range in the $\mathrm{Re}(\delta^d_{23})_{LL}$-$\mathrm{Im}(\delta^d_{23})_{LL}$ (*left*), $\mathrm{Re}(\delta^d_{23})_{RR}$-$\mathrm{Im}(\delta^d_{23})_{RR}$ (*center*) and $\mathrm{Re}(\delta^d_{23})_{LL=RR}$-$\mathrm{Im}(\delta^d_{23})_{LL=RR}$ (*right*) planes. In the plots on the top (*bottom*), the lower bound (measurement) on Δm_s is used. See the text for details

For definiteness, we choose an average squark mass of 350 GeV, a gluino mass of 350 GeV, $\mu = -350$ GeV and $\tan\beta = 3$. In Fig. 4, we present the allowed range in the $\mathrm{Re}(\delta_{23}^d)_{AA}$-$\mathrm{Im}(\delta_{23}^d)_{AA}$ plane, for $A = L, R$, using the previous lower bound (top) or the present measurement (bottom) of Δm_s. For the other mass insertions $(\delta_{23}^d)_{LR,RL}$, the constraint from Δm_s is irrelevant since the main effect comes from $\Delta F = 1$ processes. This leaves open the possibility of sizable deviation from the SM prediction in the CP asymmetries in $b \to s$ penguin decays [14]. Finally, we notice that there is still plenty of room to observe SUSY-generated CP violation in the B_s system.

We warmly thank the organizers for the very pleasant and stimulating atmosphere of the workshop, and the contributors to the flavour physics session for their interesting presentations. This work has been supported in part by the EU network "The quest for unification" under the contract MRTN-CT-2004-503369.

References

1. These proceedings, contributions by E. Baracchini, T. Cartaro, M. Dreucci, R. Ferrandes, P. Gambino, P. Massarotti, F. Mescia, M. Morello, N. Neri, P. Paradisi, G. Passaleva, V. Porretti, G. Salamanna and P. Santorelli.
2. N. Cabibbo, Phys. Rev. Lett. **10**, 531 (1963); M. Kobayashi and T. Maskawa, Prog. Theor. Phys. **49**, 652 (1973).
3. A. Abulencia *et al.* (CDF Collaboration), FERMILAB-PUB-06-344-E, hep-ex/0609040. Submitted to Phys. Rev. Lett.
4. H. G. Moser and A. Roussarie, Nucl. Instrum. Methods Phys. Res., Sect. A **384**, 491 (1997).
5. V. Vagnoni, these proceedings. UTfit Collaboration, `http://www.utfit.org`.
6. M. Bona *et al.* [UTfit Collaboration], JHEP **0610**, 081 (2006).
7. M. Bona *et al.* [UTfit Collaboration], JHEP **0603**, 080 (2006).
8. M. Bona *et al.* [UTfit Collaboration], Phys. Rev. Lett. **97**, 151803 (2006).
9. E. Gabrielli and G. F. Giudice, Nucl. Phys. B **433**, 3 (1995) [Erratum-ibid. B **507**, 549 (1997)]; M. Misiak, S. Pokorski and J. Rosiek, Adv. Ser. Direct. High Energy Phys. **15**, 795 (1998); M. Ciuchini, G. Degrassi, P. Gambino and G. F. Giudice, Nucl. Phys. B **534**, 3 (1998); C. Bobeth *et al.*, Nucl. Phys. B **726**, 252 (2005); M. Blanke, A. J. Buras, D. Guadagnoli and C. Tarantino, arXiv:hep-ph/0604057.
10. A. J. Buras *et al.*, Phys. Lett. B **500**, 161 (2001).
11. G. D'Ambrosio *et al.*, Nucl. Phys. B **645**, 155 (2002).
12. M. Ciuchini and L. Silvestrini, Phys. Rev. Lett. **97**, 021803 (2006).
13. M. Ciuchini, E. Franco, A. Masiero and L. Silvestrini, Phys. Rev. D **67**, 075016 (2003) [Erratum-ibid. D **68**, 079901 (2003)].
14. L. Silvestrini, Int. J. Mod. Phys. A **21**, 1738 (2006).

Neutrinos and Cosmic Rays: Session Summary

Eligio Lisi[1] and Laura Patrizii[2]

[1] Istituto Nazionale di Fisica Nucleare, Sezione di Bari, Italy
   `eligio.lisi@ba.infn.it`
[2] Istituto Nazionale di Fisica Nucleare, Sezione di Bologna, Italy
   `laura.patrizii@bo.infn.it`

The IFAE 2006 parallel session on "Neutrinos and Cosmic Rays" hosted 14 presentations, which are representative of the current Italian contribution to neutrino physics (with particular attention to flavor oscillations and absolute masses) and to cosmic ray physics, from both theoretical and experimental viewpoints. In this summary we report upon the main topics presented and discussed at the session.

1 Introduction

Neutrinos and Cosmic Rays represent two very important topics in the relatively new field of "Astroparticle Physics," namely, the field of research at the frontier of both particle physics and astrophysics, which has witnessed extraordinary achievements in recent years. Just to make one example, the discovery of atmospheric neutrino oscillations has not only provided the first evidence for new physics beyond the standard electroweak model, but has also shown the synergy between theoretical calculations of cosmic ray-induced neutrino fluxes and atmospheric neutrino detection techniques. There is a long tradition of Italian contributions to neutrino and cosmic ray physics, whose most recent aspects have been extensively presented and discussed in our parallel session in 14 talks (by M. Sioli, A. Ianni, M. Cirelli, F. Gatti, M. Pavan, M. Frigerio, A. Mirizzi, D. Meloni, M. Maltoni, G. Riccobene, D. De Marco, A. Chiavassa, V. Vitale, and P. Salvini). In the following, we summarize the main highlights, within the more general context of current astroparticle physics.

2 Neutrinos

Neutrino flavor transitions have been firmly established in the last few years, thanks to many beautiful experiments using both "artificial" sources (accelerator and reactor neutrinos) and "natural" sources (solar and atmospheric neutrinos), as reviewed in the two introductory talks by Max Sioli and Aldo Ianni respectively.

Max Sioli discussed accelerator and reactor beams in a temporal perspective, starting with first-generation experiments which have provided either positive (KamLAND, K2K) or negative (CHOOZ) or controversial (LSND) results; and continuing with current second-generation experiments (MINOS, OPERA, MiniBoone) and near-future projects for the next decade (T2K, NOvA, and Double Chooz), which might be followed by far-future (>2015–2020) and challenging experiments with beta-beams, superbeams, or neutrino factories. The Italian contribution to this rich program is particularly evident in the crucial CHOOZ reactor experiment (which has set the dominant upper limit on the mixing angle θ_{13}), in the CERN-to-Gran Sasso long-baseline experiments (which should observe for the first time the direct transformation from muon to tau neutrino flavor) and in simulations and optimization studies for far-future experiments (aiming at observing leptonic CP violation, if present). The Italian commitment in next-decade projects (mainly devoted to improve the sensitivity to θ_{13}) is instead still under discussion in the scientific community, with different viewpoints that emerged also at this Meeting.

Aldo Ianni reviewed natural neutrino beams, in particular from the Sun, from supernovae, and from Earth radioactivity (geo-neutrinos), leaving atmospheric neutrinos to later talks. In this field, the Italian contribution is traditionally strong in solar neutrinos, first with the (completed) GALLEX-GNO experiment at Gran Sasso, and currently with the Borexino project (in the same laboratory), which aims at measuring the ^{7}Be solar neutrino flux. Two good news were reported for Borexino: 1) the schedule is proceeding well after the well-known stop due to accidental oil spill; and 2) preliminary studies indicate that it might be possible to push the sensitivity down to solar "pep" neutrinos – although the competition with other projects in Japan (KamLAND) and Canada (SNO) will be tough. It was also remarked that a future galactic supernova explosion would be a fantastic opportunity at Gran Sasso, which has three sensitive detectors (Borexino, LVD, and Icarus-T600) at the same location.

In this context, **Alessandro Mirizzi** emphasized that future Supernova neutrino observations in large volume detectors might provide an important tool to probe, via matter effects on neutrino oscillations: 1) the θ_{13} mixing angles; 2) the neutrino mass hierarchy; and 3) the dynamics of shock waves after core collapse. Moreover, the observation of relic neutrinos from all past supernova explosions (which seems a relatively accessible goal in the next decade) might also provide constraints on cosmological issues, such as the star formation rate.

Neutrinos of astrophysical or cosmological origin are raising increasing interest. In particular, a fast-moving front is represented by cosmic relic neutrinos, whose combined effect on the microwave background and on structure formation leads to $O(\text{eV})$ bounds on the absolute neutrino masses, already competitive with laboratory bounds. The physics and the latest results in the field were discussed by **Marco Cirelli**, together with the intriguing hypothesis that the dark energy scale and the neutrino mass scale – which are accidentally (?) comparable – might be physically related, leading to a new dynamical origin for the neutrino masses ("mass-varying neutrinos")

The discussion on absolute neutrino masses continued with two experimental talks on single beta and double beta decay searches. While cosmological data constrain the sum of neutrino masses, single beta decay searches constrain independently a weighted sum of neutrino squared masses. The status of beta-decay searches was discussed by **Flavio Gatti**, with emphasis on the most important Italian contribution in this field, namely, the use of calorimetric ^{187}Re detectors. This technique (MANU, MiBETA experiments) is rapidly becoming competitive with spectroscopic experiments (which will reach their ultimate sensitivity with KATRIN in Germany), and has very good chances to become the next frontier in this field (MARE project).

The current experimental frontier in the field of neutrinoless double beta decay is instead represented by the Cuoricino (and future CUORE) ^{130}Te experiments which, together with the ^{76}Ge GERDA project, will be in pole position to probe the controversial evidence claimed by the Heidelberg–Moscow ^{76}Ge experiment, and hopefully find an unambiguous $0\nu2\beta$ signal. **Maura Pavan** reviewed the current status of the Cuoricino, CUORE and GERDA experiments at Gran Sasso and of the other competitors in the world, as well as the prospects for the next decade. It was emphasized that future data will bring important information (or constraints) not only on the neutrino nature (Dirac or Majorana) but also on the mass spectrum hierarchy – both being crucial open problems in neutrino physics.

Can we understand theoretically the values of neutrino masses and mixings? Maybe yes, if they are associated to some symmetry group linking the three lepton families and, even better, linking the lepton and quark sectors within a grand-unified theory. **Michele Frigerio** discussed, in particular, recent results in models with A_4 family symmetry, and with SO(10) grand unification, emphasizing that discrimination of models will require further exploration of related topics (quark-lepton symmetry, gauge unification, leptogenesis, rare process with lepton flavor violation), as well as greater accuracy in neutrino parameter measurements.

This brings us to the issue of how to improve significantly the current knowledge of neutrino mass-mixing parameters, trying to get the most from the difficult (and costly) experiments that might be approved in the future. **Michele Maltoni** showed, in particular, how to make the best use of a possible large-volume Cherenkov neutrino detector under discussion in the Frejus site (MEMPHYS project). On the one hand, such a detector would provide high-statistics atmospheric neutrino data; on the other hand, it might be used as a target for a dedicated neutrino beam created at CERN. The combined analysis of "natural" and "artificial" neutrino events shows that indeed a significant progress can be achieved with this setting, in particular by eliminating spurious solutions for the mass-mixing values ("degeneracy" problem).

Davide Meloni also discussed how to solve the degeneracy problem in the context of future long-baseline accelerator experiments, devoted to the search for CP violation in the neutrino sector. It was shown how the combination of different flavor oscillation channels can serve this purpose as well.

3 Cosmic rays

Cosmic ray physics is living a very exciting period, not only because many open problems (such as the origin of the very high-energy events observed) seem to point towards new unexpected phenomena, but also because new experimental techniques and projects are opening really new windows in this field, where the Italian contribution has traditionally been quite strong. The cosmic ray session at this Meeting, devoted to the latest news about gamma and neutrino astronomy and about charged cosmic ray physics, was introduced by two review talks on experiments and theory (by Andrea Chiavassa and by Daniel De Marco, respectively).

Andrea Chiavassa discussed the puzzling features of the cosmic ray energy spectrum: the "knee" (its possible sources, and their discrimination, e. g., through studies of chemical composition and of anisotropy); the "ankle" and beyond (super-GZK events, and latest results from the AUGER experiment). The talk illustrated the impressive experimental effort which is being made all over the world in order to: 1) build balloon-driven or ground-based detectors (with scintillator, Cherenkov, calorimetric, emulsion, transition radiation techniques) able to separate the chemical components; 2) realize very large-area detectors, able to collect enough statistics to test the high-energy part of the cosmic ray spectrum.

Daniel De Marco reviewed the (separate) issues of production and propagation of the highest energy cosmic rays, which might be either accelerated by shocks and/or magnetic fields (bottom-up approach) or be produced in decays of hypothetical very massive particle or topological defects (top-down approach), each mechanism having its own pros and cons. The signatures of different cosmic ray sources (and of the associated propagation effects) are not yet visible, however, in the AGASA, HiRes, and first AUGER spectra. In the future, it will be crucial not only to increase the statistics, but also to exploit the interplay between gamma ray and neutrino induced events (since both γ's and ν's are produced in decays of charged and neutral pions, created by absorption of accelerated protons).

Giorgio Riccobene emphasized that neutrino astronomy (with large volume detectors) can not only "point" towards the sources of the ultra high energy cosmic rays, but also shed light on their production mechanism (since neutrinos are produced in high-energy hadronic processes). Even the largest detector up-to-date (AMANDA), however, is not sensitive enough to see point sources – a task left to its km^3 version (ICECUBE). A large part of the talk was devoted to two underwater projects with strong Italian contributions

(ANTARES and NEMO) and to a possible km^3 evolution, which seems necessary to compete with ICECUBE and to provide independent and complementary observations.

The two final talks were devoted to gamma ray astronomy, where great experimental progress is being achieved in different directions: higher sensitivity (HESS, MAGIC), lower energy threshold (GLAST), and larger aperture plus high duty cycle (Milagro, ARGO-YBJ). **Vincenzo Vitale** discussed the difference between direct observation of gamma rays with satellite experiments and the indirect one with ground-based installations, focussing on the latter. The results of the HESS galactic plane survey were presented, together with a discussion of HESS and MAGIC sources identified at the level of several sigmas, including pulsar wind nebulae and supernova remnants. The MAGIC observations of gamma ray bursts were also analyzed. **Paola Salvini** presented other ground-based techniques for gamma ray detection, as used in ARGO-YBJ (RPC carpet) and Milagro (water-Cherenkov) to reconstruct the shower time-space pattern with full-coverage detectors. ARGO-YBJ is starting data taking, while Milagro has already detected known gamma sources, as well as a new extended source in the Cygnus region.

4 Synergy between neutrino and cosmic ray Physics

Atmospheric neutrinos represent perhaps the best known and most successful example of synergy between neutrino and cosmic ray physics: Interactions of cosmic rays in the Earth atmosphere produce showers which also contain neutrinos from pion, kaon, and muon decays – neutrinos which have carried the first indications of physics beyond the Minimal Standard Model via muon flavor disappearance (Super-Kamiokande, MACRO, Soudan-II, 1998). It has been shown in this Session that higher-statistics atmospheric neutrino searches can significantly contribute to improve our knowledge of the neutrino mass-mixing parameters; therefore, our understanding of the primary cosmic ray flux and spectrum should also be refined, in order to reduce the inherent systematics. Supernova neutrinos may also probe one of the most energetic processes in the universe (the core collapse and the subsequent shock wave) via matter effects on flavor oscillations; shedding light on the shock mechanisms (which may also generate high-energy cosmic rays) will be beneficial to both fields. Next-generation neutrino telescopes will also test neutrino properties at unprecedented energies and propagation pathlengths, where new unexpected phenomena might be revealed. For instance, pion-decay neutrinos have well-defined flavor ratios at the origin, which are modified in a predictable way by flavor oscillations: "unexpected" flavor ratios might thus signal new physics either in the astrophysical sources or in the propagation over large distances. These and other examples of the synergy between neutrino and cosmic ray physics show the usefulness of a joint Session as in this Meeting.

5 Acknowledgments

We thank all the speakers for having contributed to a very lively and interesting session. We are grateful to the organizers of IFAE 2006 for their kind hospitality in Pavia and for the excellent organization of the meeting; special thanks go to Oreste Nicrosini for his help and patience. The work of E.L. is supported in part by MIUR and INFN through the "Astroparticle Physics" research project.

6 References

This summary is based on the following talks (original titles in Italian), available in these Proceedings and in electronic format at www.pv.infn.it/~ifae2006, to which the interested reader is referred for further details on specific topics:

- Max Sioli: *Oscillazioni di neutrini con sorgenti artificiali*
- Also Ianni: *Oscillazioni di neutrini con sorgenti naturali*
- Marco Cirelli: *Cosmologia e neutrini con massa fissa e variabile*
- Flavio Gatti: *Massa del neutrino da decadimento beta*
- Maura Pavan: *Esperimenti sul decadimento doppio beta*
- Michele Frigerio: *Modelli teorici per le masse dei neutrini*
- Alessandro Mirizzi: *Fisica dei neutrini da supernova*
- Davide Meloni: *Violazione di CP nel settore leptonico*
- Michele Maltoni: *Sinergie fra ricerche con neutrini da acceleratore e atmosferici*
- Giorgio Riccobene: *Astronomia a neutrini con Km3 sott'acqua e sotto il ghiaccio*
- Daniel De Marco: *Fisica dei raggi cosmici di altissima energia: aspetti teorici*
- Andrea Chiavassa: *I raggi cosmici di alta e ultra alta energia*
- Vincenzo Vitale: *Astronomia gamma con telescopi Cherenkov*
- Paola Salvini: *Astronomia gamma con esperimenti a copertura totale*

Detectors and New Technologies

A. Cardini[1], M. Michelotto[2], and V. Rosso[3]

[1] INFN Sezione di Cagliari, Italy `alessandro.cardini@ca.infn.it`
[2] INFN Sezione di Padova, Italy `michele.michelotto@pd.infn.it`
[3] Dipartimento di Fisica "E. Fermi", Università di Pisa e Sezione INFN di Pisa, Italy `valeria.rosso@pi.infn.it`

This paper summarizes the talks given in the "Detector and New Technologies" Parallel Session. This session gives an overview of some of the research activities carried out within the *National Scientific Committee V*. As stated on the INFN web page, "The National Scientific Committee V is a precursor and incubator for new projects devoted to INFN's experiments: the researchers group develops materials, apparatus, new or improved procedures or, more generally, cutting-edge technologies for experiments in nuclear, subnuclear and astro-particle physics". The activity of this Scientific Committee develops in four main items:

- Particle Detectors
- Particle Accelerator
- Electronics and Software Development
- Interdisciplinary application of INFN cutting-edge Technologies

As conveners of this session, we had a very difficult task in selecting a few research activities within the more than a hundred excellent projects funded by the National Scientific Committee V. Developing new particle detectors has always been a very strong tradition within INFN. We selected here a few presentation on the latest development on silicon detectors for future experiments, and also on how new ideas on particle detection can be used to built state-of-the-art experiments to study very rare events. INFN has always been very active in developing custom VLSI chips for specific purposes. In this parallel session we will show that this field of research is still very active and that custom VLSI chips are definitely needed for cutting-edge experiments. Today's experiment have strong requests in hardware, but they also need adequate computing technologies. INFN partecipate in the developing of advanced software technologies for current and future experiments, and a few talks of this parallel session covered this important item. INFN is also investing in the use of particle accelerators to provide high energy hadron beams for cancer treatment. A few talks were also given on this very important research activity and on related items. We would like to underline here the fact that the Centro Nazionale di Adroterapia Oncologica (CNAO) is currently under construction here in Pavia.

1 New Detector Technologies

New detectors are under development to fulfill the requirements of future high energy physics experiments. In this session we will focus on new solid state detector technologies, in particular on CMOS Monolithic Active Pixel Sensors (MAPS) and on three-dimensional (3D) silicon detectors. These new detectors are foreseen to be used in light multi-layer vertex detectors, and in order to operate under heavy radiation fluxes a lot of the R&D work was addressed to the enhancement of their radiation tolerance.

Development of 3D Silicon Detectors at ITC-IRST

The 3D detectors have the characteristic of being fully depleted at very low voltages by introducing vertical electrodes. This property allows for very fast signal response and enhanced radiation tolerance. ITC-IRST in Trento is currently developing these very promising devices.

Monolithic Active Pixel Sensors in 130 nm CMOS Technology

CMOS MAPS incorporate on the same substrate the readout electronics and an extremely thin active radiation detector, few tens of a micron thick. The use of commercial CMOS technology allows either low power consumption and low fabrication costs. Furthermore, the MAPS sensor readout electronics will benefit from the high degree of radiation hardness offered by modern deep submicron CMOS technology.

2 New Apparatus

After a successfull research and development phase, new detector technologies are used to build important parts of an experiment. INFN has a longstanding leadership position in the developement of liquified noble gas detectors, and in this section we had the contribution of two experiment, MEG and WARP, which use this technology in their searches for very rare events.

The Liquid Xenon Calorimeter of the MEG Experiment

The aim of the MEG experiment is to search for the lepton flavour violating decay $\mu \to e\gamma$. While this decay mode is forbidden in the Standard Model and allowed at a negligeable rate in the case of massive neutrinos, it is usually enhanced by all Standard Model Extension. This makes this decay a very interesting channel when looking for physics beyond the Standard Model. The experiment is performed at the Paul Scherrer Institute where a high intensity muon beam is available. Muons come to rest in a target and their decays are studied by means of a high precision spectrometer – to reconstruct the electron trajectory – and a Liquid Xenon Calorimeter – to reconstruct the photon energy. A high intensity beam is needed to reach the desired sensitivity in a few years of data taking, but it also set very strong requirements on the detector performances. The MEG calorimeter needs to accurately measure the photon energy, direction and event time. It consist of a $0.8\,\mathrm{m}^3$ liquid xenon volume readout by 846 photomultiplier. The photomultiplier signals are digitized at $2\,\mathrm{GHz}$ to reject pile-up events. A lot of R&D work was performed to optimize the photomultiplier behaviour at low temperature and at the high illumination level at which they will work. The liquid xenon purity has also been a concern because the excellent energy resolution would have been degraded by an important scintillation light absorption. Measurements performed on a 70 litres prototype have shown that this type of detector fulfills the experiment requirements, and the final calorimeter is currently under construction.

WARP: a two-phase Argon Detector for a Direct Search of Dark Matter

The WARP experiment is looking for the elastic scattering of weakly interacting massive particles (WIMPs) in a liquid argon active target. A WIMP elastic scattering produces nuclear recoils that have to be identified against a huge background of cosmics and radioactive decays, and this is why WARP is located inside the Gran Sasso Underground Laboratory. The WIMP active target is seen by photomultipliers which have the purpose of measuring both the excitation and the ionization signal produced by the WIMP interaction. The excitation is measured by looking at the prompt scintillation light produced at the WIMP elastic scattering site. Ionization electrons produced during the WIMP collision drift to the liquid argon surface and are successively extracted in the gaseous argon where they are collected by a set of anodic wires after electron multiplication. The secondary excitation generated in the electron multiplication avalanche produces a delayed scintillating light emission which is also seen by the photomultipliers. To discriminate between WIMP recoils and background events two different criteria are used: the ratio between the prompt and the delayed light emission intensity and the pulse shape of the light emission. The experiment estimate a rejection power of the order of 10^7 for electron-like events by combining the two criteria. Data taking is currently underway.

3 Electronics for the Experiments

INFN has always played an important role in the design of experiment electronics system. Cutting-edge experiment often need custom electronics to provide optimal performances, and inside INFN a large community of VLSI chip designers exist. In this session we had a report from two experiments, CDF and LHCb, which have solved the complexity of their detector systems by means of special purpose custom chips developed by INFN researchers.

The CDF SVT Upgrade

The purpose of the CDF Silicon Vertex Trigger (SVT) is to provide an identification of large impact parameter tracks in an event – the signature of production and decay of heavy quark hadrons – at the second level trigger stage. The SVT receives data from the Silicon Vertex detector (SVX) and the Central Outer Tracker (COT) and reconstructs the particle trajectories with quasi-offline quality in less than $20\,\mu s$. The pattern recognition prior to track fitting is performed in a highly parallel way by an array of Associative Memory custom chips. In order to cope with the higher luminosity in the current phase of the CDF experiment the SVT system has been upgraded and a more powerful Associative Memory chip was designed and produced. The CDF experiment is continuously triggering on high impact parameter tracks and this is leading to significant physics results in particular in the sector of B physics.

Electronics for the Time Calibration
of the LHCb Muon Detector

The main purpose of LHCb, one of the four LHC experiments, is to study B physics at very high statistics. LHCb is a forward magnetic spectrometer equipped with a highly segmented silicon vertex detector, two RICH detectors for accurate particle identification, electron and hadron calorimetry and a complex muon detector. High transverse momentum muon are present in many heavy quark decay final states and LHCb exploits this by means of a muon system that can identify high p_t muons at $40\,\mathrm{MHz}$, the rate of collisions at LHC. The muon detector consist of about 1380 multi-wire proportional chambers located in five separate muon station, each with overall size of about $10 \times 10\,\mathrm{m}^2$, for a total of 120,000 readout channels. To provide an efficient first level muon trigger all readout channels need to be accurately time-adjusted at the nanosecond level. Two VLSI custom chips have been designed for this purpose. They provide the necessary logics and programmable delays to adjust the timing of each channel and all the functionnality needed to synchronize the full muon detector in a standalone mode. LHCb will start its data taking phase at the end of 2007.

4 Computing Technologies

The section on computing technologies has seen the contribution of IBM, a talk on the actual use of grid technologies in the WLCG Service Challenge, a talk on the new Ajax paradigm for user interaction and a report on the event filter in CMS and Atlas, the two large general purpose detectors ready to start operation next year.

High-Performances Computing Architecture and Data Processing Solutions in High Energy Physics

IBM is a major player in scientific computing, where the computing and data processing needs grow faster than the Moore's law prediction. It has the two fastest supercomputer and six of the top 10 supercomputer in the world. This result has been achieved thanks to the Power architecture, but in HEP field farms of distributed x86 worker node is more common solution. Recently the community saw an increased use of multi-core chip, where two or more processor are present in a single chip enclosure. The transition to smaller feature size (recently from 130 nm to 90 nm and down to 65 nm in 2006) allowed an increase in the number of cores instead of the maximum clock size; energy consumption had infact increased to worrying levels, causing several problems of heat removal inside the box and at the datacenter level. A complementary trend is the transition from the 1U rack mountable "pizza box" to the blade format that permits higher density and cleaner installation with a reduced number of cables and integrated management services. IBM is also involved in the Cell architecture, used mainly by game consoles but exploitable by HEP due to the 9 cores and interesting floating point performance. IBM is also pushing the Infiniband that provide a thousand MB/s bandwidth with a few microsecond latency. The GPFS is a parallel distributed filesystem that has found practical use in HEP to distribute the load of event serving to hundred of clients since it permits transparent I/O over a cluster interconnect.

An AJAX-based Client for a WEB-based Interactive Data Analysis

Ajax is a set of techniques and programming pattern, involving Javascript and XML, which allow to create web applications that give the same look and feel of traditional desktop applications. A normal web application is requested, processed and sent to the client. If the page is complex this imposes at least a few seconds of waiting on users. Ajax breaks down the interaction in small request that are processed asynchronously. The following requests involve only the refresh of a small part of the page. Iguana is the graphical analysis framework of CMS which has been ported to Ajax. The major part of the code is identical to the standard application and only the GUI part has been plugged on top of the Iguana Object Model to complement the normal

QT interface. Other example shown during the presentation are the Geant4 visualization in which users can perform navigation by mouse, panning and zooming in the client side and an Event Display for the CMS offline software showing the first cosmic events. This technology is very promising since it permits the access of physics data from a normal web browser, with no need of installing specific experiment software or even specific plugins. The use of AJAX is spreading also in various commercial application (Amazon, Google Earth, Google Mail to name a few).

Service Challanges and Grid-based Computing Architecture in the LHC Era

The computing of the LHC experiments is based on a Tiered hierarchy of Computing Centers. The Tier0 at CERN receives data from the experiments, record them on permanent MSS and distributes them to the Tier1. Tier1 provide a distributed permanent back-up of the raw data and provide the first processing of the events and data intensive analysis and reprocessing. Tier2 provide grid-enabled disk storage and concentrate on task such as simulation and end-user analysis Tier1 centers represent an integral part of the LHC experiments and they commit to undertake a number of services on a long-term basis (at least 5 years) with a high level of availability. All storage and computational services are "grid enabled", according to standard between CERN and the regional centers. The service challenges are a set of efforts towards the achievements of data services with a rate of performance and availability without precedent in the HEP history. Some of these challenges met their throughput goal (e. g. 100 MB/s to every T1 site, 500 MB/s sustained out of CERN for SC2) but failed to achieve their service goal (reliability, availability, end-to-end unattended performance). Service Challenge 3 and 4 bring the Tier2s in the loop to fully address experiments use-cases. The SC4 aim is to demonstrate that all of the offline data processing requirements expressed in the experiments Computer Model can be handled by the World-wide LHC Computing Grid at the full nominal rate of the LHC. In the first phase (until April 06) the aim is to achieve 3 weeks of sustained data rate to the targets. From May to September all the basic software components required for the initial LHC data processing must enter the loop.

The ATLAS Event Filter

The Atlas detector needs a DAQ capable of detecting about one event on a million, reducing the original 40 MHz rate to a still impressive 200 Hz 1.5 MB events. The talk concentrates on the 3rd level trigger that is the first level in the chain that operates on the complete event. The input rate is about 2 kHz and a decision must be taken in about one second. The physics algorithms are the same used in the offline code to simplify the validation of the code, while the Level2 Trigger, working only on the region of interest of the detector, uses algorithms developed for the online environment. The selection is based on a tree of chained step, where the outcome of one step is the seed of the following. The events can be rejected ad each step (early rejection). The Event Filter is organized as a set of several subfarms connected to the output links of the Event Builder. The aggregated performances are equivalent to $16\,K$ $1\,GHz$ cpu, achievable with about 1000 biprocessor dual core modern cpu. The use of memory mapped file and the decoupling between data flow and data processing provides data security and fault tolerance. The system is scalable and permits the hot swapping of processing resources that can be heterogeneously and geographically distributed. The system has been validated in several conditions and is now in the final commissioning phase.

The CMS High Level Trigger

The CMS High Level Trigger must cope with $100\,kHz$ of $1\,MB$ events coming from the first trigger level. The number of channels and bandwidth of $1000\,Gbps$ in similar to the Atlas DAQ but here the HLT is implemented in a single step only at software level on a farm of commercial PC. The software trigger is very flexible and permits to easily add new physics channels, giving a great freedom on which data one can use and to choose algorithms complexity exploiting the maximum of the benefit from technology advancements. The HLT operates on full events but applyes the algorithms only to region of interest in the sub-detectors and reconstructs the physics object only for what is needed to select the event. The algorithms are divided in sub-levels in order to discard events as soon as possible. The selected events then travel through all the selection processes and are divided in streams. In the example shown of the muon trigger, the first processing step, using the candidates coming from L1 trigger, interpolates the trajectory from internal to external muon detectors fitting a path down to the interaction region. The second sublevel, using the candidates from the previous step, reconstructs the muon trajectory in the tracker. The cpu time needed to process an event varies from 50 to $710\,ms$ with an average of about $300\,ms$ on a $1\,GHz$ PIII CPU. The physics startup need about 15000 of these reference CPU. The 2007 cpu technology using dual processor with dual core and higher clock could offer enough performance to process an events in $40\,ms$ on a 1000 dual cpu box Filter Farm. The DAQ model has been test and validated both on algorithms and on the computing side but some problems are still open: new selection of combined objects, beam halo effects, the exact determination of events occupancy, access time to the online raw data among the others.

5 Interdisciplinary Applications

The applications of nuclear techniques, innovative instrumentations and of the know-how used in the field of High Energy Physics (HEP) can produce relevant developments in different fields like Biology, Medicine and Cultural Heritage. In the following some examples will be given.

Hadron Radiobiology and Its Implications in Hadrotherapy and Radiation Protection

The therapeutic effectiveness of conventional X-ray and electron radiation in tumors treatment is intrinsically limited by its physical and radiobiological properties. In fact, the dose-depth distribution and the lateral scattering limit the physical selectivity for high precision irradiation of deep-seated tumors. On the contrary high-energy ionizing radiation has proved to be effective in the treatment of tumors. In fact many studies have been taken to evaluate the relative biological effectiveness of hadrons and the survival rate have been evaluated; these studies have been conduced also at cellular level and a micro-beam facility that can deliver targeted and counted particles to individual cells has been realized.

The CNAO Project

The results of the studies presented in the previous section, still ongoing, have lead to the realization of a dedicated facility for tumor treatment that will be located in Pavia (Italy), the Centro Nazionale di Adroterapia Oncologica Foundation (CNAO).

At CNAO an innovative accelerator that will allow proton and carbon ions cancer therapy using high precision active scanning is under construction. The synchrotron will offer the flexibility needed for dual-species operation and the variable energy needed for active scanning.

The Firenze Proton Microbeam: Setup and Applications

Protons are used not only in medical field, but also in the filed of Cultural Heritage. In fact Ion Beam Analysis (IBA) is a unique tool of non-destructive analysis, leading to obtain the complete composition of materials in works of art, or of archaeological or historical interest. The knowledge of their composition is important, on one hand, to solve problems of chronology, of art-historical interpretation, of authentication or discovery of fakes; on the other hand, it is a key point to take decisions concerning conservation procedures. In Florence a dedicated laboratory has been realized: the laboratory is equipped with a modern Tandem accelerator entirely dedicated to these purposes. The community of scholars in the field of the Cultural Heritage will thus get access to a "complete" offer of nuclear analyses and will rely on the specific competence of experienced nuclear physicists, which is of basic importance for a good running of such type of laboratories.

Theoretical progress in the muon g-2

M. Passera

Dipartimento di Fisica, Università di Padova and INFN, Via Marzolo 8,
35131 Padova, Italy
passera@pd.infn.it

Summary. Recent progress in the determination of the Standard Model prediction
for the anomalous magnetic moment of the muon is briefly reviewed.

During the last few years, in a sequence of increasingly precise measurements, the E821 Collaboration at Brookhaven has determined $a_\mu =$
$(g_\mu - 2)/2 = 116\,592\,080\,(63) \times 10^{-11}$ with a fabulous relative precision of 0.5
parts per million (ppm) [1], allowing us to test all sectors of the Standard
Model (SM) and to scrutinize viable alternatives to this theory [2]. A new
experiment, E969, has been approved (but not yet funded) at Brookhaven in
2004 [3]. Its goal is to reduce the present experimental uncertainty (still limited by statistical errors) to about 0.2 ppm. A letter of intent for an even more
precise g-2 experiment was submitted to J-PARC with the proposal to reach
a precision below 0.1 ppm [4]. But how precise is the theoretical prediction?

The SM prediction a_μ^{SM} is usually split into three parts: QED, electroweak
and hadronic (see [5] for recent reviews). The QED one arises from the subset
of SM diagrams containing only leptons and photons. The leading (one-loop)
contribution was first computed by Schwinger more than fifty years ago [6].
Also the two- and three-loop QED terms are known analytically (the latter
is mainly due to Remiddi and his collaborators [7]), while the numerical
evaluation of the four-loop term (about six times larger than the present
experimental uncertainty of a_μ), was first accomplished by Kinoshita and his
collaborators in the early 1980s. The latest results appeared in [8]. The leading
five-loop QED terms were recently evaluated in [9], and estimates obtained
with the renormalization-group method agree with these results [10]. The
total QED prediction currently stands at $a_\mu^{\mathrm{QED}} = 116\,584\,718.09\,(14)\,(08) \times$
10^{-11} [11]. The first error is determined by the uncertainty of the five-loop
QED coefficient, while the second is caused by the tiny uncertainty of the
new value of the fine-structure constant α [11, 12].

For the electroweak (EW) contribution, complete one- and two-loop calculations have been carried out. The two-loop part turned out to be unexpectedly large (about -22% of the one-loop) [13], but three-loop leading

logarithms are negligibly small. After several small refinements [14, 15], the EW contribution can be confidently given by $a_\mu^{\rm EW} = 154(1)(2) \times 10^{-11}$ [15]. The first error is due to hadronic loop uncertainties, while the second one corresponds to an allowed range of $M_H \in [114, 250]$ GeV, to the current top mass uncertainty, and to unknown three-loop effects.

The hadronic leading-order contribution $a_\mu^{\rm HLO}$, due to the hadronic vacuum polarization correction to the one-loop diagram, can be computed from hadronic e^+e^- annihilation data via the dispersion integral $a_\mu^{\rm HLO} = \frac{1}{4\pi^3} \int_{4m_\pi^2}^{\infty} ds\, K(s)\sigma^{(0)}(s)$ [16], where $\sigma^{(0)}(s)$ is the total cross section for e^+e^- annihilation into any hadronic state, with extraneous QED corrections subtracted off. A prominent role among all e^+e^- annihilation measurements is played by the precise data collected by the CMD-2 detector at the VEPP-2M collider in Novosibirsk for the $e^+e^- \to \pi^+\pi^-$ cross section at values of $\sqrt{s}$ between 0.37 and 1.39 GeV [17,18]. Also the SND Collaboration (at the VEPP-2M collider as well) released in 2005 its analysis of the $e^+e^- \to \pi^+\pi^-$ process for $\sqrt{s}$ between 0.39 and 0.98 GeV [19]. These data were recently reanalyzed, reducing the value of the measured cross-section [20]. The new SND result appears to be in good agreement with the corresponding one from CMD2. In 2004 the KLOE experiment at the DAFNE collider in Frascati presented a precise measurement of $\sigma(e^+e^- \to \pi^+\pi^-)$ via the initial-state radiation (ISR) method at the ϕ resonance [21]. This cross section was extracted for $\sqrt{s}$ between 0.59 and 0.97 GeV. There are some discrepancies between the KLOE and CMD-2 results, even if their integrated contributions to $a_\mu^{\rm HLO}$ are similar. The study of the ISR method is also in progress at BABAR [22] and Belle [23].

The latest evaluations of the dispersive integral are in good agreement: $a_\mu^{\rm HLO} = 6934\,(53)_{\rm exp}(35)_{\rm rad} \times 10^{-11}$ [24], $a_\mu^{\rm HLO} = 6948\,(86) \times 10^{-11}$ [25], $a_\mu^{\rm HLO} = 6924\,(59)_{\rm exp}(24)_{\rm rad} \times 10^{-11}$ [26], $a_\mu^{\rm HLO} = 6944\,(48)_{\rm exp}(10)_{\rm rad} \times 10^{-11}$ [27]. Further significant progress is expected from the e^+e^- collider VEPP-2000 under construction in Novosibirsk [28] and, possibly, from DAFNE-2 at Frascati [29].

The authors of [30] pioneered the idea of using vector spectral functions derived from hadronic τ decays to improve the evaluation of the dispersive integral. However, the latest analysis with ALEPH, CLEO, and OPAL data yields $a_\mu^{\rm HLO} = 7110\,(50)_{\rm exp}(8)_{\rm rad}(28)_{SU(2)} \times 10^{-11}$ [37], a value significantly higher than those obtained with e^+e^- data (see [31] for recent results from Belle). The puzzling discrepancy between the $\pi^+\pi^-$ spectral functions from e^+e^- and isospin-breaking-corrected τ data could be caused by inconsistencies in the data, or in the isospin-breaking corrections applied to the latter [32].

The hadronic higher-order contribution $a_\mu^{\rm HHO}$ can be divided in $a_\mu^{\rm HHO} = a_\mu^{\rm HHO}({\rm vp}) + a_\mu^{\rm HHO}({\rm lbl})$. The first term, $a_\mu^{\rm HHO}({\rm vp}) = -97.9\,(9)_{\rm exp}(3)_{\rm rad} \times 10^{-11}$, is the $O(\alpha^3)$ contribution of diagrams containing hadronic vacuum polarization insertions [26,33]. The second one, also of $O(\alpha^3)$, is the hadronic light-by-light contribution. As it cannot be directly determined via a dispersion relation approach using data, its evaluation relies on specific models of low-energy hadronic interactions with electromagnetic currents. Recent determination

Table 1. The *first column* shows $a_\mu^{\mathrm{SMO}} = a_\mu^{\mathrm{QED}} + a_\mu^{\mathrm{EW}} + a_\mu^{\mathrm{HLO}} + a_\mu^{\mathrm{HHO}}$. The values employed for a_μ^{HLO} are indicated by the reference on the *left*; all a_μ^{SM} values were derived with $a_\mu^{\mathrm{HHO}}(\mathrm{lbl}) = 80\,(40) \times 10^{-11}$. Errors were added in quadrature. The differences $\Delta = a_\mu^{\mathrm{EXP}} - a_\mu^{\mathrm{SM}}$ are listed in the *second column*, while the numbers of "standard deviations" (σ) appear in the *third* one. Lower discrepancies, shown in *angle brackets*, are obtained if $a_\mu^{\mathrm{HHO}}(\mathrm{lbl}) = 136\,(25) \times 10^{-11}$ is used instead of $80\,(40) \times 10^{-11}$

$a_\mu^{\mathrm{SM}} \times 10^{11}$	$\Delta \times 10^{11}$	σ
[24] 116 591 788 (76)	292 (98)	3.0 $\langle 2.5 \rangle$
[25] 116 591 802 (95)	278 (114)	2.4 $\langle 2.0 \rangle$
[26] 116 591 778 (76)	302 (98)	3.1 $\langle 2.6 \rangle$
[27] 116 591 798 (63)	282 (89)	3.2 $\langle 2.7 \rangle$
[37] 116 591 961 (70)	119 (95)	1.3 $\langle 0.7 \rangle$

vary between $a_\mu^{\mathrm{HHO}}(\mathrm{lbl}) = 80\,(40) \times 10^{-11}$ [34, 35] and $a_\mu^{\mathrm{HHO}}(\mathrm{lbl}) = 136\,(25) \times 10^{-11}$ [36]. This contribution may become the ultimate limitation of the SM prediction of the muon g-2.

The SM prediction for a_μ deviates from the present experimental value by 2–3 σ (or $\sim 1\sigma$), if data from e^+e^- collisions (or τ-decay) are employed to evaluate the leading-order hadronic term (see Table 1). The puzzling discrepancy between the $\pi^+\pi^-$ spectral functions from e^+e^- and isospin-breaking-corrected τ data could be caused by inconsistencies in the e^+e^- or τ data, or in the isospin-breaking corrections applied to the latter. Indeed, while some disagreements occur between data sets, requiring further detailed investigations, the connection of τ data with the leading hadronic contribution to a_μ is less direct, and one wonders whether all possible isospin-breaking effects have been properly taken into account.

If the E969 experiment will be funded, the present experimental uncertainty may be reduced by a factor of 2.5, or even less. While the QED and EW terms are ready to rival these precisions, much effort will be needed to reduce the hadronic uncertainty by a factor of two. This effort, challenging but possible, is certainly well motivated by the excellent opportunity provided by the muon g-2 to unveil (or constrain) "new physics" effects.

References

1. H.N. Brown *et al.*, Phys. Rev. D **62** (2000) 91101; Phys. Rev. Lett. **86** (2001) 2227; G.W. Bennett *et al.*, Phys. Rev. Lett. **89** (2002) 101804; **89** (2002) 129903 (E); **92** (2004) 161802; Phys. Rev. D **73** (2006) 072003.
2. A. Czarnecki, W.J. Marciano, Phys. Rev. D **64** (2001) 013014.
3. R.M. Carey *et al.*, Proposal of the BNL Experiment E969, 2004; B.L. Roberts, hep-ex/0501012; Nucl. Phys. Proc. Suppl. **155** (2006) 372.
4. J-PARC L.o.I. L17, B.L. Roberts contact person.
5. M. Passera, J. Phys. G **31** (2005) R75; Nucl. Phys. Proc. Suppl. **155** (2006) 365; M. Davier, W.J. Marciano, Ann. Rev. Nucl. Part. Sci. **54** (2004) 115;

M. Knecht, hep-ph/0307239; A. Nyffeler, Acta Phys. Polon. B **34** (2003) 5197; A. Czarnecki, W.J. Marciano, Nucl. Phys. Proc. Suppl. **76** (1999) 245; V.W. Hughes, T. Kinoshita, Rev. Mod. Phys. **71** (1999) S133.

6. J.S. Schwinger, Phys. Rev. **73** (1948) 416.

7. S. Laporta, E. Remiddi, Phys. Lett. B **379** (1996) 283; **301** (1993) 440.

8. T. Kinoshita, M. Nio, Phys. Rev. D **70** (2004) 113001; D **73** (2006) 013003.

9. T. Kinoshita, M. Nio, Phys. Rev. D **73** (2006) 053007.

10. A.L. Kataev, Phys. Lett. B **284** (1992) 401; Nucl. Phys. Proc. Suppl. **155** (2006) 369; hep-ph/0602098; hep-ph/0608120; A.L. Kataev, V.V. Starshenko, Phys. Rev. D **52** (1995) 402.

11. M. Passera, hep-ph/0606174.

12. G. Gabrielse *et al.*, Phys. Rev. Lett. **97** (2006) 030802; B. Odom *et al.*, Phys. Rev. Lett. **97** (2006) 030801.

13. T.V. Kukhto *et al.*, Nucl. Phys. B**371** (1992) 567; A. Czarnecki, B. Krause, W.J. Marciano, Phys. Rev. D **52** (1995) 2619; Phys. Rev. Lett. **76** (1996) 3267; S. Peris, M. Perrottet, E. de Rafael, Phys. Lett. B **355** (1995) 523.

14. M. Knecht *et al.*, JHEP **0211** (2002) 003; S. Heinemeyer, D. Stöckinger, G. Weiglein, Nucl. Phys. B**699** (2004) 103; T. Gribouk, A. Czarnecki, Phys. Rev. D **72** (2005) 53016; G. Degrassi, G.F. Giudice, Phys. Rev. D **58** (1998) 053007.

15. A. Czarnecki, W.J. Marciano, A. Vainshtein, Phys. Rev. D **67** (2003) 073006.

16. C. Bouchiat, L. Michel, J. Phys. Radium 22 (1961) 121; M. Gourdin, E. de Rafael, Nucl. Phys. B **10** (1969) 667.

17. R.R. Akhmetshin *et al.*, Phys. Lett. B **578** (2004) 285.

18. V.M. Aulchenko, JETP Lett. **82** (2005) 743; I. Logashenko, Proc. Sci. **HEP2005** (2005).

19. M.N. Achasov *et al.*, J. Exp. Theor. Phys. **101** (2005) 1053.

20. M.N. Achasov, hep-ex/0604051; hep-ex/0604052; hep-ex/0605013.

21. A. Aloisio *et al.*, Phys. Lett. B **606** (2005) 12; F. Nguyen, these proceedings.

22. V.Druzhinin, hep-ex/0601020; M.Davier, Nucl.Phys.Proc.Suppl.**144**(2005)238.

23. B.A. Shwartz, Nucl. Phys. Proc. Suppl. **144** (2005) 245.

24. A. Höcker, hep-ph/0410081.

25. F. Jegerlehner, Nucl. Phys. Proc. Suppl. **131** (2004) 213; J. Phys. G**29**(2003)101.

26. K. Hagiwara *et al.*, Phys. Rev. D **69** (2004) 093003.

27. J.F. de Trocóniz, F.J. Ynduráin, Phys. Rev. D **71** (2005) 073008.

28. Yu.M. Shatunov *et al.*, Proceedings of EPAC 2000, Vienna, June 2000, p.439.

29. F. Ambrosino *et al.*, hep-ex/0603056.

30. R. Alemany, M. Davier, A. Höcker, Eur. Phys. J. C **2** (1998) 123.

31. K. Abe *et al.*, hep-ex/0512071; H. Hayashii, PoS **HEP2005** (2006) 291.

32. W.J. Marciano, A. Sirlin, Phys. Rev. Lett. **61** (1988) 1815; V. Cirigliano, G. Ecker, H. Neufeld, Phys. Lett. B **513** (2001) 361; JHEP **0208** (2002) 002; S. Ghozzi, F. Jegerlehner, Phys. Lett. B **583**, 222 (2004); M. Davier, Nucl. Phys. Proc. Suppl. **131** (2004) 123; K. Maltman, Phys. Lett. B **633** (2006) 512; K. Maltman and C. E. Wolfe, Phys. Rev. D **73** (2006) 013004.

33. B. Krause, Phys. Lett. B **390** (1997) 392.

34. M. Knecht, A. Nyffeler, Phys. Rev. D **65** (2002) 73034; M. Knecht *et al.*, Phys. Rev. Lett. **88** (2002) 71802; A. Nyffeler, Acta Phys. Polon. B **34** (2003) 5197.

35. M. Hayakawa, T. Kinoshita, Phys. Rev. D **57** (1998) 465; **66** (2002) 019902(E); J.Bijnens, E.Pallante, J.Prades, Nucl. Phys. B**474** (1996) 379; **626** (2002) 410.

36. K. Melnikov, A. Vainshtein, Phys. Rev. D **70** (2004) 113006.

37. M. Davier *et al.*, Eur. Phys. J. C **31** (2003) 503.

Beyond leading-log approximation in the Parton Shower approach to Bhabha process

G. Balossini[1], C. M. Carloni Calame[2,1], G. Montagna[1,2],
O. Nicrosini[2], and F. Piccinini[2,1]

[1] Dipartimento di Fisica Nucleare e Teorica, via A. Bassi 6, I-27100 Pavia, Italy
[2] INFN Sezione di Pavia, via A. Bassi 6, I-27100 Pavia, Italy

Summary. An original matching procedure of the Parton Shower approach to Bhabha process with the complete exact $\mathcal{O}\left(\alpha\right)$ matrix elements is discussed, in relation to high-precision luminosity determination at flavour factories. For this purpose, a new version of **BABAYAGA** event generator has been released, with a theoretical error reduced down to 0.1%, as estimated by comparison with independent calculations.

1 Introduction

The physics programme at flavour factories has among its goals the measurement of the ratio $R = \sigma\left(e^+e^- \to \text{hadrons}\right)/\sigma\left(e^+e^- \to \mu^+\mu^-\right)$ in the energy range below about $10\,\text{GeV}$. In fact, this quantity provides a stringent test of the Standard Model predictions, i. e. a precise determination of the hadronic contribution, by means of R, required for reducing the error on the anomalous magnetic moment of the muon $(g-2)_\mu$ and the QED coupling constant $\alpha_{\text{QED}}\left(Q^2\right)$ [1].

To fix R accurately, the knowledge of the machine luminosity is needed. Nowadays the uncertainty on this parameter is quoted by KLOE collaboration at 0.6%: 0.3% experimental and 0.5% theoretical [2], i. e. the main part is yielded by the errors on the reference Bhabha cross section. So, an improvement in the high-precision computations, exploited by the already developed event generators (MCGPJ [3], BHAGENF [4], BHWIDE [5] and BABAYAGA [6]), is demanded.

In the following, we will discuss an original matching of the Parton Shower (PS) approach with the complete $\mathcal{O}\left(\alpha\right)$ matrix element and its implementation in a new version of **BABAYAGA**.

2 Matching next-to-leading order corrections with PS

Parton Shower is a Monte Carlo algorithm which solves exactly the DGLAP equation and one of its most important advantages is that all the events are generated exclusively. In the PS approach, the Bhabha cross section with the emission of an arbitrary number of photons takes the form

$$d\sigma_{\mathrm{LL}}^{\infty} = \Pi\left(Q^2, \varepsilon\right) \sum_{n=0}^{\infty} \frac{1}{n!} |\mathcal{M}_{n,\mathrm{LL}}|^2 \, d\Phi_n \, , \tag{1}$$

where $\Pi\left(Q^2, \varepsilon\right)$ is the Sudakov form-factor accounting for soft photon emission and virtual corrections, ε is an infrared separator between the soft and the hard radiation and Q^2 is the energy scale of the process. The square amplitude $|\mathcal{M}_{n,\mathrm{LL}}|^2$ describes, in LL approximation, the process with the emission of n hard photons, while $d\Phi_n$ is the exact phase-space element of the process with the emission of n additional photons with respect to the Born-like final-state configuration. Notice that in (1) the photonic radiative corrections are resummed up to all the orders of α.

The one-photon expansion of the squared matrix element is:

$$|\mathcal{M}_{1,\mathrm{LL}}|^2 = P\left(z\right) I\left(k\right) |\mathcal{M}_0|^2 \frac{8\pi^2}{E^2 z\left(1 - z\right)} \, , \tag{2}$$

where $1 - z$ is defined as the energy fraction carried by the emitted photon. $P\left(z\right)$ is the Altarelli–Parisi splitting function, while $I\left(k\right)$ is the photon angular spectrum including the initial-final state interference. Equation (2) allows to write the $\mathcal{O}\left(\alpha\right)$ expansion of (1):

$$d\sigma_{\mathrm{LL}}^{\alpha} = [1 + C_{\alpha,\mathrm{LL}}] |\mathcal{M}_0|^2 \, d\Phi_0 + |\mathcal{M}_{1,\mathrm{LL}}|^2 \, d\Phi_1 \, , \tag{3}$$

which is not coincident with the exact NLO result. In fact, it is:

$$d\sigma^{\alpha} = [1 + C_{\alpha}] |\mathcal{M}_0|^2 \, d\Phi_0 + |\mathcal{M}_1|^2 \, d\Phi_1 \, , \tag{4}$$

where C_{α} and $C_{\alpha,\mathrm{LL}}$ have the same logarithmic structure, $|\mathcal{M}_{1,\mathrm{LL}}|^2$ and $|\mathcal{M}_1|^2$ the same singular behaviour.

In order to match LL and NLO calculations, while preserving the resummation of the higher orders, avoiding the double counting of the LL contributions and guaranteeing the independence of the cross section on the ε fictictious parameter, two correction factors, by construction infrared safe and free of collinear logarithms, are needed. They are, for the emission of a virtual or soft photon,

$$F_{\mathrm{SV}} = 1 + (C_{\alpha} - C_{\alpha,\mathrm{LL}}) \, , \tag{5a}$$

and, for hard bremsstrahlung,

$$F_{\mathrm{H}} = 1 + \frac{|\mathcal{M}_1|^2 - |\mathcal{M}_{1,\mathrm{LL}}|^2}{|\mathcal{M}_{1,\mathrm{LL}}|^2} \, . \tag{5b}$$

Now the exact $\mathcal{O}\left(\alpha\right)$ cross section of (4) can be expressed, up to $\mathcal{O}\left(\alpha^2\right)$, in term of its LL approximation as:

$$\mathrm{d}\sigma_{\mathrm{LL}}^{\alpha} = F_{\mathrm{SV}}\left[1 + C_{\alpha,\,\mathrm{LL}}\right]\left|\mathcal{M}_0\right|^2 \mathrm{d}\Phi_0 + F_{\mathrm{H}}\left|\mathcal{M}_{1,\,\mathrm{LL}}\right|^2 \mathrm{d}\Phi_1 \qquad (6)$$

and, dealing in the same way, (1) becomes:

$$\mathrm{d}\sigma_{\mathrm{matched}}^{\infty} = F_{\mathrm{SV}}\,\Pi\left(Q^2, \varepsilon\right) \sum_{n=0}^{\infty} \frac{1}{n!}\left(\prod_{i=0}^{n} F_{\mathrm{H}i}\right)\left|\mathcal{M}_{n,\,\mathrm{LL}}\right|^2 \mathrm{d}\Phi_n, \qquad (7)$$

with $F_{\mathrm{H}i}$ following from the definition (5b) for each photon emission.

Details about the matching procedure can be found in [7].

3 Checks on the theoretical uncertainty

The matching procedure discussed in Sect. 2 has been implemented in the new BABAYAGA code [8]. This generator has been tested in a number of ways and, particularly, it has been compared with independent calculations (LABSPV [9], BHWIDE [5] and the old BABAYAGA 3.5 release [6]) to check the physical reliability, considering that LABSPV and BHWIDE calculate the Bhabha cross section with different approaches from BABAYAGA. The comparisons have been performed neglecting vacuum polarization effects in typical event selection criteria for luminosity at flavour factories.

As a result of the tests, only very slight discrepancies with BHWIDE and LABSPV were noticed, the latter attributable to the strictly collinear approximation of the higher orders corrections in the LABSPV code. Furthermore, also differential cross sections have been examined: in Fig. 1 and in Fig. 2 the acollinearity distribution is presented, in comparison with the ones obtained with the old BABAYAGA and at $\mathcal{O}\left(\alpha\right)$ (Fig. 1) and BHWIDE (Fig. 2). In the first plot, the effects of the higher orders corrections are clearly visible, while, in

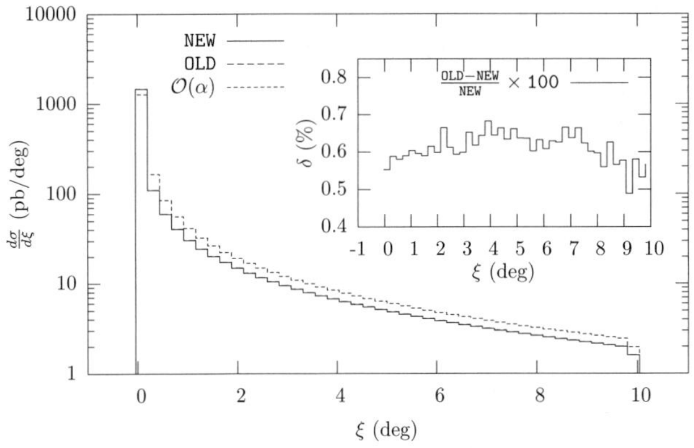

Fig. 1. Acollinearity distribution, the new BABAYAGA compared with the old version

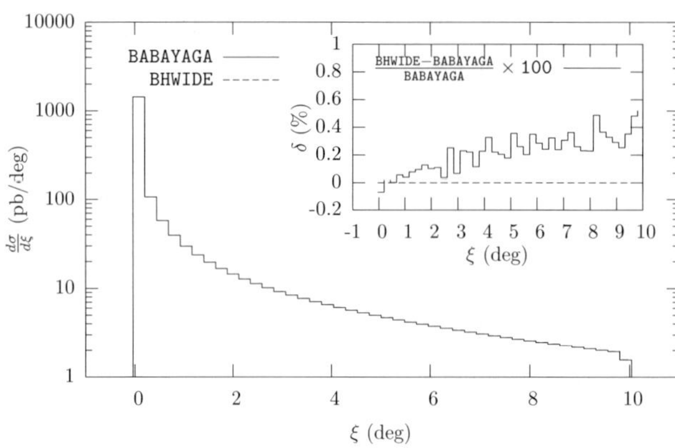

Fig. 2. Acollinearity distribution, the new BABAYAGA compared with BHWIDE

the second one, it is possible to notice how small the differences between the two codes are.

In conclusion, the matching procedure succeeds in including the missing $\mathcal{O}\left(\alpha\right)$ contributions in the PS algorithm while preserving the resummation of all the higher orders corrections. The theoretical error of BABAYAGA is shifted to the two-loop level and, by comparison with recent $\mathcal{O}\left(\alpha^2\right)$ calculations [10], it is possible to infer that the missing contributions have an impact not larger than the 0.1% on the integrated cross section, for typical event selection criteria.

See [7] for a systematic discussion of the theoretical errors.

4 Conclusions

The experimental collaborations at the flavour-factories require high-precision calculations of the Bhabha process cross section to improve their knowledge of the machine luminosity and, thus, to obtain a very good measurement of the R ratio. The recent reduction in the experimental uncertainty at DAΦNE calls for an improvement of the existent event generators. To this end, an original way to match NLO correction with a PS algorithm has been implemented in a new version of the BABAYAGA code. This procedure reduces the theoretical error on the integrated cross section down to 0.1%.

References

1. S. Eidelman and F. Jegerlehner. *Z. Phys.*, C67:585-602, 1995; F. Jegerlehner. *J. Phys.*, G29:101-110, 2003; *Nucl. Phys. Proc. Suppl.*, 131:213-222, 2004.
2. A. Aloisio et al. *Phys. Lett.*, B606:12–24, 2005; A. Denig and F. Nguyen. *KLOE note*, 202, July 2005; F. Ambrosino et al. [KLOE Collaboration]. *Eur. Phys. J.*, C47:589-596, 2006; F. Nguyen. *These proceedings.*

3. A. B. Arbuzov, G. V. Fedotovich, F. V. Ignatov, E. A. Kuraev and A. L. Sibidanov. *BUDKER-INP-2004-70*, Apr. 2005 (arXiv:hep-ph/0504233).

4. E. Drago and G. Venanzoni. *INFN-AE-97-48*, Jul. 1997.

5. S. Jadach, W. Placzek and B. F. L. Ward. *Phys. Lett.*, B390:298–308, 1997.

6. C. M. Carloni Calame, C. Lunardini, G. Montagna, O. Nicrosini and F. Piccinini. *Nucl. Phys.*, B584:459–479, 2000; C. M. Carloni Calame. *Phys. Lett.*, B520:16–24, 2001; C. M. Carloni Calame, G. Montagna, O. Nicrosini and F. Piccinini. *Nucl. Phys. Proc. Suppl.*, 131:48–55, 2004.

7. G. Balossini, C. M. Carloni Calame, G. Montagna, O. Nicrosini and F. Piccinini. arXiv:hep-ph/0607181, accepted for pubblication in Nucl. Phys. B, 2006.

8. http://www.pv.infn.it/hepcomplex/babayaga.html

9. derived from M. Cacciari, G. Montagna, O. Nicrosini and F. Piccinini. *Comput. Phys. Commun.*, 90:301–310, 1995.

10. A. A. Penin. *Phys. Rev. Lett.*, 95:010408, 2005; R. Bonciani, A. Ferroglia, P. Mastrolia, E. Remiddi and J. J. van der Bij. *Nucl. Phys.*, B701:121–179, 2004 (erratum arXiv:hep-ph/0405275v2); *Nucl. Phys.*, B716:280–302, 2005 (erratum arXiv:hep-ph/0411321v2); R. Bonciani and A. Ferroglia. *Phys. Rev.*, D72:056004, 2005.

Status and prospects
of the $\sigma_{e^+e^-\to\text{hadrons}}$ measurement

Federico Nguyen

Università and Sezione INFN "Roma TRE"
nguyen@fis.uniroma3.it

Summary. The present status and the near and far future prospects of the $\sigma_{e^+e^-\to\text{hadrons}}$ measurements are briefly reviewed.

Introduction

Precision tests of the Standard Model require an accurate measurement of the hadronic cross section, $\sigma_{\text{had}} \equiv \sigma_{e^+e^-\to\text{hadrons}}$.

Hadronic contributions [1,2] to the running of the fine structure constant, $\Delta\alpha_{\text{had}}$, and to the muon magnetic anomaly, a_μ^{had}, are evaluated from a dispersion integral over σ_{had} data points.

The kernel functions of the two integrands are such that the $\pi\pi$ final state accounts for 80% of the value and for 40% of the uncertainty in a_μ^{had}, while the largest uncertainty in $\Delta\alpha_{\text{had}}$, 35%, is provided by hadron final states measured in the energy range between 1 and 2 GeV.

An approach [3], alternative to varying the e^+e^- beam energies, is to extract σ_{had} from the hadrons invariant mass, M_{had}, measured in Initial State Radiation (ISR) events:

$$M^2_{\text{had}} \left.\frac{\mathrm{d}\sigma_{e^+e^-\to\text{hadrons }\gamma}}{\mathrm{d}M^2_{\text{had}}}\right|_{\text{ISR},\,\theta_{\min}} = H(M^2_{\text{had}}, \theta_{\min})\,\sigma_{\text{had}}(M^2_{\text{had}}) \qquad (1)$$

where H is the radiation function for a photon emitted by the $e^\pm$ with minimum polar angle $\theta_{\min}$. The main advantage is that uncertainties related to

Table 1. Hadronic mass ranges covered by recent experiments

Experiment	Mass range (GeV)	Technique
CMD-2 & SND	$0.4 < M_{\text{had}} < 1.4$	energy scan
KLOE	$0.59 < M_{\text{had}} < 1.02$	ISR events
BABAR	$0.61 < M_{\text{had}} < 10.6$	ISR events

beam energy and luminosity are the same for each M^2_{had} value. Table 1 shows the regions covered by present experiments and the experimental techniques.

The pion form factor: a comparison

The pion form factor, F_π[1], has been recently measured by CMD-2, SND and KLOE.

Table 2 shows the statistical and systematic accuracy related to the most recent published F_π measurements.

Figure 1 shows the present situation for the F_π contribution to a_μ, $a_\mu^{\pi\pi}$: there are local disagreements between KLOE and the VEPP-2M experiments, which however compensate in the dispersion integral. Above the ρ peak, the relative difference between F_π extracted using the conserved vector current hypothesis and isospin corrections from $\tau^- \to \nu_\tau \pi^- \pi^0$ decays and that obtained from $\sigma_{\pi\pi}$ is $\sim 10\%$. This result is consistent among the e^+e^- experiments quoted in Table 2.

Table 2. Number of events and relative systematic error of the most recent F_π measurements

experiment	n. of events	$\dfrac{(\delta F_\pi)_{\text{sys}}}{F_\pi}$
CMD-2 [4] ($\sqrt{s} \in [0.6, 1]$ GeV)	$10^5\,\pi^+\pi^-$	0.6%
CMD-2 [5] ($\sqrt{s} > 1$ GeV)	$3 \times 10^4\,\pi^+\pi^-$	1.2–4.2%
SND [6] ($\sqrt{s} < 0.98$ GeV)	$4.5 \times 10^6\,\pi^+\pi^-$	1.3–3.2%
KLOE [7] ($\sqrt{s} = M_\phi$, ISR)	$1.6 \times 10^6\,\pi^+\pi^-\gamma$	1.3%

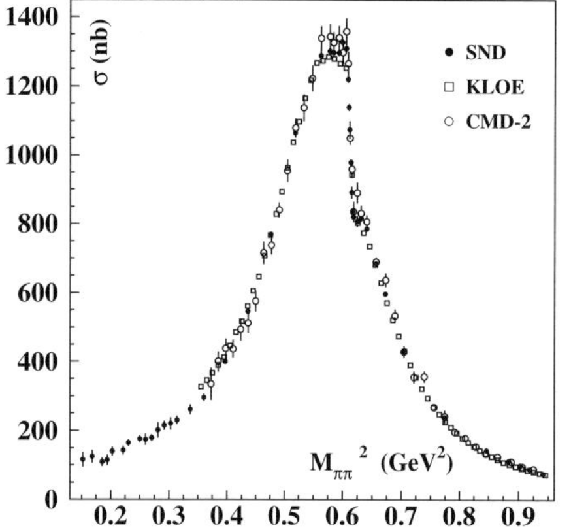

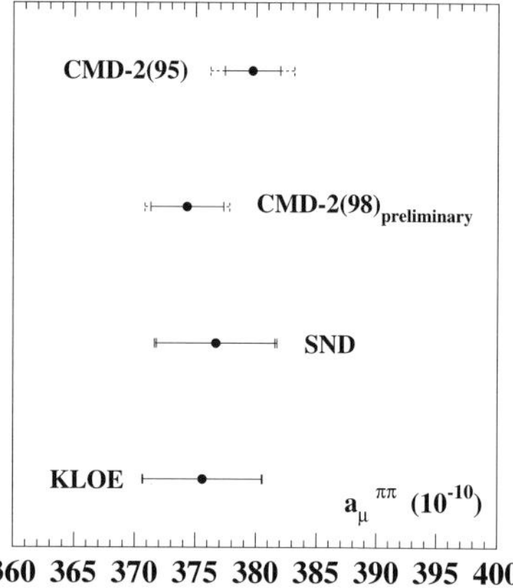

Fig. 1. *Left*: $\sigma_{\pi\pi}$ measurements from CMD-2, SND and KLOE. *Right*: comparison of the a_μ contributions from the $M^2_{\pi\pi} \in [0.35, 0.95]$ GeV^2 part of the spectra shown in the left plot. The CMD-2(98) [8] result is preliminary

---

[1] Defined via $\sigma_{\pi\pi} \equiv \sigma_{e^+e^- \to \pi^+\pi^-} \propto s^{-1}\beta_\pi^3(s)|F_\pi(s)|^2$.

KLOE is analyzing the remaining data (about 2 fb^{-1}) aiming to cover the threshold region [$2m_\pi$,0.5 GeV] and also measuring the ratio $\sigma_{\pi^+\pi^-\gamma}/\sigma_{\mu^+\mu^-\gamma}$, to reduce some of the systematic uncertainties.

The [1-2] GeV energy region

After the above precise F_π measurements, this becomes the most relevant region for both $\Delta\alpha_{\text{had}}$ and a_μ^{had}. Recently the BABAR experiment published results in the three [9] and four pions [10] studying ISR events from an integrated luminosity of 89 fb^{-1}. In the latter case the whole mass range has been covered with unprecedented precision by the same experiment.

In future, BABAR aims to collect an integrated luminosity of ~ 1 ab^{-1} and is going to perform extensive studies of exclusive channels, semi-inclusive ϕ, η, K productions and an inclusive analysis.

Future chances: KLOE-2

This is the continuation and extension of the KLOE physics program, including also an energy scan for σ_{had} measurements in [1-2] GeV energy range.

Figure 2 shows the relative statistical error of the σ_{had} data points published by BABAR [10] in the channels $2\pi^+2\pi^-$ and $\pi^+\pi^-K^+K^-$, compared with that improved by an order of magnitude, and with that obtained with an energy scan. This case assumes an integrated luminosity^2 of 20 pb^{-1} per point, rather than that given by the radiation function at the Υ peak.

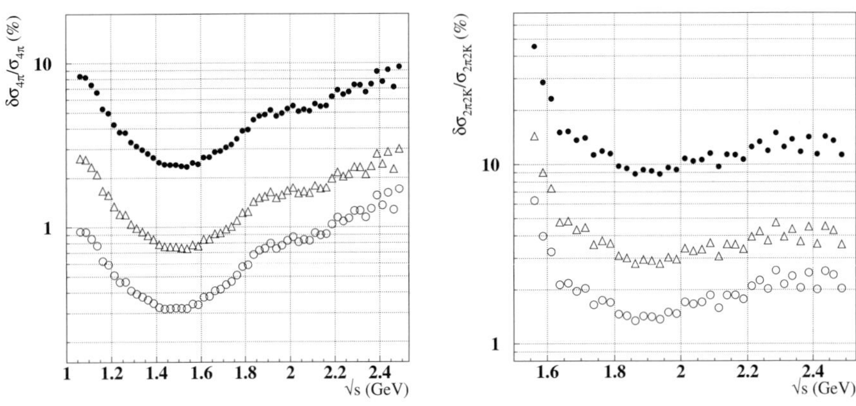

Fig. 2. Comparison of the statistical accuracy in the cross-section among an energy scan with 20 pb^{-1} per point (○), published BABAR results (●), BABAR with full statistics (△) for the $2\pi^+2\pi^-$ (*left*) and $\pi^+\pi^-K^+K^-$ (*right*) channels

---

^{2} Corresponding to 3-4 days of data taking with 10^{32} cm^{-2} s^{-1} of instant luminosity.

Status of Monte Carlo generators

In ISR analyses, an important tool is the Monte Carlo code PHOKHARA [11], able to generate the channels $\pi^+\pi^-$, $\mu^+\mu^-$, $\bar{N}N$, 3π, 4π, $\bar{K}K$ with ISR corrections at the next-to-leading order and Final State Radiation at the lowest order, with a precision of 0.5% in the radiation function.

The normalization of the measured σ_{had} is often provided by the ratio of observed Bhabha events to the theoretical cross section. In KLOE the experimental analysis of Bhabha events with $e^\pm$ emitted at large angle ($\theta > 55°$) is known at the 0.3% level [12]. The theoretical accuracy reached by the recent version of the BABAYAGA [13] Monte Carlo code is 0.1%.

Conclusions

Given its relevance to a_μ, F_π has been measured by three experiments: there is good agreement among $a_\mu^{\pi\pi}$ values, but there are systematic differences in the mass spectra. In future, new results using ISR events will be on schedule from KLOE and BABAR in the most interesting regions for both a_μ and $\Delta\alpha$.

More and more acccurate σ_{had} measurements in [1-2] GeV energy range can be performed with an energy scan at KLOE-2.

References

1. F. Ambrosino *et al.*, arXiv:hep-ex/0603056, submitted to Eur. Phys. J. C
2. F. Jegerlehner, arXiv:hep-ph/0608329
3. S. Binner, J. H. Kühn and K. Melnikov, Phys. Lett. B **459** (1999) 279
4. R. R. Akhmetshin *et al.* [CMD-2], Phys. Lett. B **578** (2004) 285; *Ibidem*, Phys. Lett. B **527** (2002) 161
5. V. M. Aulchenko *et al.* [CMD-2], JETP Lett. **82** (2005) 743
6. M. N. Achasov *et al.* [SND], arXiv:hep-ex/0605013; *Ibidem*, J. Exp. Theor. Phys. **101** (2005) 1053
7. A. Aloisio *et al.* [KLOE], Phys. Lett. B **606** (2005) 12
8. I. Logashenko, talk given at the Workshop "e^+e^- collisions from ϕ to ψ", Novosibirsk (Russia), February–March 2006
9. B. Aubert *et al.* [BABAR], Phys. Rev. D **70** (2004) 072004
10. B. Aubert *et al.* [BABAR], Phys. Rev. D **71** (2005) 052001
11. G. Rodrigo, H. Czyż, J. H. Kühn and M. Szopa, Eur. Phys. J. C **24** (2002) 71
12. F. Ambrosino *et al.* [KLOE], Eur. Phys. J. C **47** (2006) 589
13. G. Balossini *et al.*, arXiv:hep-ph/0607181

Recent Results from HERA

Andrea Parenti

Università degli Studi di Padova and INFN Sezione di Padova,
via Marzolo 8, 35131 Padova, Italy
parenti@pd.infn.it

1 Introduction

The HERA collider located in Hamburg, Germany, is a unique facility which collides protons of $920\,\mathrm{GeV}$[1] and electrons (or positrons) of $27.5\,\mathrm{GeV}$.

The $e^{\pm}p$ interactions proceed via a γ/Z^0 exchange in the neutral current (NC) reaction, or via a $W^{\pm}$ exchange in the charged current (CC) interaction.

HERA is taking data since 1992; after a shutdown in 2000, data taking has restarted in 2002 with a five-fold increase in the luminosity. The first running period is referred to as HERA-I, whereas the second one is referred to as HERA-II.

In this report I will review some recent results obtained by the H1 and ZEUS experiments at the HERA collider. Many different processes can be studied at HERA; this selection reflects my personal taste and many other relevant results have been omitted.

2 Results

Structure Functions and Parton Distribution Functions

The cross section for the $ep \rightarrow eX$ process with unpolarised lepton beams depends on the proton structure functions F_2, F_L and F_3; the prominent contribution comes from F_2. HERA is the ideal machine to study proton structure functions, since the accessible kinematical region is much larger than in hadronic colliders and fixed target experiments (see Fig. 1).

F_2 was measured by using HERA-I data [1,2] in the region $6.32 \times 10^{-5} < x < 0.65$ and $1 < Q^2 < 30\,000\,\mathrm{GeV}^2$ with an uncertainty as little as 2–3% (see Fig. 1). The longitudinal structure function F_L has been measured as well, though with worse precision [3]. F_3, the parity violating part of the

---

[1] Until 1998, the energy of the proton beam was $820\,\mathrm{GeV}$.

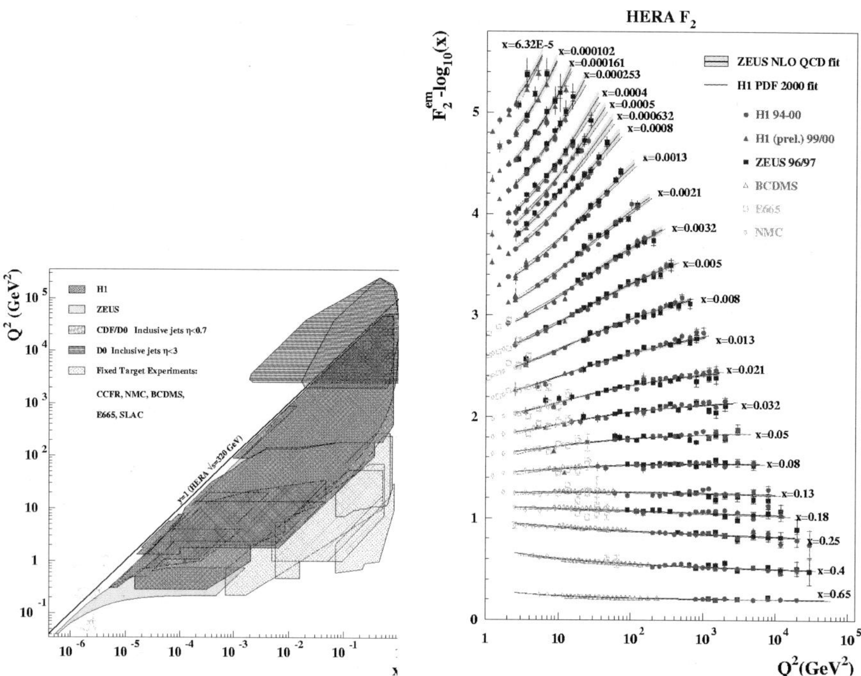

Fig. 1. The $x-Q^2$ kinematical plane and the accessible regions at hadronic colliders, in fixed target experiments and at HERA (*left*). The structure function F_2 versus Q^2 at fixed values of x, compared with a fit based on DGLAP equations (*right*)

interaction, is measurable as a difference in e^-p and e^+p NC differential cross sections. A measurement from ZEUS [4] – using e^-p collisions from HERA-II and e^+p from HERA-I – has been presented at DIS 2006; this measurement has a much lower statistical uncertainty than with HERA-I data only.

The charm and beauty contributions to F_2 have been extracted from HERA-I data by H1 [5, 6] and ZEUS [7]. H1 used the impact parameter significance in order to evaluate the heavy quark fraction in the sample; in the region $Q^2 > 150\,\mathrm{GeV}^2$ and $0.1 < y < 0.7$ the charm and beauty contributions to F_2 are $18.3 \pm 3.0\%$ and $2.72 \pm 0.74\%$, respectively. ZEUS at the time did not have any vertex detector and therefore identified only charm by the decay chain $D^* \to D^0\pi_s$, $D^0 \to K\pi$. The charm contribution to F_2 goes up to $\sim 30\%$ at $Q^2 = 500\,\mathrm{GeV}^2$ and $x = 0.012$. A vertex detector was installed in the ZEUS detector before HERA-II phase, therefore ZEUS will be able to evaluate $F_2^{b\bar{b}}$ with the new data.

The proton parton distribution functions (PDFs) were extracted by ZEUS [8] from the NC and CC cross sections, jet production in e^+p collisions, dijet production in γp collisions, by using only HERA-I data and no other data from different experiments (the so-called ZEUS-JET fit). The results are already competitive to MRST and CTEQ fits. The importance of the ZEUS-JET fit is that the uncertainty is dominated by statistics and will be reduced significantly by the use of HERA-II data, whereas in MRST and CTEQ the uncertainty comes mainly from systematics.

First Results with Polarised Lepton Beam

The leptons at HERA acquire spontaneously a transverse polarisation due to the synchrotron light emission. In 2003 spin rotators were installed before H1 and ZEUS interaction points, transforming the transverse polarisation in longitudinal. The CC cross section depends linearly on the degree of polarisation of the beam, P_e: $\sigma_{CC}^{e^{\pm}p} = (1 \pm P_e)\, \sigma_{CC,\mathrm{unpol}}^{e^{\pm}p}$.

H1 and ZEUS have measured [9, 10] the total cross section of the CC process for e^+p collisions and different values of polarisation and found a nice agreement with the expected $(1 + P_e)$ dependence.

The NC dependence on P_e is more complicated; ZEUS has measured [10] the $\mathrm{d}\sigma/\mathrm{d}Q^2$ in e^+p collisions with positive and negative values of the polarisation and found again a nice agreement with the Standard Model prediction.

Isolated Leptons at High P_T

Events with missing P_T and the production of an isolated, high-P_T lepton come mainly from the process $ep \to eWX$, $W \to l\nu$ ($l = e,\ \mu,\ \tau$); investigations of the process with $l = e,\ \mu$ have been performed by both the H1 and ZEUS experiments [11]. H1 observed an excess of high-P_T isolated leptons in HERA-I e^+p data with respect to the expectation from W production; this excess was not seen by ZEUS. The analysis of HERA-II data confirms the HERA-I results: H1 observed 15 events in 1994–2004 e^+p data having hadronic $p_T > 25\,\mathrm{GeV}$ (4.6 ± 0.8 expected), whereas no excess was seen in e^-p collisions and by ZEUS (in both $e^{\pm}p$). These results are not yet understood.

Measurement of α_S at HERA

The strong coupling constant, α_S, appears in all QCD processes: α_S can therefore be extracted from many observables.

A precise estimation of α_S at M_Z scale has been made [12] by averaging the measurements made at HERA (e. g. jet cross section, event shape, jet multiplicity, etc.); the energy dependence of α_S was measured, too. The uncertainty of HERA determination is dominated by the theoretical part, whereas the experimental part is already better than the world average total uncertainty.

Search for Pentaquarks at HERA

A number of experiments observed a narrow baryon resonance with positive strangeness, mass around 1530 MeV, decaying to nK^+ or $p(\bar{p})K_S^0$. The signals are consistent with an exotic state having quark content $uudd\bar{s}$, called Θ^+. Other experiments did not observe any signal.

ZEUS observed a peak [13] in the $p(\bar{p})K_S^0$ invariant mass spectrum in HERA-I data at $M = 1522 \pm 6\,\mathrm{MeV}$; 221 ± 48 events above the background were seen. In a similar search [14] H1 did not observe any excess over the background.

Since the Θ^+ belongs to an hypotetical antidecuplet of pentaquarks with spin-$\frac{1}{2}$, two more exotic states are foreseen: $\Xi_{3/2}^{--}$ and $\Xi_{3/2}^+$ with quark content $ddss\bar{u}$ and $uuss\bar{d}$, respectively. NA49 at CERN SPS reported the observation of $\Xi_{3/2}^{--}$ and $\Xi_{3/2}^0$, whereas searches made by other experiments were negative. ZEUS performed such a search but did not observe any signal [15].

The anti-charmed pentaquark Θ_c^0 ($uudd\bar{c}$) has been observed by H1 [16]: a peak of 50.6 ± 11.2 events was seen in the $D^{*-}p$ and $D^{*+}\bar{p}$ decay channels. However, ZEUS has not observed any excess over the background [17].

No final conclusion can be drawn on pentaquarks.

3 Conclusion

In this report a review of recent results from HERA has been presented.

The most relevant result is probably the measurement of the proton struction functions and the extraction of parton distribution functions. Almost all the results were obtained by using HERA-I data only, so we expect a big improvement in the precision when HERA-II data will be included.

References

1. C. Adloff et al. *Eur. Phys. J.*, C30: 1–32, 2003.
2. S. Chekanov et al. *Eur. Phys. J.*, C21: 443–471, 2001.
3. E. M. Lobodzinska. hep-ph/0311180.
4. V. Chekelian, C. Gwenlan, and R. S. Thorne. hep-ph/0607116.
5. A. Aktas et al. *Eur. Phys. J.*, C45: 23–33, 2006.
6. A. Aktas et al. *Eur. Phys. J.*, C40: 349–359, 2005.
7. S. Chekanov et al. *Phys. Rev.*, D69: 012004, 2004.
8. S. Chekanov et al. *Eur. Phys. J.*, C42: 1–16, 2005.
9. A. Atkas et al. *Phys. Lett.*, B634: 173–179, 2006.
10. S. Chekanov et al. *Phys. Lett.*, B637: 210–222, 2006.
11. J. Ferrando. *PoS*, HEP2005: 319, 2006.
12. Claudia Glasman. *AIP Conf. Proc.*, 792: 689–692, 2005.
13. S. Chekanov et al. *Phys. Lett.*, B591: 7–22, 2004.
14. Sergei Chekanov. *PoS*, HEP2005: 086, 2006.
15. S. Chekanov et al. *Phys. Lett.*, B610: 212–224, 2005.
16. A. Atkas et al. *Phys. Lett.*, B588: 17, 2004.
17. S. Chekanov et al. *Eur. Phys. J.*, C38: 29–41, 2004.

Twistors and Unitarity

Pierpaolo Mastrolia

Institute of Theoretical Physics, University of Zürich, CH-8057 Switzerland

Introduction. Motivated by the demand for a higher accuracy description of the multi-particle processes that will take place at the upcoming LHC experiment, the theoretical efforts to develop new computational techniques for perturbative calculation have received a strong boost, stemmed from *on-shell* techniques and unitarity-based methods.

Spinors and Twistors. The spinor-helicity representation of QCD amplitudes, that was developed in the 1980s, has been an invaluable tool in perturbative computations ever since. Accordingly, one can express any on-shell massless vector as the tensor product of two spinors carrying different flavour under $SU(2)$, say λ and $\widetilde{\lambda}$. Instead of Lorentz inner products of momenta, amplitudes can be expressed in terms of spinor products. The tree-level amplitudes for the scattering of n gluons vanish, when the helicities of the gluons are either a) all the same, or b) all the same, but one of opposite helicity. The first sequence of nonvanishing tree amplitudes is called maximally helicity-violating (MHV), formed by two gluons of negative helicity, and the rest of positive helicity ones, whose simple form [1,2] involves only spinor products of a single flavour, λ. Witten observed that under the twistor transform [4], MHV amplitudes, due to their *holomorphic* character, appear to be supported on lines in twistor space, thus revealing an intrinsic structure of local objects, once transformed back to Minkowski space [3]. Amplitudes with more negative-helicity gluons are represented by intersecting segments [5,6], each representing a MHV arrangement of particles.

CSW Construction. This representation turned out in a novel diagrammatic interpretation, in the form of the Cachazo–Svrček–Witten (CSW) construction [6], which can be used alternatively to Feynman rules. To compute an amplitude for n gluons, out of which m gluons carry negative helicity, one has to: draw all the possible graphs connecting the n external gluons, with at least three legs in each vertex (node); assign the helicities, $\pm$, to the internal legs; keep the graphs having only MHV-configuration nodes; consider any internal line as a scalar propagator; sum over all and only MHV-diagrams.

The prescription for the spinor products involving an off-shell momenta is realized by the introduction of a light-cone *reference* spinor [6,7] which disappears after adding all the contributions up. Because of the efficiency of the MHV rules for gluon tree amplitudes in QCD, they were soon generalized to other processes, like scattering of massless fermions [8], amplitudes with a Higgs boson [9], and more general objects carrying a Lorentz index, like the fermionic currents, to compute amplitudes involving electroweak vector bosons [10].

BCFW On-Shell Recurrence Relation. In parallel with the extension of tree-level MHV rules to several processes, the twistor structure of one-loop amplitudes began to be investigated – a long list of referencs should be added here! Unexpectedly, these one-loop computations led to a new, more compact representation of tree-amplitudes (appearing in the infrared-divergent parts of the one-loop amplitudes) [15]. These results could be reinterpreted as coming from a quadratic recursion, the so called BCFW on-shell recurrence relation [16,17], depicted diagrammatically in Fig. 1. The amplitude is represented as a sum of products of lower-point amplitudes, evaluated on shell, but for *complex* values of the shifted momenta (denoted by a hat). There are two sums. The first is over the helicity h of an internal gluon propagating between the two amplitudes. The second sum is over an integer k, which labels the partitions of the set $\{1, 2, \ldots, n\}$ into two consecutive subsets (with minimum 3 elements), where the labels of the shifted momenta, say 1 and n, belong to distinct subsets. Britto, Cachazo, Feng and Witten (BCFW) showed that these on-shell recursion relations have a very general proof, relying only on factorization and complex analysis [17]. Accordingly, the same approach can be applied to amplitudes in General Relativity [18], to scattering with massive theories [19], and to the rational coefficients of the integrals that appear in (special helicity configuration) one-loop amplitudes [20]. In the contest of the on-shell formalism, the CSW construction has been interpreted as a variant of the BCFW relation, obtained by a very peculiar analytic continuation of momenta [21].

Unitarity-based Methods. Unitarity-based methods are an effective tool to compute amplitudes at loop-level [22–25]. The analytic expression of any one-loop amplitude, written, by Passarino–Veltman reduction, in terms of a basis of scalar integral functions (boxes, triangles, and bubbles), may contain a polylogarithmic structure and a pure rational term. To compute the

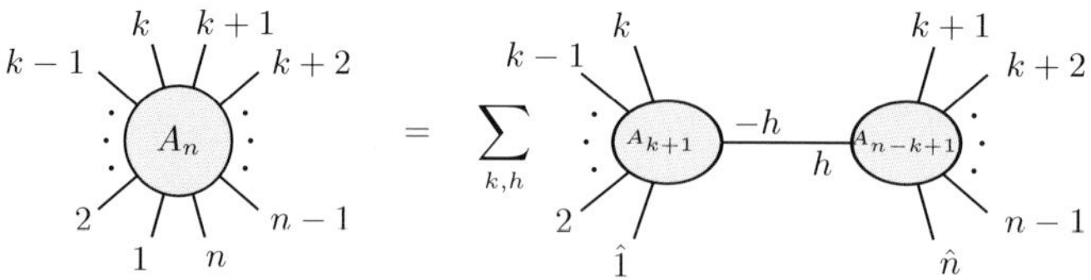

Fig. 1. BCFW Recurrence Relation for tree amplitudes

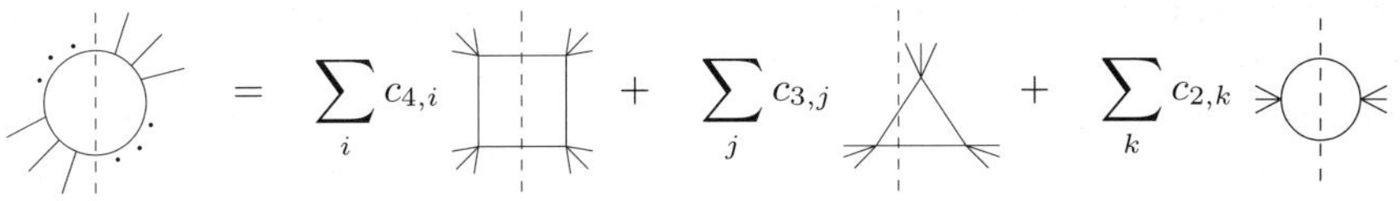

$$= \sum_i c_{4,i}\ \Box\ +\ \sum_j c_{3,j}\ \triangle\ +\ \sum_k c_{2,k}\ \bigcirc$$

Fig. 2. Double-cut of a one-loop amplitude in terms of the master-cuts.

amplitude, it is sufficient to compute the (rational) coefficients of the linear combination separately, and the principle of the unitarity-based method is to exploit the unitarity cuts of the scalar integrals to extract their coefficients, see Fig. 2. Unitarity in four-dimension (4D), requiring the knowledge of the four-dimensional on-shell tree amplitudes which are sewed in the cut, is sufficient to compute the polylogarithmic terms and the trascendenal constants of one-loop amplitudes. By exploiting the analytic continuation of tree-amplitudes to complex spinors, and the properties of the complex integration, new techniques have generalized the cutting rules. On the one side, the *quadruple-cut* technique [11] yields the immediate computation of boxes' coefficient. On the other side, the polylogarithmic structure related to box-, triangle- and bubble-functions can be detected by a *double-cut* and computed by a novel way of performing the cut-integral [26,27], which reduces the integration to the extraction of residues in spinor variables. However, on general grounds, amplitudes in nonsupersymmetric theories, like QCD, suffer of rational ambiguities that are not detected by the four-dimensional dispersive integrals, and D-dimensional unitarity [24,28–30] can be used to determine these terms as well. Alternatively, according to the combined *unitarity-bootstrap* approach, after computing the cut-containing terms by 4D-unitarity, the use of a BCFW-like recurrence relation yields the reconstruction of the rational part [31]. In the very recent past – after the IFAE 2006 workshop – an optimized tool has been developed by tailoring the Passarino–Veltman reduction on the integrals that are responsible of the rational part of scattering amplitudes [35]. That has given rise to further refinements and new developments of algorithms for the tensor reduction of Feynman integrals [36,37].

I conclude with the Table 1, which collects the efforts of the last 12 years, invested in the analytic computation of the six-gluon amplitude in QCD, numerically evaluated not so long ago [42].

Amplitude	$\mathcal{N}=4$	$\mathcal{N}=1$	$\mathcal{N}^{\mathrm{cut}}=0$	$\mathcal{N}^{\mathrm{rat}}=0$
$--++++$	[22]	[23]	[23]	[32]
$-+-+++$	[22]	[23]	[38]	[34,35]
$-++-++$	[22]	[23]	[38]	[34,35]
$---+++$	[23]	[39]	[20,27]	[33]
$--+-++$	[23]	[26,40,41]	[27]	[35]
$-+-+-+$	[23]	[26,40,41]	[27]	[35]

Table 1. The analytic computation of the one-loop six-gluon amplitude in QCD

References

1. S. J. Parke and T. R. Taylor, Phys. Rev. Lett. **56**, 2459 (1986).
2. F. A. Berends and W. Giele, Nucl. Phys. B **294**, 700 (1987);
   M. L. Mangano, S. J. Parke and Z. Xu, Nucl. Phys. B **298**, 653 (1988).
3. E. Witten, Commun. Math. Phys. 252:189 (2004).
4. R. Penrose, J. Math. Phys. **8**, 345 (1967); Rept. Math. Phys. **12**, 65 (1977).
   V. P. Nair, Phys. Lett. B214:215 (1988).
5. R. Roiban, *et al.*, JHEP **0404**, 012 (2004); Phys. Rev. D **70**, 026009 (2004);
   R. Roiban and A. Volovich, Phys. Rev. Lett. **93**, 131602 (2004).
6. F. Cachazo, P. Svrček and E. Witten, JHEP 0409:006 (2004).
7. I. Bena, Z. Bern and D. A. Kosower, Phys. Rev. D **71**, 045008 (2005).
8. G. Georgiou and V. V. Khoze, JHEP **0405**, 070 (2004); J. B. Wu and C. J. Zhu,
   JHEP **0409**, 063 (2004); G. Georgiou, *et al.*, JHEP **0407**, 048 (2004).
9. L. J. Dixon, *et al.*, JHEP **0412**, 015 (2004); S. D. Badger, *et al.*, JHEP **0503**,
   023 (2005).
10. Z. Bern, *et al.*, Phys. Rev. D **72**, 025006 (2005).
11. R. Britto, F. Cachazo and B. Feng, Nucl. Phys. B **725**, 275 (2005).
12. F. Cachazo, P. Svrček and E. Witten, JHEP 0410:074 (2004).
13. A. Brandhuber, B. Spence and G. Travaglini, Nucl. Phys. B 706:150 (2005).
14. F. Cachazo, P. Svrček and E. Witten, JHEP 0410:077 (2004); F. Cachazo,
    hep-th/0410077; I. Bena, *et al.*, Phys. Rev. D **71**, 106010 (2005).
15. R. Roiban, M. Spradlin and A. Volovich, Phys. Rev. Lett. **94**, 102002 (2005).
16. R. Britto, F. Cachazo and B. Feng, Nucl. Phys. B **715**, 499 (2005).
17. R. Britto, *et al.*, Phys. Rev. Lett. **94**, 181602 (2005).
18. J. Bedford, *et al.*, Nucl. Phys. B **721**, 98 (2005); F. Cachazo and P. Svrcek,
    hep-th/0502160.
19. S. D. Badger, *et al.*, JHEP **0507**, 025 (2005); S. D. Badger, *et al.*, JHEP **0601**,
    066 (2006); D. Forde and D. A. Kosower, hep-th/0507292.
20. Z. Bern, *et al.*, JHEP **0511**, 027 (2005).
21. K. Risager, JHEP **0512**, 003 (2005).
22. Z. Bern, *et al.*, Nucl. Phys. B425:217 (1994).
23. Z. Bern, *et al.*, Nucl. Phys. B435:59 (1995).
24. Z. Bern and A. G. Morgan, Nucl. Phys. B467:479 (1996).
25. Z. Bern, *et al.*, Ann. Rev. Nucl. Part. Sci. 46:109 (1996); Nucl. Phys. Proc.
    Suppl. 51C:243 (1996); JHEP 0001:027 (2000); JHEP 0408:012 (2004).
26. R. Britto, *et al.*, Phys. Rev. D **72**, 065012 (2005).
27. R. Britto, B. Feng, and P. Mastrolia, Phys. Rev. D **73**, 105004 (2006).
28. Z. Bern, *et al.*, Phys. Lett. B **394**, 105 (1997).
29. A. Brandhuber, *et al.*, JHEP **0510**, 011 (2005).
30. C. Anastasiou, *et al.*, arXiv:hep-ph/0609191.
31. Z. Bern, *et al.*, Phys. Rev. D **71**, 105013 (2005); Phys. Rev. D **72**, 125003
    (2005).
32. Z. Bern, L. J. Dixon and D. A. Kosower, hep-ph/0507005.
33. C. F. Berger, Z. Bern, L. J. Dixon, D. Forde and D. A. Kosower, hep-ph/0604195.
34. C. F. Berger, Z. Bern, L. J. Dixon, D. Forde and D. A. Kosower, hep-ph/0607014.
35. Z. G. Xiao, *et al.*, hep-ph/0607015, hep-ph/0607016, hep-ph/0607017.

36. G. Ossola, C. G. Papadopoulos and R. Pittau, hep-ph/0609007.
37. T. Binoth, J. P. Guillet and G. Heinrich, hep-ph/0609054.
38. J. Bedford, *et al.*, Nucl. Phys. B **712**, 59 (2005).
39. S. J. Bidder, *et al.*, Phys. Lett. B **606**, 189 (2005).
40. S. J. Bidder, *et al.* Phys. Lett. B **608**, 151 (2005).
41. S. J. Bidder, *et al.*, Phys. Lett. B **612**, 75 (2005).
42. R. K. Ellis, W. T. Giele and G. Zanderighi, JHEP **0605** (2006) 027.

Resummations in QCD: recent developments

Andrea Banfi

Università degli Studi di Milano Bicocca and INFN, Sezione di Milano, Italy
`andrea.banfi@mib.infn.it`

There are processes which involve two widely separated scales Q and Q_0, where Q is a hard scale, and real emissions characterised by momentum scales larger than Q_0 are not measured, so that only their virtual counterpart contributes to physical observables. Since both real and virtual corrections are infrared and collinear (IRC) divergent, but their sum is finite, incomplete real-virtual cancellations give rise to large logarithms $L = \ln Q/Q_0$ in perturbative (PT) expansions, which spoil their convergence.

PT predictions should therefore be improved by resumming logarithmic enhanced terms to all orders. In the luckiest cases, a resummed cross section $\sigma(L)$ can be written in the exponentiated form $\sigma(L) = \exp\{Lg_1(\alpha_{\mathrm{s}}L) + g_2(\alpha_{\mathrm{s}}L) + \alpha_{\mathrm{s}}g_3(\alpha_{\mathrm{s}}L) + \ldots\}$, where g_1, g_2, g_3, etc. contain leading logarithms (LL, $\alpha_{\mathrm{s}}^n L^{n+1}$), next-to-leading logarithms (NLL, $\alpha_{\mathrm{s}}^n L^n$) next-to-next-to-leading logarithms (NNLL, $\alpha_{\mathrm{s}}^n L^{n-1}$) and so on.

The basis of resummation is all-order factorisation of IRC divergences [1]. Any IRC singular contribution to a Feynman graph $\mathcal{G}$ (a.k.a. leading region $\mathcal{G}_L$) can be written schematically as the following convolution:

$$\mathcal{G}_L = H \otimes \prod_{\ell=1}^{n_\ell} J_\ell \otimes S \,, \tag{1}$$

where:

1. H is the hard vertex containing lines whose virtuality is of order of the hard scale of the process;
2. n_ℓ jet functions J_ℓ, one for each hard massless leg ℓ, containing all collinear singularities;
3. a soft function S embodying infrared singularities.

Since the contribution of $\mathcal{G}$ to a physical cross section is obtained by summing over all possible final-state cuts, it is clear that virtual corrections are observable independent, while the contribution of real emissions is what discriminates an observable from the other, and understanding it is the key to resummation. In particular there is a considerable difference between inclusive observables, like total cross sections, in which one is not interested in the structure of the final state, and final-state observables, like event-shape distributions, where one puts a direct veto on real emissions.

For inclusive observables, real contributions are factorised via an integral transform, and the transformed cross section $\tilde{\sigma}(L)$ exponentiates:

$$\tilde{\sigma}(L) \simeq C(\alpha_{\mathrm{s}}) \, \mathrm{e}^{E(\alpha_{\mathrm{s}},L)} \,. \tag{2}$$

The best known examples of inclusive resummations are threshold [2] and transverse momentum [3] resummations. Here we briefly present results [4,5] for Higgs production in hadron-hadron collisions, the knowledge of which is fundamental for the LHC. Results for other processes can be found in refs. [2] and [3].

Threshold resummation is needed for the total Higgs cross section when the available partonic energy $\sqrt{\hat{s}}$ is close to the Higgs mass M_H. In this limit only soft emissions are allowed and the cross section develops large logarithms $\ln(1-M_H^2/\hat{s})$. After a Mellin transform these become logarithms of N, the variable conjugated to $M_H^2/\hat{s}$, and at threshold $N \to \infty$. The exponent $E(\alpha_{\mathrm{s}}, \ln N)$ is given by

$$E\left(\alpha_{\mathrm{s}}\left(M_H^2\right), \ln N\right) = \int_0^1 \mathrm{d}z \, \frac{z^{N-1} - 1}{1 - z}$$
$$\cdot \left[2 \int_{M_H^2}^{(1-z)^2 M_H^2} \frac{\mathrm{d}q^2}{q^2} A\left(\alpha_{\mathrm{s}}\left(q^2\right)\right) + D\left(\alpha_{\mathrm{s}}\left((1-z)^2 M_H^2\right)\right) \right]. \tag{3}$$

The function $A(\alpha_{\mathrm{s}})$ contains soft and collinear contributions (double logarithms), while single logarithms are embodied in the function $D(\alpha_{\mathrm{s}})$. Both functions have an expansion in powers of α_{s}. Our knowledge of $E(\alpha_{\mathrm{s}}, \ln N)$ extends at NNLL level, i. e. we know A_3 [6] and D_2 [4].[1] The numerical results of [4] show one of the main benefits of resummation, the considerable reduction of renormalisation and factorisation scale dependence.

Large logarithms $\ln(q_{\mathrm{T}}/M_H)$ in the Higgs transverse momentum distribution $\mathrm{d}\sigma/\mathrm{d}q_{\mathrm{T}}$ are resummed via a Fourier transform. In the space of impact parameter b (conjugated to q_T) the resummed exponent is [5]

$$E(\alpha_{\mathrm{s}}(M_H^2), \ln b) = - \int_{b_0^2/b^2}^{M_H^2} \frac{\mathrm{d}q^2}{q^2} \left[A\left(\alpha_{\mathrm{s}}\left(q^2\right)\right) \ln \frac{M_H^2}{q^2} + B\left(\alpha_{\mathrm{s}}\left(q^2\right)\right) \right], \quad (4)$$

where $A(\alpha_{\mathrm{s}})$ is the same as for threshold resummations, while single logarithms build up the function $B(\alpha_{\mathrm{s}})$. The exponent in eq. (4) is known up to NNLL accuracy, i. e. we know the second order coefficient B_2 [5]. Not only does resummation reduce renormalisation and factorisation scale uncertainties, but also ensures that the resulting q_{T} spectrum vanishes linearly at small q_{T}, as is expected on physical grounds.

---

[1] Actually also D_3 is known [7]. However, since A_4 is still unknown, we cannot push resummations beyond NNLL accuracy.

In recent years a method to combine threshold and transverse momentum resummation has been developed. This procedure, called joint resummation, has been exploited to extend transverse momentum resummations to the threshold region [8], where eq. (4) breaks down, and to improve the description of the q_{T} spectra of prompt photons [9] and heavy quarks [10] .

There are also processes, involving low momentum scales, where a fixed logarithmic resummation is not sensible, since the factorial divergence of the PT exponent (always present in QCD) cannot be neglected. For instance in the decay $b \to s\gamma$, after resummation of all terms responsible for the divergence, the N-th moment $F_N(M_B)$ of the photon energy spectrum can be written as

$$F_N(M_B) = F_N^{\mathrm{PT}}(m_b) \times \mathrm{e}^{(N-1)\frac{M_B - m_b}{M_B}} \times F_N^{\mathrm{NP}}\left((N-1)\frac{\Lambda_{\mathrm{QCD}}}{M_B}\right) ,$$

$$F_N^{\mathrm{PT}}(m_b) = \exp\left\{ \frac{C_F}{\beta_0} \int_0^\infty \frac{du}{u} T(u) \left(\frac{\Lambda_{\mathrm{QCD}}}{m_b}\right)^u \right. \tag{5}$$

$$\left. \cdot [\mathcal{B}_S(u)\Gamma(-2u)(N^{2u} - 1) + \mathcal{B}_J(u)\Gamma(-u)(N^u - 1)] \right\} .$$

The functions $\mathcal{B}_J(u)$ and $\mathcal{B}_S(u)$ are the Borel transforms of the jet and soft function respectively (see (1)). The factorial divergence of the PT exponent reflect in poles away from $u = 0$ in the integrand in (5). Regulating the u-integral with a principal value prescription and using the pole b-quark mass m_b, one finds that all leading power corrections depend only on the mass difference $M_B - m_b$ between the meson and the quark, while higher power corrections, contained in the function F_N^{NP}, turn out to be not very important, so that one can even attempt a measure of m_b and α_{s} [11].

For final-state observables, such as event-shapes or jet-resolution parameters (see [12] for a recent review), a general statement concerning exponentiation of resummed distributions does not exist. This is mainly due to the fact that veto conditions on real emissions differ from one variable to the other, and in many cases a full analytical resummation is unfeasible. However there is a class of variables v, which share the properties of globalness and recursive infrared-collinear (rIRC) safety [13], for which one can resum the corresponding rate $\Sigma(v)$ at NLL level. More precisely, one can show that for rIRC safe variables all leading logarithms (and part of the NL logarithms) exponentiate, and the remaining NLL contributions factorise:

$$\Sigma(v) = \mathrm{e}^{-R(v)} \mathcal{F}(R') , \qquad R' = -v\frac{\mathrm{d}R}{\mathrm{d}v} . \tag{6}$$

Both the exponent $R(v)$ and the correction factor $\mathcal{F}(R')$ can be computed via a general master formula. This makes it possible to perform the resummation in a fully automated way, as implemented in the program CAESAR [13].

For other variables, for instance the so-called non-global variables [14], the situation is less clear. Non-global variables measure radiation only in a restricted phase space region, a typical example being the energy flow away from hard jets. In this case some approximations for multi-parton matrix elements that are valid for global rIRC variables do not hold any more, and a resummation can be performed only numerically and in the large N_c limit [14, 15]. Furthermore, large logarithms for these observables come only from soft emissions at large angles, so that one naively expects leading logarithms to be of the form $\alpha_s^n L^n$. However there has been recently a claim that in hadron-hadron collisions super-leading logarithms $\alpha_s^n L^{n+1}$ arise at higher orders [16]. Properties of these new logarithms have not been fully investigated yet.

References

1. J. C. Collins, D. E. Soper and G. Sterman, Adv. Ser. Direct. High Energy Phys. **5** (1988) 1.
2. R. Bonciani, S. Catani, M. L. Mangano and P. Nason, Nucl. Phys. B **529** (1998) 424; S. Catani, M. L. Mangano, P. Nason, C. Oleari and W. Vogelsang, JHEP **9903** (1999) 025;
   N. Kidonakis, G. Oderda and G. Sterman, Nucl. Phys. B **525** (1998) 299;
   N. Kidonakis and J. F. Owens, Phys. Rev. D **63** (2001) 054019;
   R.Bonciani, S.Catani, M.L. Mangano and P.Nason, Phys.Lett. B **575**(2003)268;
   T. O. Eynck, E. Laenen and L. Magnea, JHEP **0306** (2003) 057;
   S. Moch, J. A. M. Vermaseren and A. Vogt, Nucl. Phys. B **726** (2005) 317.
3. J. C. Collins, D. E. Soper and G. Sterman, Nucl. Phys. B **250** (1985) 199;
   J. C. Collins and D. E. Soper, Nucl. Phys. B **284** (1987) 253;
   D. de Florian and M. Grazzini, Nucl. Phys. B **704** (2005) 387; Nucl. Phys. B **616** (2001) 247.
4. S. Catani, D. de Florian, M. Grazzini and P. Nason, JHEP **0307** (2003) 028.
5. G. Bozzi, S. Catani, D. de Florian and M. Grazzini, Nucl. Phys. B **737**(2006)73.
6. S. Moch, J. A. M. Vermaseren and A. Vogt, Nucl. Phys. B **688** (2004) 101.
7. S. Moch and A. Vogt, Phys. Lett. B **631** (2005) 48;
   E. Laenen and L. Magnea, Phys. Lett. B **632** (2006) 270;
   A. Idilbi, X. d. Ji and F. Yuan, hep-ph/0605068;
   V. Ravindran, J. Smith and W. L. van Neerven, hep-ph/0608308.
8. E. Laenen, G. Sterman and W. Vogelsang, Phys. Rev. D **63** (2001) 114018.
9. E. Laenen, G. Sterman and W. Vogelsang, Phys. Rev. Lett. **84** (2000) 4296.
10. A. Banfi and E. Laenen, Phys. Rev. D **71** (2005) 034003.
11. J. R. Andersen and E. Gardi, JHEP **0506** (2005) 030; E. Gardi, JHEP **0404** (2004) 049.
12. M. Dasgupta and G. P. Salam, J. Phys. G **30** (2004) R143, and references therein.
13. A. Banfi, G. P. Salam and G. Zanderighi, JHEP **0503** (2005) 073.
14. M. Dasgupta and G. P. Salam, Phys. Lett. B **512** (2001) 323.
15. M. Dasgupta and G. P. Salam, JHEP **0203** (2002) 017;
    R. B. Appleby and M. H. Seymour, JHEP **0212** (2002) 063;
    A. Banfi and M. Dasgupta, Phys. Lett. B **628** (2005) 49.
16. J. R. Forshaw, A. Kyrieleis and M. H. Seymour, JHEP **0608** (2006) 059.

Resummation
of Drell–Yan rapidity distributions

Paolo Bolzoni

Dipartimento di Fisica, Università di Milano and INFN, Sezione di Milano,
Via Celoria 16, I-20133 Milano, Italy
paolo.bolzoni@mi.infn.it

The threshold resummation of Drell–Yan rapidity distributions was first considered in 1992 [1]. At that time, it was suggested a resummation formula for the case of zero rapidity. Very recently, thanks to the analysis of the full NLO calculation of the Drell–Yan rapidity distribution, it has been shown [2], that the result given in [1] is valid at NLL for all rapidities.

Here, we present an all-order threshold resummation formula valid for all values of rapidity. In particular, resummation can be reduced to that of the rapidity-integrated process, which is given in terms of a dimensionless universal function for both DY and $W^{\pm}$ and Z^0 production, and has been largely studied [3–5] even to all logarithmic orders [6]. Finally, we give predictions of the full rapidity-dependent NLL Drell–Yan cross section for the case of the fixed-target E866/NuSea experiment. We find that the agreement of resummation at the NLL level is better than the NNLO calculation [7].

We consider the general Drell–Yan process in which the collisions of two hadrons (H_1 and H_2) produce a virtual photon γ^* (or an on-shell vector boson V) and any collection of hadrons (X). In particular, we are interested in the differential cross section $\frac{d\sigma}{dQ^2\,dY}(x, Q^2, Y)$, where Q^2 is the invariant mass of the photon or of the vector boson, x is defined as usual and Y is the rapidity of $\gamma^*(V)$ in the hadronic center-of-mass:

$$
x \equiv \frac{Q^2}{S}, \quad S = (P_1 + P_2)^2, \quad Y \equiv \frac{1}{2}\ln\left(\frac{E + p_{\mathrm{z}}}{E - p_{\mathrm{z}}}\right), \tag{1}
$$

where E and p_{z} are the energy and the momentum along the collisional axis of $\gamma^*(V)$ respectively. At the partonic level, a parton 1(2) in the hadron H_1 (H_2) carries a longitudinal momentum $p_1 = x_1 P_1$ ($p_2 = x_2 P_2$). Thus, the rapidity in the partonic center-of-mass (y) is obtained performing a boost of Y between the two frames:

$$
y = Y - \frac{1}{2}\ln\left(\frac{x_1}{x_2}\right). \tag{2}
$$

In the threshold limit the gluon–quark channels are suppressed and, so, in order to study resummation, we will consider only the quark–anti-quark contributions. According to the standard factorization of QCD, the Y-dependent cross quark–anti-quark cross section has the form:

$$\frac{\mathrm{d}\sigma}{\mathrm{d}Q^2\,\mathrm{d}Y}(x,Q^2,Y) = \sum_{q,\bar{q}'} \sigma_{q\bar{q}'}^{(0)}(x,Q^2) \int_{x_1^0}^1 \frac{\mathrm{d}x_1}{x_1} \int_{x_2^0}^1 \frac{\mathrm{d}x_2}{x_2} F_q^{H_1}(x_1,\mu^2) F_{\bar{q}'}^{H_2}(x_2,\mu^2)$$

$$\cdot\, C\left(z,\frac{Q^2}{\mu^2},\alpha_{\mathrm{s}}(\mu^2),y\right)\,, \tag{3}$$

where F_q and $F_{\bar{q}'}$ are quark (or anti-quark) parton densities in the hadron H_1 and H_2 respectively, $z = x/(x_1 x_2)$ and y depends on x_1 and x_2 through (2). The coefficients $\sigma_{q\bar{q}'}^{(0)}(x,Q^2)$ are naive dimensional factors for the different Drell–Yan processes reported in [8].

Usually resummation is performed in the space of the variable N, which is the Mellin conjugate of x, since Mellin transformation turns convolution products into ordinary products. In the case of the rapidity distribution, however, this is not sufficient. In fact, the Mellin trans form with respect to x does not diagonalize the double integral in (3), because the partonic center-of-mass rapidity y depends on x_1 and x_2. The ordinary product in Mellin space can be recovered performing the Mellin transform with respect to x of the Fourier transform with respect to Y. It can be shown that, up to terms suppressed by factors $1/N$ in the large-N limit (which corresponds to the threshold limit), the resummed Mellin–Fourier transform of the cross section is given by:

$$\sigma_{q\bar{q}'}^{\mathrm{res}}(N,Q^2,M) = F_q^{H_1}(N + iM/2,\mu^2) F_{\bar{q}'}^{H_2}(N - iM/2,\mu^2)$$

$$\cdot\, C_{\mathrm{I}}^{\mathrm{res}}\left(N,\frac{Q^2}{\mu^2},\alpha_{\mathrm{s}}(\mu^2)\right)\,, \tag{4}$$

for each quark–anti-quark channel. Here we have omitted the dimensional prefactors for brevity, M is the Fourier conjugate of Y and $C_{\mathrm{I}}^{\mathrm{res}}(N,Q/\mu^2,\alpha_{\mathrm{s}}(\mu^2))$ is the rapidity-integrated Drell–Yan coefficient function resummed to the desired logarithmic accuracy. This shows that, near threshold, the Mellin–Fourier transform of the coefficient function does not depend on the Fourier moments and that this is valid to all orders of QCD perturbation theory. Furthermore this result remains valid for all values of hadronic center-of-mass rapidity, because we have introduced a suitable integral transform over rapidity.

To show the importance of this resummation, we want obtain a NLO determination of the cross section improved with NLL resummation for proton-proton collisions at the Fermilab fixed-target experiment E866/NuSea [9]. In order to do this, we must keep the resummed part of the cross section, add the full NLO cross section and subtract the double-counted logarithmic enhanced contributions. This matching has to be done in the x and Y spaces,

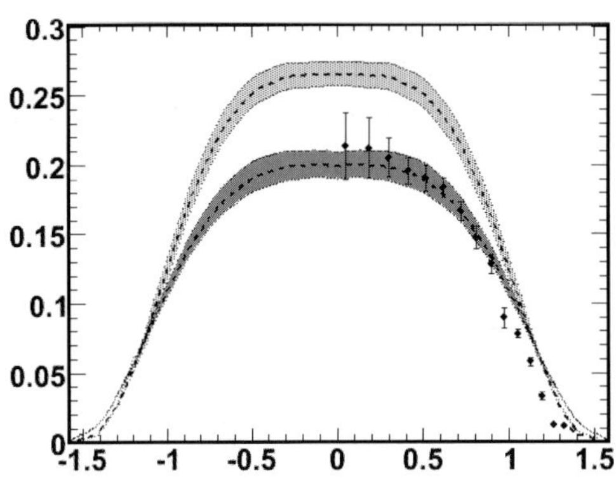

Fig. 1. Y dependence of $\mathrm{d}^2\sigma/(\mathrm{d}Q^2 dY)$ in units of pb/GeV^2. The curves are, from top to bottom, the NLO result and the NLO+NLL resummation together with the E866/NuSea data. The bands are obtained varying the factorization scale between $\mu^2 = 1/2Q^2$ and $\mu^2 = 2Q^2$

because we are not able to calculate the Mellin–Fourier moments of the full NLO cross section analytically. To perform the inverse Mellin transform, we adopt the "Minimal Prescription" proposed in [10]. The center-of-mass energy has been fixed at $\sqrt{S} = 38.76\,\mathrm{GeV}$ and the invariant mass of the virtual photon γ^* has been chosen to be $Q^2 = 64\,\mathrm{GeV}^2$ in analogy with [7]. We have evolved up the MRST 2001 parton distributions taken at $\mu^2 = 1\,\mathrm{GeV}^2$. However, results obtained using more modern parton sets should not be very different. The LO and NLO parton sets are given in [11, 12]. The evolution of parton densities at the scale μ^2 has been performed in the variable flavor number scheme considering the quarks massless.

In Fig. 1, we report the experimental data of [9] converted to the Y variable, together with our NLO and NLL resummed predictions. The effect of the NLL resummation in the central rapidity region is almost as large as the NLO correction, but it reduces the cross section instead of enhancing it for not large values of rapidity. The agreement with data is good and a great improvement for not large rapidity is obtained with respect to the NLO calculation. We note also that the NLL resummation gives better results than the NNLO calculation performed in [7]. This suggests that NLL resummation is more important than a high-fixed-order calculation.

To summarize, we have shown a resumation formula for the Drell–Yan rapidity distributions to all logarithmic accuracy and valid for all values of rapidity. Furthermore, we have analyzed the impact of NLL resummation for the fixed-target experiment E866/NuSea, showing a better agreement with data than NNLO calculation.

References

1. Eric Laenen and George Sterman. Resummation for Drell-Yan differential distributions. Presented at Particles & Fields 92: 7th Meeting of the Division of Particles Fields of the APS (DPF 92) (NOTE: Dates changed from Oct 13-17), Batavia, IL, 10-14 Nov 1992.

2. Asmita Mukherjee and Werner Vogelsang. Threshold resummation for W-boson production at RHIC. *Phys. Rev.*, D73:074005, 2006.

3. George Sterman. Summation of large corrections to short distance hadronic cross-section. *Nucl. Phys.*, B281:310, 1987.

4. S. Catani and L. Trentadue. Resummation of the QCD perturbative series for hard processes. *Nucl. Phys.*, B327:323, 1989.

5. S. Catani and L. Trentadue. Comment on QCD exponentiation at large x. *Nucl. Phys.*, B353:183–186, 1991.

6. Stefano Forte and Giovanni Ridolfi. Renormalization group approach to soft gluon resummation. *Nucl. Phys.*, B650:229–270, 2003.

7. Charalampos Anastasiou, Lance J. Dixon, Kirill Melnikov, and Frank Petriello. Dilepton rapidity distribution in the Drell-Yan process at NNLO in QCD. *Phys. Rev. Lett.*, 91:182002, 2003.

8. T. Gehrmann. QCD corrections to double and single spin asymmetries in vector boson production at polarized hadron colliders. *Nucl. Phys.*, B534:21–39, 1998.

9. J. C. Webb *et al.* Absolute Drell-Yan dimuon cross sections in 800-Gev/c p p and p d collisions. 2003.

10. Stefano Catani, Michelangelo L. Mangano, Paolo Nason, and Luca Trentadue. The resummation of soft gluon in hadronic collisions. *Nucl. Phys.*, B478:273–310, 1996.

11. A. D. Martin, R. G. Roberts, W. J. Stirling, and R. S. Thorne. NNLO global parton analysis. *Phys. Lett.*, B531:216–224, 2002.

12. Alan D. Martin, R. G. Roberts, W. J. Stirling, and R. S. Thorne. MRST2001: Partons and alpha(s) from precise deep inelastic scattering and Tevatron jet data. *Eur. Phys. J.*, C23:73–87, 2002.

Recent jet measurements at the Tevatron

Sofia Vallecorsa

University of Geneva
sofia.vallecorsa@cern.ch
on behalf of the CDF collaboration

The Run II physics program started at the Tevatron in spring 2001, introducing a new level of QCD precision measurement at hadron colliders. The higher center-of-mass energy ($\sqrt{s} = 1.96\,\mathrm{TeV}$ compared to Run I $\sqrt{s} = 1.8\,\mathrm{TeV}$) corresponds to a larger jet production rate, which together with an improved acceptance of CDF [1] and D0 [2] detectors enables stringent tests to pQCD over a wide range in jet energy and rapidity. More than $1\,\mathrm{fb}^{-1}$ of data is already available to analysis: in this contribution some new jet physics measurements are described focusing on high pt jet inclusive production, multi jets final states in association to weak bosons (W and Z) and heavy flavor jet production.

1 Inclusive jet production

Run I inclusive jet cross section measurement [3] had risen great interest for the apparent transverse energy excess, later explained within the Standard Model by the poor knowledge of gluon PDFs in the high x range [4]. In Run II the extended rapidity range constrains those PDFs while the high p_T tail of the distribution is sensitive to new physics.

CDF has recently performed two independent measurements of the inclusive jet cross section using a cone-based (Midpoint) [5] algorithm and a longitudinally invariant k_T algorithm [6]. D0 has made a similar measurement using a cone based algorithm: details can be found in [10].

Figure 1 shows the CDF inclusive cross section for K_T jets with a D parameter equal to 0.7 (it roughly accounts for the cone size of the jet). The measurement has been performed in five intervals over a large rapidity region $0.1 < |Y| < 2.1$, it is corrected for energy smearing effects. NLO prediction by JETRAD [7] (CTEQ6.1M PDFs and renormalization and factorization scales set to $p_\mathrm{T}^{\mathrm{jet}}/2$), also corrected to account for underlying event and hadronization effects, is superimposed to data distribution. In the most forward region ($1.7 < |Y| < 2.1$) theoretical uncertainty (mainly from PDFs)

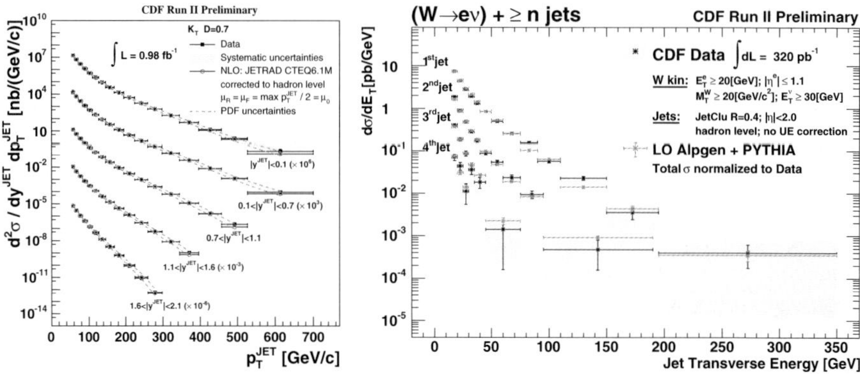

Fig. 1. *Left* Inclusive jet cross section for k_T-jets in 5 rapidity region (CDF). Data is compared to NLO JETRAD prediction corrected for underlying event and hadronization effects. Experimental systematics (yellow band) include jet energy scale uncertainty. *Right* $W \to e\nu + n$ jets differential cross sections as a function of jet E_T. Measurement is overlapped to LO Alpgen + Pythia events normalized to data

is far larger than the experimental value, which proves the important role played by CDF data in tightly constraining gluon distribution at high x. The result is in agreement with the cross section measured by CDF for central Midpoint-cone 0.7 jets $(0.1 < |Y| < 0.7)$, using $1\,\mathrm{fb}^{-1}$ of data. A more detailed description can be found in [11].

2 Bosons + jets production

The production of high energy jets in association to W and Z bosons represents the main source of background to many top quark physics as well as Higgs and SUSY searches, and the presence of the heavy bosons ensures that the hard scattering has occurred at the pQCD energy scale.

Figure 1 shows CDF result for the $W \to e\nu +$ jet cross section as a function of the first, second, third and fourth jet E_T. The measurement is performed in the central region (electron $|\eta| < 1.1$, jet $|\eta| < 2$) using the Run I cone0.4 algorithm. The plot also shows the LO prediction calculated by Alpgen [9] together with Pythia (MonteCarlo is normalized to data). The absolute jet energy scale represents the main source of systematic uncertainty on data (yellow shaded area on the plot) while MonteCarlo uncertainties are mainly due to PDFs.

D0 measurement of $Z \to e^+ e^- +$ jets cross section for about $343\,\mathrm{pb}^{-1}$ of data is described in [12], where a full comparison to LO Jetrad prediction is carried out and the ratio $R = \frac{\sigma(Z \to e^+ e^- + n\mathrm{jets})}{\sigma(Z \to e^+ e^-)}$ is measured as a function of jet multiplicity.

3 Heavy flavour jets

Heavy flavour jets are identified using different methods: searching for B decay muons inside the jets or directly reconstructing B hadron decay vertexes. Correspondingly background contribution from charm, gluon and light quarks jets can be separated from the b-quark signal using a fit to the muon momentum measured with respect to the jet axis or a fit to the secondary vertex mass distribution. Using the first method D0 has measured the inclusive cross section for μ tagged jets [10], CDF has used secondary vertex tagging to measure the inclusive b-jet and the $b\bar{b}$ jet cross section.

The inclusive b-jet cross section is measured for cone 0.7 jet in the central rapidity region ($|Y| < 0.7$). The result, corrected for energy smearing effects, is shown in Fig. 2 compared to a pQCD NLO prediction [8] computed using CTEQ6M PDFs and $\mu_R = \mu_F = \mu_0 = 1/2\sqrt{p_t^2 + m_b^2}$ and corrected for underlying event and hadronization effects. Shaded band represents the total systematics uncertainty on data dominated by jet energy scale uncertainty.

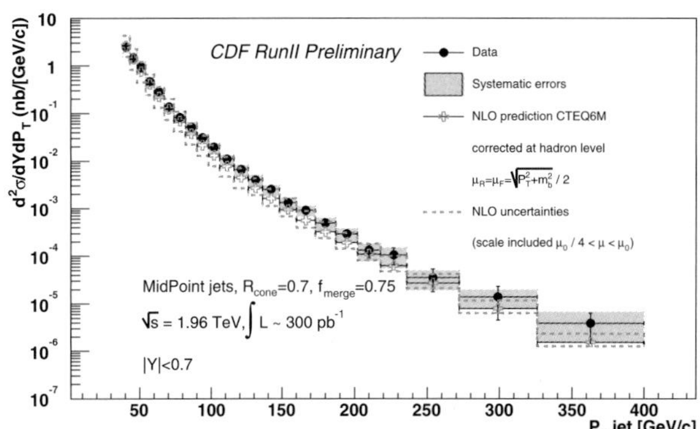

Fig. 2. Inclusive b-jet cross section compared to NLO prediction computed using a bquark bare mass of 4.75 GeV$/c^2$. Experimental systematic uncertainties (shaded band) includes jet energy scale, b-tagging efficiency and tagged jets b-purity uncertainty

Figure 3 shows a preliminary $b\bar{b}$ cross measurement (about $70\,\mathrm{pb}^{-1}$) using jets in $|\eta| < 1.2$. Cross section as a function of the leading jet E_T, di-jet invariant mass and of the azimuthal angle between the two jets ($\Delta\phi$, figure 3) are compared to LO Pythia and Herwig and to NLO MC@NLO. The $\Delta\phi$ distribution peaks out the leading order flavour creation processes however at small opening angles the effect of NLO corrections is apparent: the cross section clearly deviates from pythia.

4 Conclusions

Few examples of the very intense physics program carried out at the Tevatron by CDF and D0 experiments have been described in this contribution. The inclusive cross section measurements shows a good agreement to NLO

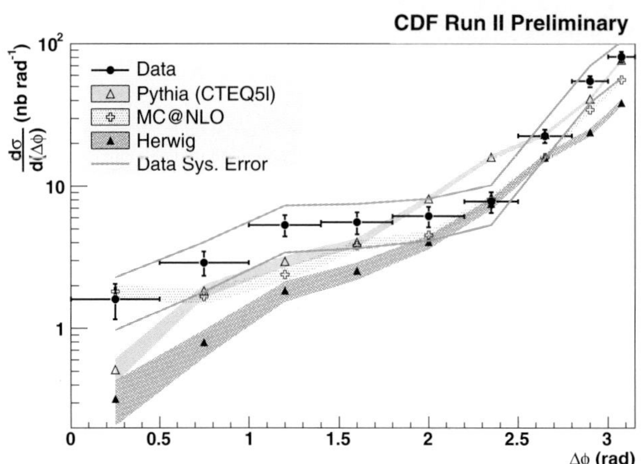

Fig. 3. Azimuthal angle between $b\bar{b}$ jets. Data systematic uncertainty is dominated by jet energy scale and b-tagging uncertainty. The measurement is compared to LO Pythia and Herwig and NLO MC@NLO prediction.

pQCD predictions. It represents an important constraint to gluon PDFs in the high x region (large jet rapidity) and it is a good test of the longitudinally invariant jet reconstruction algorithm for the first time at a hadron collider. Jet production associated to weak bosons also represents a key test to LO and NLO Matrix Elements and Parton Shower predictions in a multi-jet environment. The $b\bar{b}$ measurement shows a preliminary agreement with NLO prediction while the inclusive b jet cross section goes further suggesting a significant contribution from next-to-NLO calculation.

References

1. The CDF collaboration, D. Acosta *et al.*: Phys. Rev. D **71**, 33 (2001)
2. T. LeCompte, H.T. Diehl: Annu. Rev. Nucl. Part. Sci **50**, 71 (2000)
3. The CDF collaboration, T. Affolder *et al.*: Phys. Rev. D **64**, 032001 (2001)
   The CDF collaboration, B. Abbot *et al.*: Phys. Rev. Lett. **82**, 2451 (1999)
4. J. Houston *et al.*: Phys. Rev. Lett. D **77**, 444 (1996)
5. C.G. Blasey *et al.*: hep-ex/0005012
6. S. Catani *et al.*: Nucl. Phys. B **406**, 187 (1993)
7. W.T. Giele, W.N. Glover, D.A. Kosower: Nucl. Phys. B **403**, 633 (1993)
8. S. Frixione, M. Mangano: Nucl. Phys. B **483**, 321 (1997)
9. M. Mangano *et al.*, CERN-TH/2002-129, hep-ex/0206293.
10. http://www-d0.fnal.gov/Run2Physics/WWW/results/qcd.html
11. http://www-cdf.fnal.gov/physics/new/qcd/QCD.html
12. V. Abazov *et al.*, D0 Collaboration, hep-ex/0608052

Vector Boson Production Associated with Jets @ LHC (ATLAS)

Monica Verducci on behalf of ATLAS Collaboration

CERN 1211 Geneva 23, Switzerland
monica.verducci@cern.ch

The current uncertainty on the proton content (PDFs) affects the potential discovery of new physics at LHC. The vector boson production associated with an inclusive jet is one of the most interesting candidate to constraint this uncertainty over the whole LHC kinematic regime in the ATLAS experiment. In particular, the Standard Model process such as Z plus one jet production, described here in details, is a process of a particular interest not only for PDF's study but for the Higgs and SUSY physics discovery.

1 Introduction

At the LHC, every measurement will be deeply affected by the uncertainty related to the parametrization of the partonic content of the proton (PDFs). Every cross section calculation in a proton–proton collision is the convolution of the partonic cross sections with the PDF's, where the parameters of these distribution functions are obtained by fit to data using the DGLAP evolution equation since the QCD does not predict the parton content of the proton. In LHC, the kinematic region accessible will be much broader than the other existing colliders, this will allow the possibility to test the QCD at very high (TeV scale) and low (electroweak mass scale) momentum fraction of the parton involved in the hard process. In addition, some regions will be accessible both by LHC and by the other experiments like HERA, representing a good cross check for the first results at LHC.

2 Vector Boson Production with Jets

Using the processes γ, Z and W with inclusive jet production, LHC will be able to constrain the gluon and b quark PDF. The most interesting results are reported here, with particular emphasis to the $Z + $ jet results.

The direct photon production, at LO due to Compton scattering ($\approx 90\%$) and annihilation ($\approx 10\%$) processes, is particular sensitive to the gluon distribution at mid-to-high x range. The photon and jet p_{T} distributions are extremely sensitive to the different sets of PDFs, these differences can be of the order of 16–18%. The only constrain to this measurement comes from the efficiency for the photon selection in the ATLAS detector.

Another interesting channel is the Z and W boson production, where the cross sections are dominated by the PDF uncertainty. These are very clean signals at the LHC, the relative background is less than 1%, while from the rapidity distributions one can estimate an uncertainty on the PDFs of about 8%. This permits, if the detector systematics can be controlled to 4% level, to achieve a precision of about 4% (see [1] for more details).

Finally, the $gb \rightarrow Zb$ process, sensitive to the b content of the proton, is another useful channel to study the PDF's in ATLAS both for the high statistics then for the relative small and reducible background, see [2] and [3].

In addition, this channel is very important because it is a possible background for $gb \rightarrow hb$, where the Z boson and the Higgs boson decay to the same final state ($b\bar{b}, \tau^{+}\tau^{-}, \mu^{+}\mu^{-}$) [4], and for SUSY events characterized by multiple jets, leptons, and missing transverse energy. Moreover, Z+jet events will be used to calibrate the calorimetric jet energy measurements, performing an "in situ calibration" with reconstructed jets [5].

3 Z + jets Analysis

The Z + bjet analysis, in first place, selects events with Z and one jet, then identifies in these events a Z boson associated with a jet from a b quark. Signal events must contain a Z boson decaying into a couple of muons and a b jet with a $p_{\mathrm{T}} \geq 15\,\mathrm{GeV}$ and $|\eta| \leq 2.5$.

The sample of Z + jet events was selected requiring two high p_{T} muons satisfying the following kinematic cuts, and at least one jet:

- two muons of opposite charge,
- both muons with $p_{\mathrm{T}} \geq 20\,\mathrm{GeV}$,
- di-muon invariant mass in the range $70 \leq M^{\mu\mu} \leq 110\,\mathrm{GeV}$,
- and one jet.

The number of events is estimated using the formula (1), where the following values have been used for the cross sections at the NLO: for b quark $\sigma_b = 1040\,\mathrm{pb}$, for c quark $\sigma_c = 1390\,\mathrm{pb}$ and for the other background $\sigma_x = 15\,870\,\mathrm{pb}$ (see [3] for details on the calculations). The selection efficiencies used in (1) are the following: ϵ_{acc} due to the geometric acceptance (defined as two muons in $|\eta| \leq 2.5$, and one jet in $|\eta| \leq 2.5$ and with a $pT > 15\,\mathrm{GeV}$), ϵ_{cuts} for the cuts applied for the di-muon invariant mass and finally ϵ_{flav} for the selection of the sample with the jet of a particular flavor. The efficiency for the trigger and the reconstruction algorithm are of order of 95% and are included in the ϵ_{cuts}.

$$N_{\text{flav}} = \sigma_{\text{flav}}^{\text{table}} \cdot BR(Z \to \mu\mu) \cdot \mathcal{L} \cdot \delta t \cdot \epsilon_{\text{acc}} \cdot \epsilon_{\text{cuts}} \cdot \epsilon_{\text{flav}} \tag{1}$$

An estimation of the number of events expected for $30\,\text{fb}^{-1}$ of integrated luminosity has been provided.

3.1 Z + bjet Selection

Two different and independent b-jet identification algorithms have been used: the inclusive b-tagging and the soft muon tagging.

The inclusive b-tagging algorithm is used to identify the jet of b flavor with the Cone algorithm with $\Delta R = \sqrt{(\Delta\phi)^2 + (\Delta\eta)^2} = 0.7$, see [6], while the soft muon tagging provides an independent way to identify the b-jet using the muon track, reconstructed in the tracking detectors, for the muon from the semileptonic decay of a b-meson. The obtained purity (ratio between the number of events with a b-quark and the number of selected signal events) and the expected number of events are reported in Table 1, the P_{T} distributions of the jets (a) and of the third muons (b) are reported in the Fig. 1.

Given the large statistics of the available data samples, the measurement will be limited by systematic effects which will be mainly due to the knowledge of the selection efficiency and background estimation. The efficiency of b selection has been estimated in a previous analysis to be less than 5%. The efficiency of mis-tagging, estimated with a sample of W + jet events [2], is of the order of few percent in all the muon p_{T} range. A preliminary estimation of the systematics due to the calibrations of the detectors [5], is of the order of 1%.

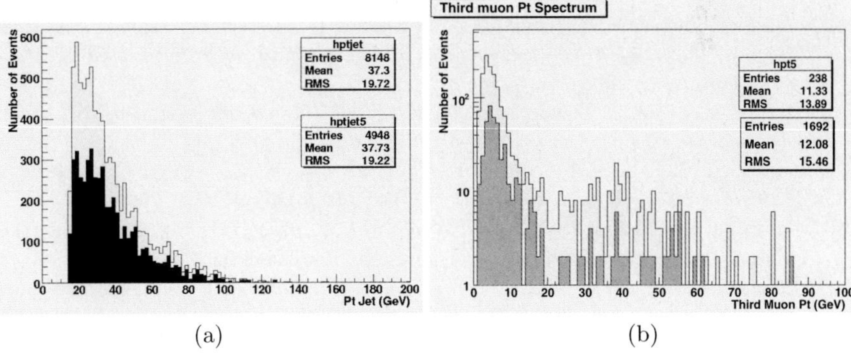

(a)   (b)

Fig. 1. a P_{T} spectrum of jets obtained after the cuts on the di-muon invariant mass. The dotted area represents the tagged b-jets. **b** Third muon P_{T} spectrum obtained after all cuts and requiring at least one jet. The muon P_{T} is reconstructed in the Muon Spectrometer using the MUID reconstruction algorithm

Table 1. The efficiencies, the purity and the number of estimated events with the full simulation for the b-tagging algorithm and the soft muon tagging algorithm. The number of events is calculated for $30\,\mathrm{fb}^{-1}$ of integrated luminosity. The b, c jets selected satisfied the cuts on p_T and $\eta : p_\mathrm{T} > 15\,\mathrm{GeV}$ and $|\eta| \leq 2.5$

CUTS	b-Tag Efficiency	Soft Muon Tag Efficiency
μ in $|\eta| < 2.5$	49.9%	49.9%
$p_\mathrm{T} > 20\,\mathrm{GeV}$ and $70 < \mathrm{Mass} < 110\,\mathrm{GeV}$	59.5%	59.5%
B-tagging Algo	56.2%	7.2%
Purity	60.7%	37.2%
Number of b Events	176 642	22 630
Number of Background Events	204 265	68 088

4 Conclusion

The Standard Model channels as direct photon, Z, W and inclusive jet productions are optimal examples to constrain PDF's at LHC and in particular in the ATLAS experiment. Given the very high statistics of these processes the uncertainties on the PDF's will be dominated by the systematic effects. For the $Z + b$ cross section measurement, the possible sources of systematic uncertainties and their impact on the measurement have been evaluated. The systematics due to the b-tagging algorithm have been estimated to be of the order of few percents as well as the systematics related to the jet and energy resolution.

References

1. A. Cooper-Sarkar, M. Dittmar, D.C. Gwenlan, H. Stenzel, A. Tricoli *LHC final states and their potential experimental and theoretical accuracies*, CERN-2005-014 (14 December 2005)
2. S.Diglio, A.Farilla, A.Tonazzo, M.Verducci, *Associated production of a Z boson and a b-jet in ATLAS*, ATL-COM-PHYS-2006-051, 2006
3. J.Campbell, R.K.Ellis, F.Maltoni and S.Willenbrock, *Associated production of a Z Boson and a single heavy quark jet*, Phys.Rev.D69:074021, 2004
4. J.Campbell, R.K.Ellis, F.Maltoni, S.Willenbrock, *Higgs-Boson production in association with a single bottom quark* Phys.Rev.D67:095002, 2003
5. R.Lefevre, C.Santoni *In situ determination of the scale and resolution of the jet energy measurements using Z^0 +jets events*, ATLAS Internal Note, ATL-PHYS-2002-026, 2002
6. S. Correard *et al. b-tagging with DC1 data*, ATL-PHYS-2004-006, (2004)

Recent developments
on precise electroweak observables

Sandro Uccirati

Dipartimento di Fisica Teorica, Università di Torino, Italy
INFN sezione di Torino, Italy
uccirati@to.infn.it

Abstract

I briefly review some tests of the Standard Model in the light of the improvements of the last year in experimental and theoretical electroweak physics. In particular I focus my attention on the new values of the masses of the W boson and top quark and on the new theoretical computation of the two-loop electroweak corrections to $\sin^2\theta_{\mathrm{eff}}$.

The global fit

According to the ElectroWeak Working Group (EWWG) reports [1], the experimental improvements during winter 2006 concern essentially three variables. We have a new value for the W width, due both to Tevatron and to Lep2, so that the average changes from $\Gamma_W = 2.133(69)\,\mathrm{GeV}$ in winter 2005 to $\Gamma_W = 2.115(58)\,\mathrm{GeV}$ in winter 2006. The most interesting improvements come from the measurements of the W-mass at Lep2 and of the top-quark mass at Tevatron.

Winter 2005	Winter 2006
$m_W = 80.452(34)\,\mathrm{GeV}$	$m_W = 80.404(30)\,\mathrm{GeV}$
$m_t = 178.0(4.3)\,\mathrm{GeV}$	$m_t = 172.5(2.3)\,\mathrm{GeV}$

Figures 1 and 2 represent some plots taken from the EWWG [1]. We can easily see from them that the new values for m_W and m_t give a more stringent limit for the Higgs boson mass and a narrower window for the validity of the Standard Model (SM). In particular a little Higgs is more and more expected.

Besides the global fit, which could hide some local discrepancies among data, also specific analysis on some observables can be done. Indeed not all

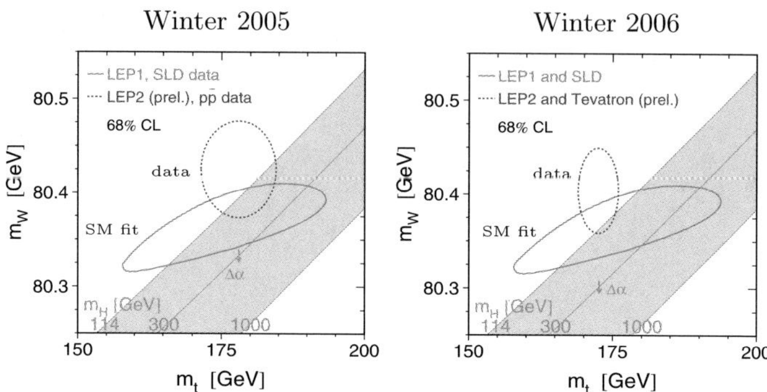

Fig. 1. The contour curves represent a probability of 68% CL, coming from the data and from the SM fit for $m_W - m_t$. The green band comes from the m_W dependence on m_t in the SM at different values of m_H and with $\Delta\alpha_{\mathrm{had}}^{(5)} = 0.02758\,(35)$

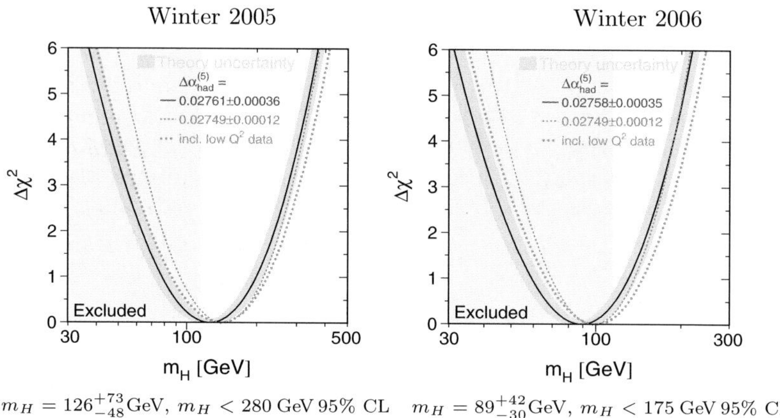

$$m_H = 126^{+73}_{-48}\,\mathrm{GeV},\ m_H < 280\ \mathrm{GeV}\ 95\%\ \mathrm{CL} \qquad m_H = 89^{+42}_{-30}\,\mathrm{GeV},\ m_H < 175\ \mathrm{GeV}\ 95\%\ \mathrm{CL}$$

Fig. 2. The solid line is the global $\Delta\chi^2$ obtained by varying m_H and fixing the fitting variables to their best-fit values. The associated band is an estimate of the current theoretical uncertainty, obtained by varying all variables within their one sigma ranges and using different renormalization schemes

physical parameters are sensitive to the Higgs mass and it is worth discussing the behaviour of the most interesting ones: the W boson mass and the effective leptonic mixing angle.

The mass of the W boson m_W

By analyzing the muon decay, the W boson mass can be related to the precisely known Fermi constant G_μ via the W–Z interdependence:

$$m_W^2\left(1 - \frac{m_W^2}{m_Z^2}\right) = \frac{\pi\alpha}{\sqrt{2}G_{\mathrm{F}}}(1 + \Delta r)\,. \tag{1}$$

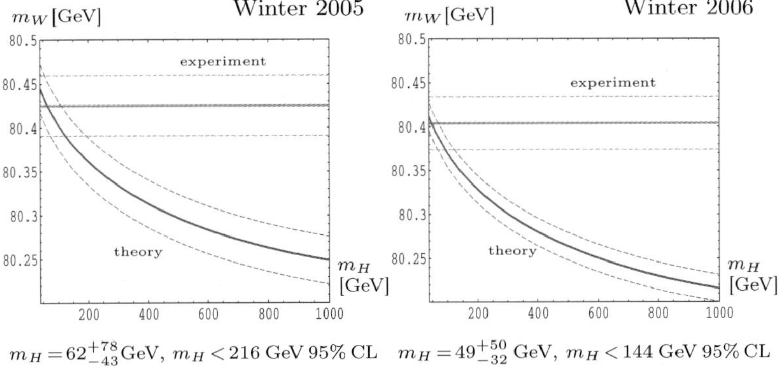

Fig. 3. Comparison theory-experiment for m_W. The estimation of theoretical error is: 6 MeV from $\Delta\alpha^{(5)}_{\text{had}}$, 14 MeV from m_t (new value) and 4 MeV from higher order corrections

Δr contains the higher order corrections (for the two-loop ones see [2]). It turns out that m_W has a logarithmic dependence on the Higgs mass, giving us the possibility to put some constraints on m_H and on the SM directly by comparing with the experimental measurements (Fig. 3). The partial fit on m_W, obtained using the fitting programs of [3], gives a central value for m_H much smaller than the lower bound of 11 4.4 GeV from direct search.

The effective leptonic mixing angle $\sin^2\theta_{\text{eff}}^{\text{lept}}$

Another important experimental observable which can be computed theoretically is $\sin^2\theta_{\text{eff}}^{\text{lept}}$. It is defined in terms of the ratio of the dressed vector and axial-vector couplings $g_{V,A}$ of the Z boson to leptons:

$$\sin^2\theta_{\text{eff}} = \frac{1}{4}\left(1 - \text{Re}\,\frac{g_V}{g_A}\right), \qquad \bar{u}_l\,\gamma_\mu(g_V + g_A\gamma_5)\,v_l\,\epsilon_Z^\mu = Z_{\bar{l}}^l \qquad (2)$$

It is related to the vector boson mass ratio via the κ factor:

$$\sin^2\theta_{\text{eff}} = \left(1 - \frac{m_W^2}{m_Z^2}\right)\kappa = \sin^2\theta_{\text{eff}}^{(0+1)} - 2\frac{m_W}{m_Z^2}\Delta m_W^{(2)} + \left(1 - \frac{m_W^2}{m_Z^2}\right)\Delta\kappa^{(2)}. \quad (3)$$

m_W is computed perturbatively (Eq. (1)), as well as κ. The last two terms correspond to the two-loop electroweak contributions due to m_W and to κ, recently computed in [4]. The fermionic corrections from m_W dominate, while the contributions from m_W and κ in the bosonic sector largely cancel (Table 1).

Measurements of the various asymmetries of the Z resonance allow to determine the effective leptonic mixing angle $\sin^2\theta_{\text{eff}}^{\text{lept}}$ with high accuracy. The current experimental value is 0.23153 (16) [1]. In Fig. 4, the programs developed in [3] with the new results of [4] have been used. Again the new m_t shrinks the theoretical error bar and lowers the preferred value for m_H.

Table 1. Fermionic and bosonic two-loop electroweak corrections to $\sin^2\theta_{\text{eff}}$ corresponding to the last two terms in (3). m_H is given in GeV and the normalization factor for $\Delta\sin^2\theta_{\text{eff}}$ is 10^{-5}

m_H	$\Delta\sin^2\theta_{\text{eff}}^{\text{fer}}(\Delta m_W)$	$\Delta\sin^2\theta_{\text{eff}}^{\text{fer}}(\Delta\kappa)$	$\Delta\sin^2\theta_{\text{eff}}^{\text{bos}}(\Delta m_W)$	$\Delta\sin^2\theta_{\text{eff}}^{\text{bos}}(\Delta\kappa)$
100	93.89	2.38	1.93	−1.65
200	98.51	−0.73	0.97	−1.05
600	106.89	−6.43	0.19	0.40
1000	105.36	−5.83	−1.16	2.47

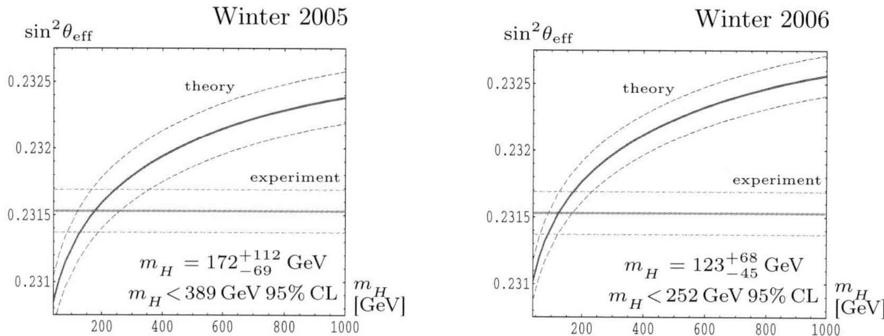

Fig. 4. Comparison theory-experiment for m_W. The theoretical error has three sources: 1.0×10^{-4} from $\Delta\alpha_{\text{had}}^{(5)}$, 0.7×10^{-4} from m_t (new value) and 0.5×10^{-4} from the truncation of the perturbation series

Conclusion

The comparison of the Standard Model with the new experimental analysis shows a slightly new scenario for the expected Higgs boson mass, much smaller then in winter 2005 (from the global fit: $m_H = 89^{+42}_{-30}$ GeV, upper limit of 175 GeV at 95%). New values for m_t and m_W are now expected, in order to reduce the theoretical uncertainties and make these bounds more stringent.

Acknowledgements

I would like to thank Giovanni Ossola for his help in making all fits and plots.

References

1. The LEP Collaborations, the LEP Electroweak Working Group, the SLD Electroweak and Heavy Flavour Groups, hep-ex/0509008.
2. A. Freitas, W. Hollik, W. Walter and G. Weiglein, Phys. Lett. B **495** (2000) 338, E: ibid. B **570** (2003) 260 and Nucl. Phys. B **632** (2002) 189, E: ibid. B **666** (2003) 305; M. Awramik and M. Czakon, Phys. Lett. B **568**, 48 (2003) and

Phys. Rev. Lett. **89** (2002) 241801; A. Onishchenko and O. Veretin, Phys. Lett. B **551** (2003) 111; M. Awramik, M. Czakon, A. Onishchenko and O. Veretin, Phys. Rev. D **68**, 053004 (2003).

3. A. Ferroglia, G. Ossola and A. Sirlin, arXiv:hep-ph/0406334. and Eur. Phys. J. C **35** (2004) 501; A. Ferroglia, G. Ossola, M. Passera and A. Sirlin, Phys. Rev. D **65** (2002) 113002.

4. M. Awramik, M. Czakon, A. Freitas and G. Weiglein, Phys. Rev. Lett. **93** (2004) 201805; W. Hollik, U. Meier and S. Uccirati, Nucl. Phys. B **731** (2005) 213, Phys. Lett. B **632** (2006) 680 and "The effective electroweak mixing angle $\sin^2\theta_{\mathrm{eff}}$ with two-loop bosonic contributions," work in preparation; M. Awramik, M. Czakon and A. Freitas, hep-ph/0605339 and hep-ph/0608099.

W and Z bosons physics at LHC at low luminosity

Sara Bolognesi[1,2]

[1] INFN, Sezione di Torino, Italy
[2] Dipartimento di Fisica Sperimentale, Universita' di Torino, Italy
   `bolognes@to.infn.it`

The W and Z bosons production cross sections at LHC will be huge: $\sigma(pp \to W \to l\nu) \sim 20\,\mathrm{nb}$, $\sigma(pp \to Z \to l^+l^-) \sim 2\,\mathrm{nb}$. Moreover the W and Z production and decay processes have been measured with high accuracy in previous experiments. Thus the W and Z bosons will play a key role during the first data taking at LHC allowing to test the detector performances and to tune the Monte Carlo generators. In fact, with only $1\mathrm{fb}^{-1}$ of integrated luminosity, the processes involving EW bosons will be used to calibrate Electromagnetic Calorimeters and to align Muon and Tracker Systems. With $10\,\mathrm{fb}^{-1}$ of integrated luminosity, the study of Z and W events will improve our knowledge of the Parton Distribution Functions (PDF) and it will provide a raw luminosity monitoring. In the following we will focus on these EW bosons studies achievable already at low luminosity. These analysis hold the key for all the future searches because they provide the way to control many of the main experimental and theoretical systematics at LHC.

1 Calibration using W and Z events

1.1 The Muon and the Tracker System

The software alignment of the Muon and Tracker System will be performed mainly using single isolated muon tracks mostly from W decay. In [1] some techniques to align the CMS Muon system are described.

The calibration of the Muon System using $Z \to \mu\mu$ events is an interesting possibility presented by ATLAS in [2]. In this analysis the energy loss of the muon in the calorimeter system and the inaccurate knowledge of the magnetic field are taken into account. The energy loss is split into the momentum-dependent and the θ-dependent part $\Delta p(p,\theta) = \epsilon(p) \times \Delta p(\theta)$. An angle-dependent scale factor for the magnetic fields $B = \frac{B}{\alpha(\theta,\phi)}$ is allowed. In order to obtain the desired accuracy of $30\,\mathrm{MeV}$ for the energy loss and of 5×10^{-4} for the magnetic field scale factors, about $6000\ Z \to \mu\mu$ events have to be processed for each pair of parameters $\alpha_i, \Delta p_j$ which have to be calibrated.

1.2 The Electromagnetic Calorimeter

The Z mass constraint in $Z \to e^+e^-$ events is a powerful tool to calibrate the Electromagnetic Calorimeter, as explained in [3] for the CMS ECAL. Effects due to partial containment, incomplete bremsstrahlung energy recollection, energy lost in the material and also simple crystal-to-crystal variation of scintillation light yield have to be taken into account in order to relate the cluster energy to the particle's initial energy. Considering these corrections, the energy measurement of a certain calorimetric object can be written down as

$$E_{\text{ele},\gamma}(GeV) = \mathcal{F} \times \mathcal{G}(GeV/ADC) \times \sum_i c_i \times A_i(ADC) \qquad (1)$$

where $\mathcal{F}$ is the correction function depending on the particle hypothesis (electron or photon), the clustering algorithm and the position and momentum of the reconstructed particle, $\mathcal{G}$ is a global ADC/GeV conversion factor, while c_i and A_i are respectively the inter-calibration coefficients and the signal amplitudes and they are summed over the clustered crystals. The inter-calibration coefficients can be computed using the nominal Z mass to constraint the e^+e^- invariant mass in an iterative event-by-event procedure. The capability of this method can be tested by comparing artificial injected mis-calibration with respect to the computed calibration coefficients. Only Z's with both legs identified as electrons with very low bremsstrahlung and far away from the module borders are considered. The achievable inter-calibration precision is about 0.6% with $2\,\text{fb}^{-1}$.

Another interesting method of ECAL calibration is based on the comparison between the measured momentum in the tracker and the energy deposited in the calorimeter for single isolated electrons, mainly from W decay [4]. However, this calibration procedure can't be applied during the first data taking because it requires the Tracker System already aligned.

2 Measurements achievable with $10\,\text{fb}^{-1}$ of data taking

2.1 PDF measurement and luminosity monitor from W production

Thanks to the very accurate theoretical prediction for the W boson production cross section [5–8], the rate of W boson events at LHC can be used to estimate the PDF or, alternatively, as a raw luminosity monitor. Obviously this will be possible only with a good control of the experimental systematics, mainly the detector acceptance (ϵ) and the background contamination:

$$\frac{N_{\text{events}} - N_{\text{backgr.}}}{\epsilon} = \sigma(q_1 q_2 \to W/Z + X) \times \text{PDF}(x_1, x_2, Q^2) \times L_{pp} \,. \qquad (2)$$

In the last years HERA data have dramatically improved our knowledge of the PDF (mainly of the gluon structure functions). As a consequence the

PDF uncertainties on predictions for the W/Z rapidity spectra have reached a precision of about 8% [9]. This may be good enough to consider exploiting these processes as luminosity monitors.

On the other side the W/Z production events can be used to extract information on the PDF, assuming that the proton-proton luminosity is measured with a different experimental procedure. This analysis is particularly interesting seeing that LHC can extend the PDF measurement to a big unexplored region of x and Q^2. In fact, the proton's momentum fractions of the scattering partons are given by $x_{1,2} = \frac{Q}{\sqrt{s}} \exp(\pm y)$ where Q is the center of mass energy of the partonic process, $\sqrt{s}$ is the center of mass energy of the hadronic reaction (14 TeV at LHC) and y is the rapidity of the parton. In the W/Z production processes the center of mass energy Q is about equal to the boson mass, so over the measurable rapidity range ($|y| < 2.5$) the x values remain in the range $10^{-4} < x < 0.1$. This ensures that the gluon is the dominant parton in the EW bosons production processes.

As a consequence the uncertainties on the gluon PDF are the major contributors to the uncertainty on the W^+, W^- and Z bosons spectra. Therefore there is a big correlation between these uncertainties and they can be strongly reduced by taking the ratios:

$$A_W = \frac{W^+ - W^-}{W^+ + W^-}\,, \qquad A_{ZW} = \frac{Z}{W^+ + W^-}\,. \tag{3}$$

However, whereas the Z rapidity distribution can be fully reconstructed from its decay leptons, we actually measure the W's decay lepton rapidity spectrum rather than the W one, because of the neutrino. Nevertheless the cancellation of the uncertainties, due to the gluon PDF, is still effective in the lepton asymmetry $A_l = (l^+ - l^-)/(l^+ + l^-)$, thereby the final PDF uncertainties on A_l are about 4%. In [9] the analysis of the $W \to e\nu$ channel has been performed using the ATLAS fast simulation. The spread between PDF sets in the electron asymmetry suggests that measurements of this variable could discriminate between PDF sets if the systematics will be kept under 5%.

2.2 Measurement of the *W* mass

The W mass has been measured with an accuracy of 30 MeV at LEP and Tevatron. The goal of LHC is measuring it with an accuracy of 15 MeV using 10 fb^{-1} of integrated luminosity. This will enable to strengthen the constraint on the SM validity range by means of the EW global fit. With an accuracy of 15 MeV on the W mass, and of 2 GeV on the top mass, the range mass allowed for the Higgs boson will be reduced to $\frac{\delta M_H}{M_H} \sim 30\%$.

Since the W mass can't be experimentally completely reconstructed because of the missing neutrino, the analysis is based on the measurement of the transverse mass

$$M_\mathrm{T}^W = \sqrt{2 p_\mathrm{T}^l p_\mathrm{T}^\nu (1 - \cos \Delta\phi)} \tag{4}$$

where the neutrino transverse momentum can be reconstructed from the hadronic recoiling system transverse momentum ($p_{\mathrm{T}}^{\mathrm{hadr}}$): $\boldsymbol{p}_{\mathrm{T}}^{\nu} = -(\boldsymbol{p}_{\mathrm{T}}^{l} + \boldsymbol{p}_{\mathrm{T}}^{\mathrm{hadr}})$.

A possible strategy to extract the W mass is comparing the M_{T}^{W} distribution observed from data with those obtained from samples simulated with different W mass. This method requires a very high precision in the simulation of both the instrumental and the physical effects. The statistical error on M_W will decrease to $2\,\mathrm{MeV}$ with only $10\,\mathrm{fb}^{-1}$ of integrated luminosity, so a very good control of the systematics will be necessary. The Missing Transverse Energy resolution and the Pile Up effect have to be carefully depicted and a very accurate physical model will be necessary to describe the hadronic recoiling system and the transverse momentum distribution of the W. In [8] the magnitude of the main expected systematics has been extrapolated from the Tevatron results using a fast simulation of the ATLAS detector. The dominant source of uncertainty ($\sim 15\,\mathrm{MeV}$) is the lepton energy and momentum scale which has to be known with an accuracy of 0.02%. However recently a strategy to limit this uncertainty to $4\,\mathrm{MeV}$ has been presented in [10] by the ATLAS collaboration. The lepton energy and momentum resolution and the response to the recoil system are the other main uncertainties related to the detector ($\sim 10\,\mathrm{MeV}$ each). All these uncertainties will be constrained in situ using the $Z \to ll$ data sample thanks to the hugestatistics available.

An alternative method to measure the W mass relies completely on the comparison with the $Z \to ll$ channel. The M_{T}^{W} distribution can be directly compared with the M_{T}^{Z} distribution from $Z \to ll$ events, opportunely rescaled [11]. This method has been applied in CMS [12]. The ratio between the two experimental distributions will reduce the detector systematics but the statistical error will increase because of the lower statistics available in the $Z \to ll$ channel. Additional systematics due to the scale extrapolation and the different behavior between the two channels have still to be evaluated.

References

1. A. Calderón *et al.*, CMS Note 2006/016
2. M. Aleksa, ATL-MUON-99-001
3. P. Meridiani, R. Paramatti, CMS Note 2006/039
4. L. Agostino *et al.*, CMS Note 2006/021
5. C. Anastasiou *et al.*, Phys. Rev. D **69**, 094008 (2004)
6. S. Dittmaier, M. Kramer, Phys. Rev. D **65**, 073007 (2002)
7. U. Baur, S. Keller and W.K. Sakumoto, Phys. Rev. D **57**, 199 (1998)
8. G. Azuelos *et al.*, hep-ph/0003275
9. A. Tricoli, A. Cooper-Sarkar, C. Gwenlan, hep-ex/0509002
10. N. Besson, M. Boonekamp, ATL-PHYS-PUB-2006-007.
11. W.T. Giele, S. Keller, Phys. Rev. D **57**, 4433 (1998)
12. A. Ghezzi *et al.*, CMS Internal Note 2005/034.

Electroweak corrections
to the charged-current Drell–Yan process

C.M. Carloni Calame[1,2], G. Montagna[2,1], O. Nicrosini[1], and A. Vicini[3]

[1] INFN, Sezione di Pavia, Via A. Bassi 6, Pavia (Italy)
[2] Dipartimento di Fisica Nucleare e Teorica, Via A. Bassi 6, Pavia (Italy)
[3] Dipartimento di Fisica and INFN-Sezione di Milano, Via Celoria 16,
   20133 Milano (Italy)

Introduction. The production of a high-transverse-momentum lepton pair, known as Drell–Yan (DY) process, plays an important role at hadron colliders, such as the Fermilab Tevatron and the CERN LHC, thanks to the large cross section and the clean signature of the final state, with at least one high-transverse-momentum lepton to trigger on. After the LEP and Tevatron precision measurement of the gauge boson masses, the DY processes represent standard candles which can be used for the detector calibration. The production of gauge bosons, in association with jets, is an important background to interesting physics channels, like the top quark pair production. The mass of the W boson will be measured at the LHC from the transverse mass distribution, but also from the ratio $(\mathrm{d}\sigma^W/\mathrm{d}M_\perp)/(\mathrm{d}\sigma^Z/\mathrm{d}M_\perp)$, with a foreseen final uncertainty of $\Delta m_W \sim 15\,\mathrm{MeV}$ [1]. The latter, combined with the improvement in the determination of the top quark mass (foreseen with an accuracy of 1–$2\,\mathrm{GeV}$), will put more stringent bounds in all the precision tests of the Standard Model. The DY processes provide, both in neutral and charged current channels, stringent constraints on the density functions which describe the partonic content of the proton [2]. The important progress in the calculation of the QCD corrections has reduced at the per cent level the residual theoretical uncertainties which affect the DY cross sections; as a consequence, it has been proposed to use them as a luminosity monitor of the collider [3,4]. The large mass tail of the invariant mass distribution represents an important background to the search for new heavy gauge bosons [5].

QCD calculations and generators. The accuracy in the determination of the theoretical cross-section has greatly increased over the years. Next-to-next-to-leading order (NNLO) QCD corrections to the total cross section have been computed in [6], but differential distributions with the same accuracy have been obtained only recently in [7]. A realistic phenomenological study and the data analysis require the inclusion of the relevant radiative corrections and their implementation into Monte Carlo event generators, in order to simulate

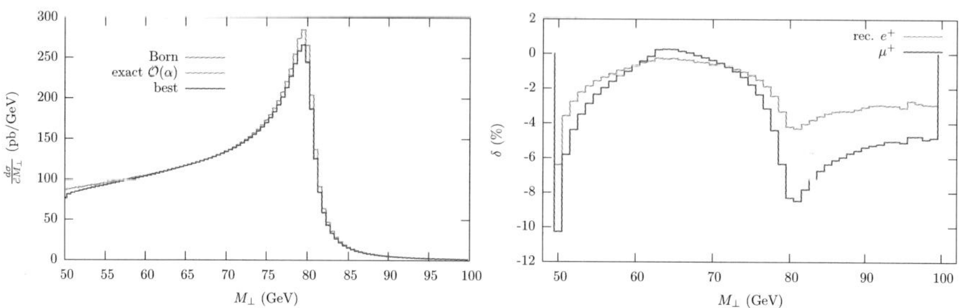

Fig. 1. Transverse mass distribution and $\mathcal{O}(\alpha)$ relative correction

all the experimental cuts and to allow, for instance, an accurate determination of the detector acceptances. The Drell–Yan processes are included in the standard QCD Parton Shower generators `HERWIG` and `PYTHIA` [8,9]. Recently there have been important progresses to improve the QCD radiation description to NLO, which has been implemented in the code `MC@NLO` [10]. Another important issue is the good description of the intrinsic transverse momentum of the gauge boson, which can be obtained by resumming up to all orders the contributions of the form $\alpha_s \log(p_\perp^W/m_W)$. The generator `RESBOS` [11], used for data analysis at Tevatron, includes these effects.

Electroweak calculations and generators. The size of the NNLO QCD corrections and the improved stability of the results against changes of the renormalization/factorization scales raises the question of the relevance of the $\mathcal{O}(\alpha)$ electroweak (EW) radiative corrections, which were computed, in the charged current channel, first in the pole approximation [12,13] and then fully in [14–16]. The generator `WGRAD` [15] includes the exact $\mathcal{O}(\alpha)$ EW corrections. The latter have been shown to induce a shift on the value of m_W, extracted at the Tevatron from the study of the transverse mass distribution (see Fig. 1[1]), of about $160\,\mathrm{MeV}$ in the muon channel [17], mostly due to final-state QED radiation. In view of the very high experimental precision foreseen at the LHC ($\Delta m_W \sim 15\,\mathrm{MeV}$), final-state higher-order (beyond $\mathcal{O}(\alpha)$) QED corrections may induce a significant shift, as shown in [18] (see Fig. 2, left panel). Some event generators can account also for multiple-photon radiation: in the published version of `HORACE` [18] final-state QED radiation was simulated by means of a QED Parton Shower; the generator `WINHAC` [19] uses the Yennie–Frautchi–Suura formalism to exponentiate final-state-like EW $\mathcal{O}(\alpha)$ corrections; finally, the standard tool `PHOTOS` [20] can be used to describe QED radiation in the W decay. A detailed series of tuned comparisons between different EW Monte Carlo generators have been done in [21], in order to check the reliability of different numerical predictions, with $\mathcal{O}(\alpha)$ accuracy, in a given setup of input parameters and cuts.

---

[1] All the figures have been obtained with the new version of `HORACE` [23].

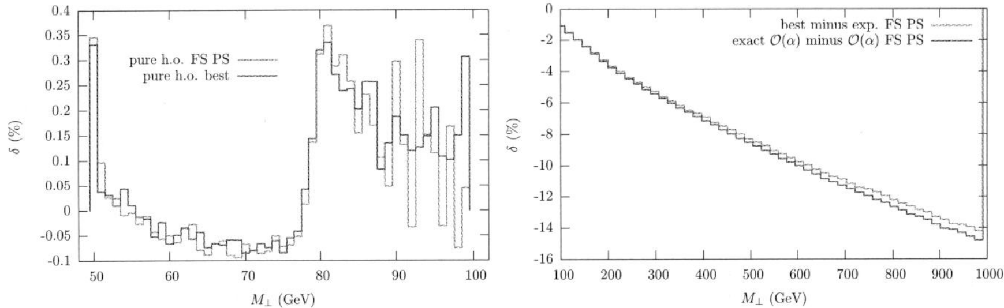

Fig. 2. Relative effect, with respect to the Born results, of the radiative corrections to the transverse mass distribution. In the *left panel* the effect of the higher-order QED effects is displayed. In the *right panel* the large negative corrections due to the $\mathcal{O}(\alpha)$ EW Sudoakov logs is shown

The new HORACE version. A precision electroweak calculation of the charged current Drell–Yan process has been recently completed [22] and includes the exact $\mathcal{O}(\alpha)$ electroweak matrix elements and leading-logarithmic QED higher-order corrections in the Parton Shower approach. The calculation is implemented in the new version of the Monte Carlo event generator `HORACE` [23], which combines, in a unique tool, the good features of the QED Parton–Shower approach with the additional effects present in the exact $\mathcal{O}(\alpha)$ EW calculation. In Fig. 2 (right panel) it is shown the increasing size of the EW $\mathcal{O}(\alpha)$ corrections, mostly due to the EW Sudakov logs. They are important above the W resonance, for a precise measurement of the W decay width and in the large transverse mass tail, to give a precise estimate of the background to new gauge boson searches. In Fig. 3 the effect of the $\mathcal{O}(\alpha)$ and higher order QED corrections on the charged lepton pseudorapidity distribution is presented. The latter is relevant to give a precise determination of the detector acceptance, which in turn is a key ingredient in the luminosity monitoring of the LHC.

Perspectives. It is important, in view of the precision aimed in the W mass measurement, in the luminosity monitoring and in the new physics searches, to study the interplay between QCD and EW corrections (a first analysis in

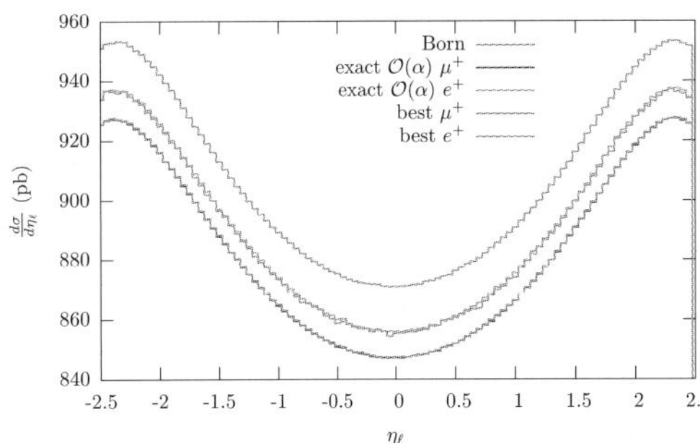

Fig. 3. Charged lepton pseudorapidity distribution. From *top* to *bottom*, Born, corrected electron, corrected muon distributions

this direction can be found in [24]). Another possible upgrade of HORACE will be achieved with the introduction of the exact EW $\mathcal{O}(\alpha)$ corrections also in the neutral current channel.

References

1. ATLAS collaboration, Technical Design Report CERN/LHCC/99-15; CMS collaboration, Technical Proposal, CERN/LHCC/94-38.
2. H. L. Lai et al., Phys. Rev. D **51** (1995) 4763 [arXiv:hep-ph/9410404]; A. D. Martin, R. G. Roberts, W. J. Stirling and R. S. Thorne, Eur. Phys. J. C **28** (2003) 455 [arXiv:hep-ph/0211080].
3. M. Dittmar, F. Pauss and D. Zurcher, Phys. Rev. D **56**, 7284 (1997)
4. S. Frixione and M. L. Mangano, JHEP **0405**, 056 (2004)
5. F. Abe et al. [CDF Collaboration], Phys. Rev. Lett. **79** (1997) 2192.
6. R. Hamberg, W. L. van Neerven and T. Matsuura, Nucl. Phys. B **359**, 343 (1991) [Erratum-ibid. B **644**, 403 (2002)].
7. C. Anastasiou, L. J. Dixon, K. Melnikov and F. Petriello, Phys. Rev. Lett. **91**, 182002 (2003); Phys. Rev. D **69**, 094008 (2004) K. Melnikov and F. Petriello, [arXiv:hep-ph/0609070].
8. G. Corcella et al., "HERWIG 6.5 release note," [arXiv:hep-ph/0210213].
9. T. Sjostrand, P. Eden, C. Friberg, L. Lonnblad, G. Miu, S. Mrenna and E. Norrbin, Comput. Phys. Commun. **135** (2001) 238 [arXiv:hep-ph/0010017].
10. S. Frixione and B. R. Webber, JHEP **0206** (2002) 029 [arXiv:hep-ph/0204244].
11. C. Balazs and C. P. Yuan, Phys. Rev. D **56** (1997) 5558 [arXiv:hep-ph/9704258].
12. D. Wackeroth and W. Hollik, Phys. Rev. D **55** (1997) 6788 [arXiv:hep-ph/9606398].
13. U. Baur, S. Keller and D. Wackeroth, Phys. Rev. D **59** (1999) 013002
14. S. Dittmaier and M. Krämer, Phys. Rev. D **65** (2002) 073007
15. U. Baur and D. Wackeroth, Phys. Rev. D **70** (2004) 073015
16. A. Arbuzov, D. Bardin, S. Bondarenko, P. Christova, L. Kalinovskaya, G. Nanava and R. Sadykov, Eur. Phys. J. C **46** (2006) 407
17. V. M. Abazov et al. [CDF Collaboration], Phys. Rev. D **70** (2004) 092008;
18. C. M. Carloni Calame, G. Montagna, O. Nicrosini and M. Treccani, Phys. Rev. D **69** (2004) 037301 [arXiv:hep-ph/0303102].
19. W. Placzek and S. Jadach, Eur. Phys. J. C **29** (2003) 325 [arXiv:hep-ph/0302065].
20. P. Golonka and Z. Was, Eur. Phys. J. C **45** (2006) 97 [arXiv:hep-ph/0506026].
21. C. Buttar et al., [arXiv:hep-ph/0604120].
22. C. M. Carloni Calame, G. Montagna, O. Nicrosini and A. Vicini, "Precision electroweak calculation of the charged current Drell–Yan process", submitted to JHEP [arXiv:hep-ph/0609170].
23. http://www.pv.infn.it/hepcomplex/horace.html
24. Q. H. Cao and C. P. Yuan, Phys. Rev. Lett. **93** (2004) 042001

Single Top at Hadron Colliders

Simona Rolli

Tufts University, Department of Physics and Astronomy, Medford, MA, 02155, USA
`rolli@fnal.gov`

1 Introduction

At hadron colliders, the strong production of $t\bar{t}$ pairs yelds large top quark samples, allowing detailed studies of many properties of top quark production and decay. However, the precise determination of the properties of the Wtb vertex, and the associated coupling strenghts, will more likely be obtained from measurements of the electroweak production of single top quarks. Single top quarks can be produced via three different reactions. These reactions are shown in Fig. 1 from left to right. The first two graphs, usually referred to as the $2 \rightarrow 2$ and $2 \rightarrow 3$ processes, respectively, both refer to the same physical W-gluon fusion process. The second production mechanism (the third graph from the left), referred to as the Wt process, is the direct production of a top quark and a W boson. This process is immeasurably small at the Tevatron, but is predicted to have a sizeable cross-section (~ 60–$110\,\mathrm{pb}$) at the LHC. The third reaction proceeds via production of an off-shell W and will be called the $W*$ process.

The primary physics interest in single top production is the ability to directly determine the coupling strength for the tWb vertex. The single top cross-section is unambiguously predicted by the SM (apart from the coupling), and it is important to cross check the W-gluon fusion, Wt, and W^* cross-sections separately. The various processes of single top production have different sensitivities to new physics. For example, the W^* channel is sensitive to an additional heavy W' boson, since new s-channel diagrams in which the W' is exchanged would occur. In contrast, additional contributions to the

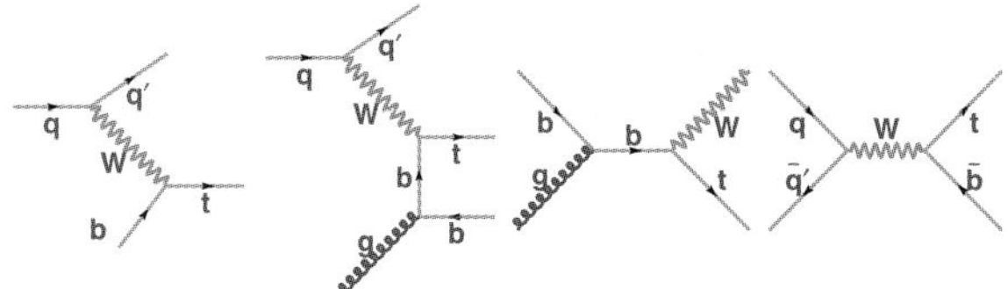

Fig. 1. Single Top generation processes at hadron colliders

W–gluon fusion process from new t–channel diagrams with a W' would be suppressed by $1/m_{W'}^2$. Therefore, existence of a W' boson would be expected to produce an enhancement in both $\sigma(W*)$ and $\sigma(W*)/\sigma(Wg)$. On the other hand, the W-gluon fusion process channel is more sensitive to modifications of the top quark's couplings to the other SM particles. Finally, because it is an inherently weak production process, the W and top quark are produced in the appropriate mixture of helicities, as unambiguously predicted by the SM. A helicity analysis of top quark decay can check for new physics, such as right handed couplings, or an unexpected admixture of the left handed and longitudinal components for the W. In the following we will report on the search for SM t-channel and s-channel single top quark production in $p\bar{p}$ collisions at a center of mass energy of 1.96 TeV, performed at the TeVatron Fermilab collider (Sect. 2) [2]. At the LHC the production of single top quarks is about a third of that of top pairs. With more than two millions single top events produced every year during a low luminosity run, a precise determination of all contributions to the total single-top cross section seems achievable. All three production channels will have a sizeable cross section, resulting in the observation of distinct final state topologies. Due to the limited space available here, the reader will find a complete description of the analysis strategies at the LHC in [1].

2 Single Top at the TeVatron

At the Tevatron the most important production mode for top quarks is via the strong interaction, with top quarks produced in pairs. Top quarks can also be produced via electroweak interactions, and in such case they are produced singly. Single top production has not been observed yet at the Tevatron. The two relevant production processes are the t- and s-channel. The cross section predicted by the SM are (1.98 ± 0.25) pb and (0.88 ± 0.11) pb at $\sqrt{s} = 1.96$ TeV respectively [2].

CDF searched for evidence of single top production in the s-channel and t-channel using Neural Networks tecniques. The networks are three layers perceptrons implemented using the NeuroBayes package [3]. There are three distinct networks used in the analysis: two for the individual s- and t-channel and one for the combined $s + t$ channels. In the training of the s-channel network, the t-channel events are treated as background and viceversa. Only events where the W from top decays leptonically are considered, giving rise to a final state signature consisting of an isolated high P_T lepton, missing transverse energy and at least one b-jet. In the s-channel a second b-jet can be observed coming from the decay of the virtual W-boson. In the t-channel the second b-jet (coming from gluon splitting) is rather forward and cannot be detected. On the other hand an additional light jet can be observed. The main backgrounds process are due to $t\bar{t}$ production, W/Z production in association with jets and QCD multijet production, where a jet is mistakenly measured

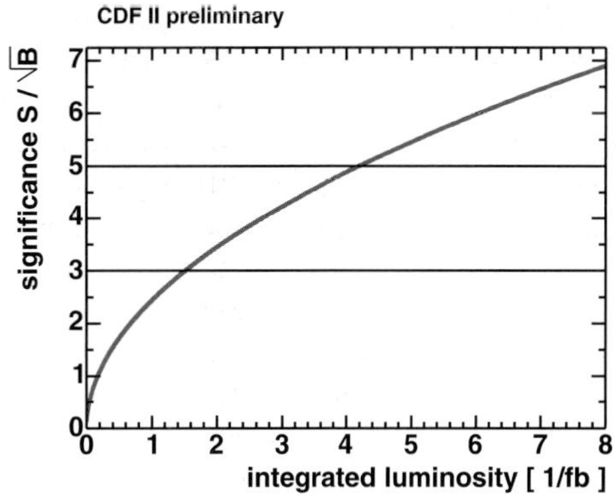

Fig. 2. CDF analysis projection for single top sensitivity

as a lepton or as giving false missing energy. Using the NN output [4] and placing an appropriate cut on it, the resulting upper limit on the $s + t$ cross section is 3.4 pb at 95% C.L. For the t- and s-channel the cross section limits are 3.1 pb and 3.2 pb, respectively.

D0 searched for evidence of single top production in the s-channel and t-channel using a likelihood discriminant method [5]. Only events where the W from top decays leptonically are considered. The data analyzed corresponds to 363 ± 24 pb^{-1} (muon channel) and 366 ± 24 pb^{-1} (electron channel). After the event selection a final discriminating likelihood variable is constructed in order to characterize the signal-type events and reject background-type ones. Up to 16 likelihood discriminant variables are used to derive upper limits on the the production cross section. The method used is bayesian: a Poisson distribution for the observed events is assumed as well as a flat prior probability for the signal cross section; the priors fir the signal acceptance and backgrounds are multivariate Gaussians centered on their estimates and described by covariance matrices taking into account correlations between different sources and bins. The upper limits on the production cross sections at 95% C.L.are: 5.0 pb for the s-channel and 4.4 pb for the t-channel.

Assuming that there will not be any improvement in the analysis technique, method and resolution, it will take about 1.5 fb^{-1} to finally observe single top at the TeVatron. Both experiments are currently taking data at sustained rate and already more than 1 fb^{-1} of data has been recorded on tape. Figure 2 shows the CDF sensitivity expressed by $S/\sqrt{(B)}$ as a function of the integrated luminosity.

3 Single Top at LHC

At the LHC single top will be produced with a total cross section which will give a yeald about a third of that relative to pair production. All three production channels will be visible. The ATLAS and CMS experiments are preparing for data taking that will start, at a center of mass energy of 14 TeV in Spring 2008. Using large samples of simulated data, for signal and background, the collaborations are defining the best analysis strategy [6], while at the same time defining the parameters of several algorithms necessary to the analysis. In the early stages of data taking, when established top quark characteristics can be used as reference points for calibration and commissioning purposes, top production (in pairs or singly) are a good place where to study b-tagging performance [7] or the effect of different jet algorithms as well as calibration effects due to calorimeter resolution.

4 Conclusions and Outlook

The Fermilab Tevatron will collect data until 2009, reaching an integrated luminosity of 4 to 6 fb^{-1}. This will allow the CDF and D0 experiments to observe for the first time top quarks singly produced via electroweak interactions. The ATLAS and CMS experiments are preparing for data taking that will start, at a center of mass energy of 14 TeV in Spring 2008. After the pioneering era at the TeVatron, which will give the first evidence for single top production, top physics will play a major role at the LHC: millions of top quarks per year will be available for precision measurements aimed at unveiling the properties of the sixth quark.

References

1. S. Rolli, `http://ncdf70.fnal.gov:8001/ifae2006/IFAE_SingleTop6.pdf`
2. B.W.Harris, E.Laenen, L.Phaf, Z.Sullivan, S.Wienzierl, Phys. Rev.D **66**, 054024(2002).
3. M.Feindt, e-print archive physics/0402093 (2004).
4. CDF Collaboration, CDF note 8185-conf, `ttp://www-cdf.fnal.gov/physics/new/top/confNotes/cdf8185_stconfnote_700pb.pdf`
5. D0 Collaboration, D0 note 4871-conf, `http://www-d0.fnal.gov/Run2Physics/WWW/results/prelim/TOP/T20/T20.pdf`
6. ATLAS Detector and Physics Performance Technical Design Report CERN/LHCC/99-14, ATLAS TDR 14, 25 May 1999.
7. S. Rolli, M. Cobal, M.P. Giordani, ATL-COM-PHYS-2006-036.

Top physics at the LHC

Andrea Dotti

INFN Sezione di Pisa, Ed. C Polo Fibonacci, Largo B. Pontecorvo, 3 – 56127 Pisa
andrea.dotti@pi.infn.it

Since its discovery at Tevatron in 1995, the top quark has been studied with
an increasing level of precision by the CDF and D0 experiments. However
most of the measurements are still statistically limited and larger top samples
are needed to open a window on new physics.

A new collider, the Large Hadron Collider, under construction at CERN,
will soon provide about 8 millions of top–antitop events in one year at low
luminosity, giving a new opportunity for precision measurements of the top
quark properties.

Prior to the precise top quark properties measurements, requiring detailed knowledge of the detector response and the understanding of the possible sources of systematic uncertainties, the top–antitop events will be used
form the very early data taking period in commissioning phase, allowing the
understanding of the detector response to different physics objects and a preliminary investigation of top quark mass measurement.

1 Introduction

The LHC is under construction in the 27 km LEP ring at CERN (Geneva),
when it will be finished (the first interactions are foreseen in 2007), will
deliver proton–proton collisions, at a center-of-mass energy of 14 TeV and
at a starting luminosity of 10^{33} cm^{-2}s^{-1} (^{1}), to the two general purpose
experiments ATLAS [1] and CMS [2].

It is assumed [3] that the integrated luminosity collected by each one of
the two experiments would be in the range 0.1–10 fb^{-1} by the end of 2008.

At LHC top quarks are mainly produced as unpolarised $t\bar{t}$ pairs via the
gluon fusion mechanism $gg \rightarrow t\bar{t}$, contributing to 90% of the total $t\bar{t}$ cross

---

^{1} The peak luminosity of 10^{34} cm^{-2}s^{-1} will be reached after three years of the low
luminosity regime.

section, the quark annihilation process accounts for the remaining 10% according to the large gluon component in the proton parton distributions. The most recent cross-section prediction at next-to-next-to leading order (NNLO) including soft-gluon corrections is $\sigma(t\bar{t}) = 873\,\mathrm{pb}$ for $m_t = 175\mathrm{GeV}/\mathrm{c}^2$ [4].

In the SM the top quarks decays before hadronization and almost exclusively into a W boson and a b-quark. The signature of the $t\bar{t}$ pair final states, "di-leptonic", "semi-leptonic" and "full hadronic", is given by the decay mode of the two W bosons, either leptonic ($W \to l\nu$) or hadronic ($W \to q\bar{q}'$). When both W bosons decay leptonically the $t\bar{t}$ signal signature is called di-leptonic, semi-leptonic when one W boson decays leptonically and the other hadronically and full hadronic when both W decay hadronically[2].

A detailed review of top quark physics potential at the LHC can be found in reference [5], here only the determination of top mass with the first data will be discussed.

2 Early top studies at LHC

The top mass is a fundamental parameter of the standard model. Together with a small set of other parameters, the top quark mass is used as an input parameter for the theoretical predictions of the electroweak precision observables. Even if the top mass is known with a better precision than other quark masses, the current precision is the dominant effect on the theoretical uncertainties of the electroweak precision observables.

In addition to the top quark measurement the first data will be used for commissioning and calibrate the detector in-situ with physics processes. The top production is an ideal laboratory for initial studies: in addition to the high production rate, the semi-leptonic channel is easy to trigger, thanks to the isolated high-p_T lepton and the high missing energy coming from the neutrino and even at low luminosity more than 50 events per day after trigger and analysis cuts should be selected. The additional presence of four jets, two of which coming from b-quarks, makes this channel interesting for many detector characteristics and performances related to lepton identification, jet reconstruction and calibration, missing energy and b-tagging. Top events will be very useful to give a prompt feedback on detector performances and as a calibration tool. The events are selected requiring an isolated high p_T lepton, missing energy and at least four jets, the physical backgrounds, mainly coming from $W + \mathrm{jets}$, $Z + \mathrm{jets}$, $WW/ZZ/WZ$ production, are reduced to a negligible level by the selection cuts and the request of one or more b-jet.

---

[2] In the di-leptoninc and semi-leptonic signature the lepton in the final state is considered to be a muon or an electron. The W boson decaying in the τ lepton, given the following preferred τ decay to hadrons, is likely to be considered more close to the hadronic mode and hence is generally considered separately.

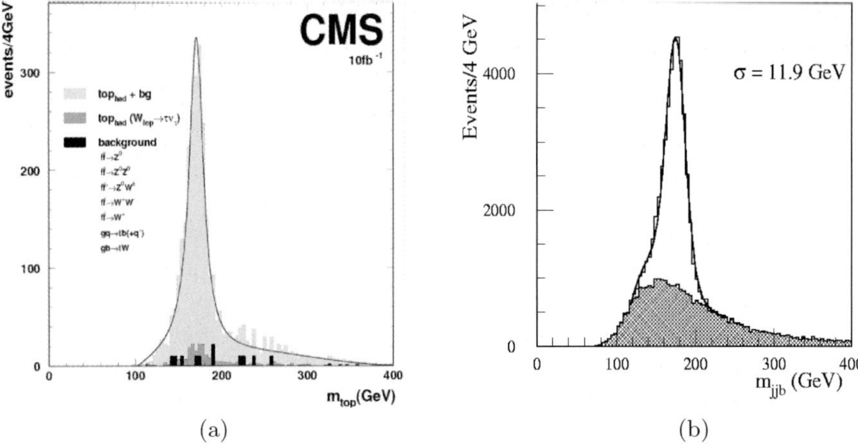

Fig. 1. Top quark reconstructed invariant mass: **a** the contribution from τ channel and non top background is visible in darker colors, **b** the combinatorial background is put in evidence in the full histogram

The W boson and top quark reconstructions are very simple, for the hadronically decaying top, through the combination of two light jets (to reconstruct the W boson) and two light jets and one b tagged jet (for the top quark reconstruction). The main background comes from wrong combinations of assigning the jets to the W boson or to the top quark. This is clearly visible in Fig. 1a where the hadronic top invariant mass distribution, reconstructed from the jets in the event, is shown. The background coming from the $t\bar{t}$ pairs with at least one τ lepton is present in the final state is presented together with the remaining backgrounds from non top events, the combinatorial background is put in evidence, in color, in the distribution presented in Fig. 1b. It has been shown [6] that even in the pessimistic case where no b-tagging can be applied, a clean sample of several thousand events will be collected in few weeks during the first data taking period.

Figure 2 shows the reconstructed top quark mass for an integrated luminosity of $300\,\mathrm{pb}^{-1}$, background from $W + 4$jets events is included. To take into account the big theoretical uncertainties in the $W + 4$jets cross-section and the the lack of the $W + N$jets (with N different from 4) events, the background is artificially increased of three times. A $W + N(< 4)$jets event can be reconstructed as a four jets event when a parton in the final state radiates a hard gluon and the jet reconstruction algorithm reconstructs two distinct jets, with a similar mechanism a $W + N(> 4)$jets can be reconstructed as a four jets event when a jet is lost because it is outside the acceptance of the detector or two jets are merged in a single one by the reconstruction algorithms. To increase the signal over background ratio events with a jet–jet invariant mass around $10\,\mathrm{GeV}$ the W boson mass are selected.

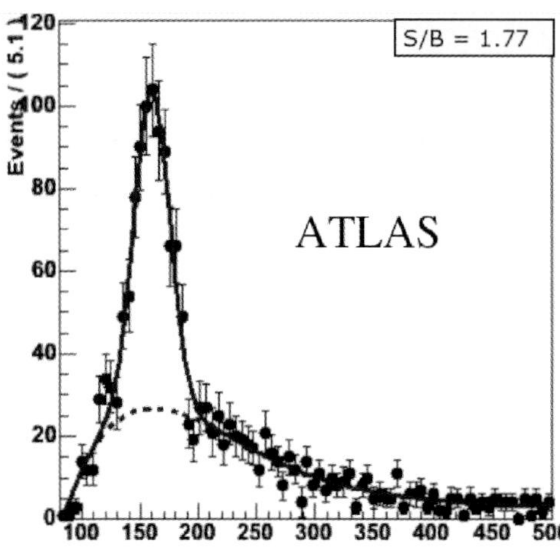

Fig. 2. Top quark mass measurement with no b-tagging, the top mass is still visible on the combinatorial and W + 4jets background

The top signal can be exploited to get, using data themselves, the light energy scale which is one of the major source of systematic error in the top mass measurement. A high purity sample (up to 80%) of $W \rightarrow jj$ decays originating from the hadronic decat of the top quark can be extracted [7] and used to improve the initial performance on the jet energy scale that is expected to be of the order of 10%. The absolute jet scale is determined by constraining the invariant mass from light jets coming from the hadronic W decay to peak at the W mass nominal value.

Calibration factors, eventually depending on jet energy, can be applied also outside top physics analyses as general calibration factors as the procedure does not require any hypothesis on the scale function.

References

1. ATLAS collaboration: *ATLAS Technical Proposal*, CERN/LHCC/94-43 (1994)
2. CMS collaboration: *CMS Technical Proposal*, CERN/LHCC/94-38 (1994)
3. F. Gianotti, M. Mangano: *LHC physics: the first one-two years*, CERN-TH-PH/2005-072, hep-ph/0504221 (2005)
4. N. Kidonakis, R. Vogt: Phys. Rev. D **68**, 114014 (2003) ; Eur. Phys. J. C **33**, s466 (2004)
5. G. Altarelli, M.L. Mangano: *Workshop on Standard Model Physics (and more) at the LHC*, CERN-2000-004 (2000)
6. S. Bentvelsen, M. Cobal: *Top studies for the ATLAS Detector Commissioning*, ATLAS Internal note : ATL-PUB-2005-024
7. J. Lu, D.M. Gingrich, H. Ahmed: *Investigation of Light-Jet Energy Calibration and Top-Quark Mass Measurement Using $t\bar{t}$* , ATLAS Internal note : ATL-COM-PHYS-2005-047

QCD corrections to Higgs physics at the LHC

Giuseppe Bozzi

Laboratoire de Physique Subatomique et de Cosmologie,
Université Joseph Fourier/CNRS-IN2P3,
53 Avenue des Martyrs, F-38026 Grenoble, France
bozzi@lpsc.in2p3.fr

We summarize the status of QCD corrections for SM Higgs boson production
at hadron colliders and briefly sketch the main search strategies at the LHC.

1 Introduction

The Higgs boson [1], which is responsible for the electroweak symmetry breaking in the Standard Model, is yet to be discovered. The experimental searches
performed at LEP allowed to put a lower bound $m_H > 114$ GeV [2] on the
mass m_H of the SM Higgs boson, whereas an upper limit $m_H < 219$ GeV at
95% CL have been obtained through global SM fits to electroweak precision
measurements [3].

The search for the Higgs boson will be one of the most important tasks
for the CERN LHC [4, 5]. A considerable effort has been devoted in recent
years to improve the theoretical predictions: the most relevant production
mechanisms and decay modes of the SM Higgs boson, together with the
main background processes, are now known to high precision.

This brief overview is mainly intended to summarize the status of the
QCD corrections to SM Higgs boson production at the LHC, focusing on the
gluon fusion and weak boson fusion processes. For a comprehensive review,
we refer the reader to [6].

2 SM Higgs production at the LHC

Gluon fusion. At the LHC, the Higgs boson will be predominantly produced
through the gluon fusion process $gg \to H$, involving a heavy-quark (mainly
top-quark) loop. The total cross Sect. at leading-order (LO) in QCD perturbation theory ($\mathcal{O}(\alpha_s^2)$) has been known for a long time [7]. The next-to-
leading order (NLO) corrections [8–10], are very large (LO is increased by
up to 100%). The complexity involved with the heavy-quark loop makes the
computation of higher-order corrections extremely difficult. In the large-m_t

limit ($m_t \gg m_H$) it is possible to introduce an effective lagrangian [11] that directly couples the Higgs to gluons:

$$\mathcal{L}_{\text{eff}} = -\frac{1}{4}\left[1 - \frac{\alpha_s}{3\pi}\frac{H}{v}(1+\Delta)\right]\text{Tr}\,\mathcal{G}_{\mu\nu}\mathcal{G}^{\mu\nu} \tag{1}$$

where the coefficient Δ is known up to $\mathcal{O}(\alpha_s^3)$ [12]. It was shown [13] that NLO calculations based on the effective lagrangian approximate the full NLO result within 10% up to $m_H=1$ TeV. The high accuracy of this approximation is due to the fact that the Higgs boson is predominantly produced in association with partons of relatively low transverse-momenta, which are unable to resolve the heavy-quark loop [14]. The next-to-next-to-leading order (NNLO) corrections have been evaluated in the large-m_t limit [14–17]. In the case of a light Higgs boson ($m_H \sim$ 100-200 GeV), the total cross section increases by about 10–25% at LHC and shows a reduced scale dependence (10–15%), thus exhibiting the features of a well-behaved perturbative series. In addition to the fixed-order results, the resummation of logarithmically-enhanced terms due to multiple soft-gluon emission has been carried out at the full next-to-next-to-logarithmic (NNLL) level [18] and at the N^3LL level [19], providing a further 7–8% increase w.r.t. NNLO and reducing scale dependence to less than 4%. Also the NLO EW contributions have been recently computed [20], showing a 5–8% effect below the $m_H = 2m_W$ threshold.

Reliable theoretical predictions for the Higgs differential (q_T and y) distributions are necessary to better understand the kinematics of the final states and compare with realistic experimental acceptances. The most advanced predictions available at present are the NNLO rapidity distribution [21], and the NLO ($\mathcal{O}(\alpha_s^4)$) transverse-momentum distribution [22]. These predictions in fixed-order perturbation theory have been supplemented with the resummation of higher-order corrections due to multiple soft-gluon emission. In the region of small transverse-momentum, q_T-resummation has been performed at the NNLL level [23], while joint (threshold and q_T) resummation has been performed at the full NLL level [24].

Weak boson fusion. This production mechanism occurs as the scattering between two (anti)quarks with weak boson (W or Z) exchange in the t-channel and with the Higgs boson radiated off the weak-boson propagator. Since the parton distribution functions of the incoming valence quarks peak at values of the momentum fractions $x \sim 0.1$ to 0.2, this process tends to produce two highly-energetic outgoing quarks. In addition the large weak boson mass provides a natural cutoff on its propagator: as a consequence, the jets are produced with a transverse energy of the order of a fraction of the weak boson mass and thus with a large rapidity interval between them (typically one at forward and the other at backward rapidity). Moreover, since the exchanged weak boson is colourless, no further hadronic production occurs in the rapidity interval between the quark jets (except for the Higgs decay products). All these phenomenological features make Higgs production via weak boson fusion (WBF) a very promising tool for precision measurements at LHC.

The LO partonic cross section can be found in [25]. As for QCD corrections, gluon radiation occurs to $\mathcal{O}(\alpha_s)$ only as bremsstrahlung off the quark legs. NLO corrections to Higgs production via WBF have been computed for the total cross Sect. [26] and for Higgs production in association with two jets [27]. They have been found to be typically modest (5–10%).

3 SM Higgs searches at the LHC

$m_H <$ **140 GeV**. In the lower mass range the Higgs particle dominantly decays into $b\bar{b}$ pairs. Because of the overwhelming QCD background the signal will be very difficult to extract, thus this decay mode is almost discarded. The same considerations apply to the leptonic decay $H \to \tau^+\tau^-$, hidden by the huge Drell–Yan lepton pair production background. Thus, the most promising channel at LHC in the case of a light Higgs is $gg \to H \to \gamma\gamma$. The decay $H \to \gamma\gamma$ is known up to 3-loop QCD [28], while the irreducible $pp \to \gamma\gamma$ background is available at NLO in the program DIPHOX [29], which also includes all relevant photon fragmentation effects. The loop-mediated process $gg \to \gamma\gamma$ contributes about 50% to the background and has been calculated at $\mathcal{O}(\alpha_s^2)$ [30].

140 GeV $< m_H <$ **180 GeV**. For intermediate Higgs masses the relevant processes are $qq \to qqV^*V^* \to qqH, (H \to \gamma\gamma, \tau^+\tau^-, V^*V^*)$ and $gg \to H \to W^+W^- \to l^+l^-\nu\bar{\nu}$. In the case of the first (WBF) channel, the decays $H \to \tau^+\tau^-, V^*V^*$ are known at 2- [31] and 3-loop [32] QCD respectively. The dominant background is Higgs production through gluon fusion in association with two jets. The full top mass dependence of the cross section at LO $(\mathcal{O}(\alpha_s^4))$ has been evaluated in [33]. Other important backgrounds to WBF are Vjj and $VVjj$ production, for which NLO corrections are available [34]. The second (gluon-initiated) channel is the most challenging one because backgrounds are of the order of the signal rate or larger. The dominant processes are $pp \to W^+W^-$, known at NLO [35] and recently supplemented with soft-gluon effects [36], and off-shell $t\bar{t}$ production, known only at LO [37].

$m_H >$ **180 GeV**. Above the ZZ threshold the $H \to ZZ \to l^+l^-l^+l^-$ process provides a very clean signature with small background [38].

The great theoretical effort made in the last years provided us with a very detailed knowledge about Higgs hadroproduction and decay. This knowledge will be essential to improve search strategies, especially in the delicate low- and intermediate-mass regions.

If the Higgs boson exists, there is no escape route for it at the LHC.

References

1. For a review on Higgs Physics, see: J. F. Gunion, H. E. Haber, G. L. Kane and S. Dawson, *The Higgs Hunter's Guide* (Addison–Wesley, Reading, Mass., 1990)

2. R. Barate et al. [LEP Collaborations], Phys. Lett. B **565**, 61 (2003)
3. LEP EW Working Group, report LEPEWWG/2005-01 [hep-ex/0511027]
4. CMS, report CERN/LHCC/94-38; ATLAS, report CERN/LHCC/99-15
5. S. Abdullin et al., report CMS-NOTE-2003-033;
6. A. Djouadi [hep-ph/0503172]
7. H.Georgi,S.Glashow,M.Machacek and D.Nanopoulos,Phys.Rev.Lett.**40**, 692 (1978)
8. S. Dawson, Nucl. Phys. B **359**, 283 (1991)
9. A. Djouadi, M. Spira and P. M. Zerwas, Phys. Lett. B **264**, 440 (1991)
10. M. Spira, A. Djouadi, D. Graudenz and P. Zerwas, Nucl. Phys. B **453**, 17 (1995)
11. J. Ellis, M. Gaillard and D. Nanopoulos, Nucl. Phys. B **106**, 292 (1976)
12. K. Chetyrkin, B. Kniehl and M. Steinhauser, Phys. Rev. Lett. **79**, 353 (1997)
13. M. Kramer, E. Laenen and M. Spira, Nucl. Phys. B **511**, 523 (1998)
14. S. Catani, D. de Florian and M. Grazzini, JHEP **0105**, 025 (2001); JHEP **0201**, 015 (2002)
15. R. V. Harlander, Phys. Lett. B **492**, 74 (2000); V. Ravindran, J. Smith and W. L. van Neerven, Nucl. Phys. B **704**, 332 (2005)
16. R. V. Harlander and W. B. Kilgore, Phys. Rev. D **64**, 013015 (2001)
17. R. V. Harlander and W. B. Kilgore, Phys. Rev. Lett. **88**, 201801 (2002); C. Anastasiou and K. Melnikov, Nucl. Phys. B **646**, 220 (2002); V. Ravindran, J. Smith and W. L. van Neerven, Nucl. Phys. B **665**, 325 (2003)
18. S. Catani, D. de Florian, M. Grazzini and P. Nason, JHEP **0307**, 028 (2003)
19. S. Moch and A. Vogt, Phys. Lett. B **631**, 48 (2005)
20. U. Aglietti, R. Bonciani, G. Degrassi and A. Vicini, Phys. Lett. B **595**, 432 (2004); G. Degrassi and F. Maltoni, Phys. Lett. B **600**, 255 (2004)
21. C. Anastasiou, K. Melnikov and F. Petriello, Nucl. Phys. B **724**, 197 (2005)
22. D. de Florian, M. Grazzini and Z. Kunszt, Phys. Rev. Lett. **82**, 5209 (1999); V. Ravindran, J. Smith and W. L. Van Neerven, Nucl. Phys. B **634**, 247 (2002); C. J. Glosser and C. R. Schmidt, JHEP **0212**, 016 (2002)
23. G. Bozzi, S. Catani, D. de Florian and M. Grazzini, Phys. Lett. B **564**, 65 (2003); Nucl. Phys. B **737**, 73 (2006);
24. A. Kulesza, G. Sterman and W. Vogelsang, Phys. Rev. D **69**, 014012 (2004)
25. R. N. Cahn and S. Dawson, Phys. Lett. B **138**, 464 (1984)
26. T. Han and S. Willenbrock, Phys. Lett. B **273**, 167 (1991)
27. T. Figy, C. Oleari and D. Zeppenfeld, Phys. Rev. D **68**, 073005 (2003)
28. M. Steinhauser, in *Proc. of the Ringberg Workshop*, World Scientific (1996)
29. T. Binoth, J. P. Guillet, E. Pilon and M. Werlen, Eur. Phys. J. C **16**, 311 (2000)
30. Z. Bern, L. J. Dixon and C. Schmidt, Phys. Rev. D **66**, 074018 (2002)
31. A. Djouadi and P. Gambino, Phys. Rev. D **53**, 4111 (1996)
32. B. Kniehl and M. Steinhauser, Phys. Lett. B **365**, 297 (1996)
33. V. Del Duca, W. Kilgore, C. Oleari, C. Schmidt and D. Zeppenfeld, Nucl. Phys. B **616**, 367 (2001)
34. C. Oleari and D. Zeppenfeld, Phys. Rev. D **69**, 093004 (2004); B. Jaeger, C. Oleari and D. Zeppenfeld, [hep-ph/0608272]
35. U. Baur, T. Han and J. Ohnemus, Phys. Rev. D **57**, 2823 (1998)
36. M. Grazzini, JHEP **0601**, 095 (2006)
37. N. Kauer and D. Zeppenfeld, Phys. Rev. D **65**, 014021 (2002)
38. J. F. Gunion, G. L. Kane and J. Wudka, Nucl. Phys. B **299**, 231 (1988)

Standard Model Higgs Boson Searches at the Large Hadron Collider

Stefano Rosati

Istituto Nazionale di Fisica Nucleare, INFN, Sezione di Roma

1 Introduction

In this paper, a brief overview of the Standard Model Higgs boson discovery potential of the ATLAS [1] and CMS [2] experiments at the Large Hadron Collider (LHC) is given. A subset of the main search channels, and of the requirements they have posed to the design of the two experiments, will be described in more detail. The full discovery potential of the two experiments, together with the possibility they will have to assess the profile of an Higgs boson in case of its discovery, will be also described.

2 Standard Model Higgs boson production and decay channels

The main Standard Model Higgs production mechanism at the LHC will be the gluon-.gluon fusion process, with the $qq \rightarrow qqH$ proccess, or Vector Boson Fusion (VBF), accounting for about 20% of the total cross section. NLO order corrections are of major relevance especially for the gluon–gluon fusion production, with K-factors ranging from 1.7 to 2.0. The K-factors for the other production mechanisms (VBF and associated production) are in the range between 1.1 and 1.3. [3]. Figure 1 shows the production cross sections and the decay branching ratios for the Standard Model Higgs boson at the LHC (from HDECAY [4]).

3 Main search channels and experimental requirements

In this section a brief review of a subset of the most relevant search channels will be given. The experimental signature and the main backgrounds will be discussed, together with the main requirements that each channel is posing on the detector performances.

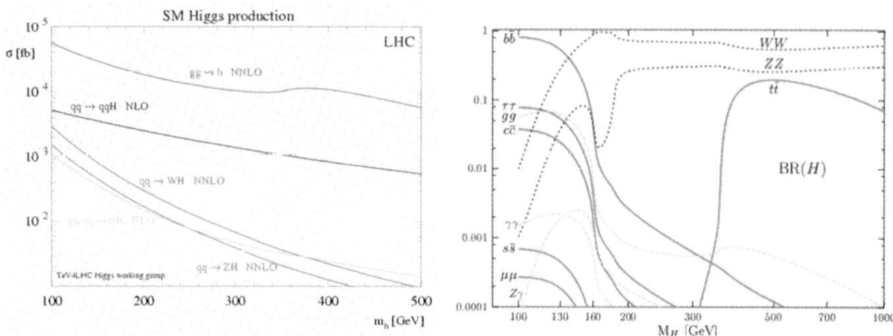

Fig. 1. Standard Model Higgs boson production cross sections and decay channels

3.1 $H \rightarrow \gamma\gamma$

The main background to this channel is the $\gamma\gamma$ continuum. Fundamental for the background rejection are the energy and the angular resolution in photon reconstruction, to allow a good separation of the signal peak with respect to the irreducible continuum. The measurement of the photon direction is relying on the identification of the primary interaction vertex, exploiting the underlying event tracks. The ATLAS experiment can take advantage also of the possibility that its EM calorimeter has, to directly reconstruct the photon direction through the measurement of the longitudinal shower developement. Both LHC experiments can obtain an experimental resolution of about 1% in the reconstruction of the Higgs mass in this channel.

3.2 $H \rightarrow ZZ^{(*)} \rightarrow$ 4-lepton

This decay channel provides a clean signature in the mass range between 120 GeV and $2m_Z$, above which it becomes the most relevant search channel, with two real Z bosons in the final state. The main backgrounds are the irreducible $ZZ^*\gamma^*$ and the reducible $t\bar{t}$ and $Zb\bar{b}$ production. Background rejection is based on di-lepton and 4-lepton mass reconstruction, decay leptons angular distributions (against $ZZ^*\gamma^*$), leptons isolation energy and impact parameter, b-jets veto (against $t\bar{t}$ and $Zb\bar{b}$).

For masses above 180 GeV the 4-lepton channel is the most reliable for a Standard Model Higgs discovery. Above 300 GeV the natural Higgs width becomes dominant above the mass resolution, and the discovery potential is mainly limited by the available integrated luminosity.

This channel is a benchmark process for leptons reconstruction performance: efficiency, p_T and impact parameter resolution have central importance for this channel, together with the lepton identification performance the high background environment foreseen at LHC.

4 Vector Boson Fusion

Higgs production through VBF, although giving an event rate smaller than for gluon–gluon fusion, can lead to a better signal/background ratio, due to the possibility of better identifying signal events through forward jet tagging [5]. Thus, in spite of the smaller cross section and the smaller K-factors, VBF channels play an important role in particular for the low Higgs mass region. One of the channels studied is the one with the Higgs decaying as $H \to \tau\bar{\tau}$. The invariant mass of the $\tau\bar{\tau}$ pair is reconstructed in the collinear approximation, the resolution obtained is about 10%.

5 Summary of the experimental discovery potential

In Fig. 2 the significances expected by each of the two experiments are shown. A significance grater than 5 is expected over the full mass range, for an integrated luminosity of $30\,\mathrm{fb}^{-1}$.

6 Measurement of the Higgs boson parameters

If a Standard Model Higgs boson will be discovered at the LHC, the measurement of its parameters (mass, partial widths, couplings, spin and CP parity) will allow an insight into the structure of the electroweak symmetry-breaking mechanism. In Fig. 3 the expected mass resolution that can be achieved by the ATLAS experiment with $300\,\mathrm{fb}^{-1}$ of integrated luminosity is shown. The procedures and the performances expected for the measurement of other Higgs parameters can be found in [6].

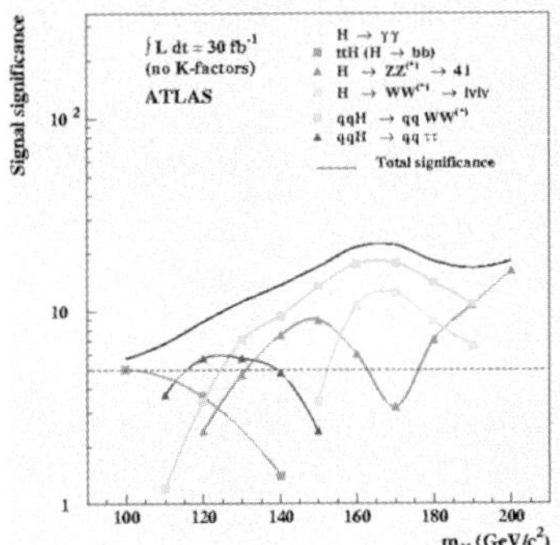
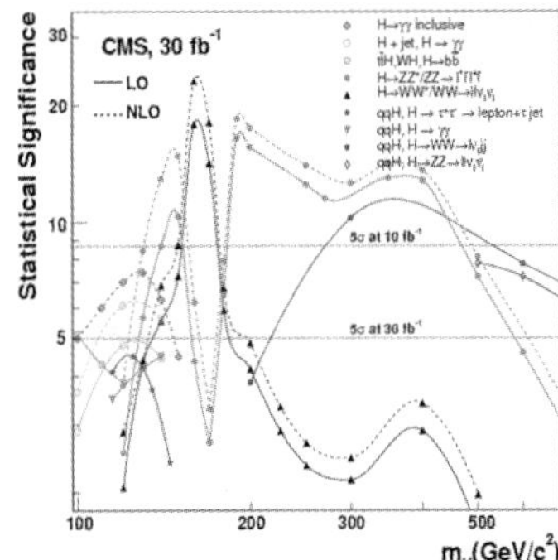

Fig. 2. Significances expected with $30\,\mathrm{fb}^{-1}$ integrated luminosity for the ATLAS and CMS experiments. The ATLAS significances have been evaluated using LO cross sections

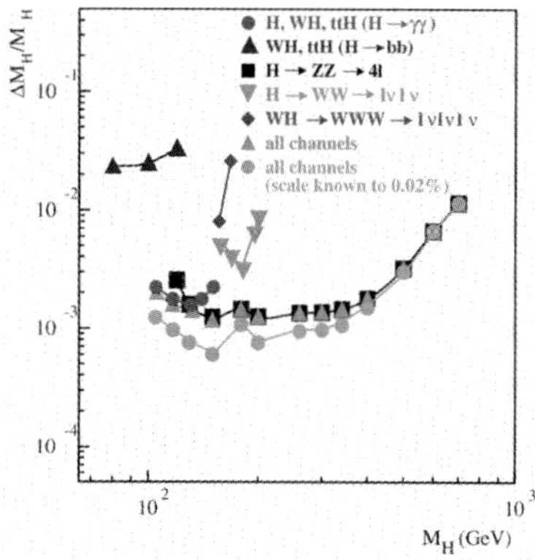

Fig. 3. Expected resolution of the ATLAS experiment in the measurement of the Higgs boson mass, for an integrated luminosity of $300 \, \text{fb}^{-1}$

7 Summary and conclusions

The prospectives for the discovery of a Standard Model Higgs boson at the Large Hadron Collider, with the experiments ATLAS and CMS, have been presented. Some of the most relevant search channels have been reviewed in more detail. A signal significance greater than 5 can be obtained by each of the two experiments with an integrated luminosity of $30 \, \text{fb}^{-1}$. The expected performance in the measurement of the Higgs boson profile has also been summarized.

References

1. ATLAS Collaboration, Detector and Physics Performance Technical Design Report, CERN/LHCC/99-14 (1999).
2. CMS Collaboration, CMS TDR 8, CERN/LHCC/2006-001 2 Feb 2006, CERN/LHCC/2006-021, 26 Jun 2006.
3. R.V. Harlander, W.B. Kilgore, Phys.Rev.Lett. **88** (2002) 201801; C.Anastasiou,K.Melnikov, Nucl.Phys. **B646** (2002) 220; S.Frixione, B.R.Webber JEP **0206** 2002 029; M.L. Mangano *et al.* JHEP **0307** (2003) 001; T.Gleisberg *et al.*, JHEP **0402** (2004) 056.
4. A. Djouadi, J. Kalinowski, M. Spira, Comput. Phys. Commun. **108** (1998) 56-74.
5. S. Asai *et al.* Prospects for the search for a Standard Model Higgs boson in ATLAS using Vector Boson Fusion, Eur. Phys. J. **C 32S2** (2004) 19.
6. C.P. Buszello, I. Fleck, P. Marquard, J.J. van der Bij, Prospective analysis of Spin and CP-sensitive variables in H→ZZ→4l at the LHC, Eur.Phys.J. **C32** (2004) 209; M.Duhrssen *et al.*, Extracting Higgs boson couplings from LHC data, Phys.Rev. **D70** (2004) 113009.

Searching for extra-SUSY signals at LHC

Lorenzo Menici

University of Rome "La Sapienza"
lorenzo.menici@roma1.infn.it

1 Search for Extra Neutral Gauge Bosons

The presence of a heavy neutral gauge bosons Z', besides the Z^0, is foreseen in various non-SUSY models. There are no reliable theoretical predictions, however, of its mass scale. Current lower limits from LEP and the Tevatron on the Z' mass [1] are 600–900 GeV, depending on the model. LHC is going to be the first opportunity to search for Z' bosons in a mass range significantly larger than 1 TeV: here we review a work on the process $Z' \to \mu^+\mu^-$, for a work concerning $Z' \to ee$ we suggest the reading of [2].

1.1 $Z' \to \mu^+\mu^-$

The CMS analysis performed in [3] involves the decay into $\mu^+\mu^-$ pairs of six Z' bosons described in the literature whose properties are representative of a broad class of extra gauge bosons: Z_{SSM} within the Sequential Standard Model [4], Z_ψ, Z_η, Z_χ, that arise in E_6 ad SO(10) groups [5]. Z_{LRM} and $Z_{\mathrm{ALRM}}, Z_\chi$, arising in the framework of left-right models described in [6]. The generation of signal events includes the full $\gamma^*/Z^0/Z'$ interference structure and the Z' bosons is assumed to decay only to three ordinary families of quarks and leptons. The dominant and irreducible background to $pp \to Z' \to \mu^+\mu^-$ is the Drell–Yan production of muon pairs, $pp \to \gamma/Z^0 \to \mu^+\mu^-$. The contribution from other background sources (ZZ, ZW, WW, tt, bb, pile-up of muons from minimum-bias collisions) is at the level of only a few percent of the Drell–Yan have been neglected. The overall reconstruction efficiency including acceptance, trigger and off-line reconstruction for $Z' \to \mu^+\mu^-$ events with a mass between 1 and 5 TeV lies in the range of 70–75%. A very low integrated luminosity, less than $0.1\,\mathrm{fb}^{-1}$ should be sufficient to discover Z' bosons at 1 TeV, a mass value which will likely be above the Tevatron reach. An integrated luminosity of $10\,\mathrm{fb}^{-1}$ is sufficient to reach 5σ significance at 3 TeV for some but not all the models considered. Depending on the model, the mass reach is in the range between 2.6 and 3.4 TeV.

With an integrated luminosity of $100\,\text{fb}^{-1}$ the mass reach lies in the region between 3.4 and $4.3\,\text{TeV}$. Such reaches do not take into account systematic uncertainties which are expected to be instrumental (alignment, magnetic field, calibration) or ascribed to the procedure of fitting the mass distributions and to theoretical uncertainties (parton distributions, K-factors).

2 Search for Extra Dimensions

Using the idea of Kaluza and Klein [7], recent different models have shown that EDs with an "observable" size might describe our world. The two basic progenitors of modern EDs theories: Compact Large and Warped EDs.

2.1 Compact Large Extra Dimensions

Arkani–Hamed, Dimopoulos and Dvali [8,9] showed that the hierarchy problem could be reformulated as a topologically if we suppose that gravity is allowed to propagate in more that four dimensions. Now the "true" gravity scale is set by the value of $4 + n$–dimensional Plank scale, M_D, defined by $M_\text{P}^2 = V_n M_\text{D}^{n+2}$, where V_n is the volume of the compactified space. Then the size R of EDs depends on both n and M_* and, trying to address the hierarchy problem assuming $M_* \sim \text{TeV}$, R can be much much larger than M_P^{-1} (e. g. $n = 2$ implies $R \sim 100\ \mu\text{m}$). Performing a 4D reduction of the linearized $(4 + n)$D Einstein theory, one obtains a massless graviton $G^{(0)}$ and a Kaluza–Klein (KK) tower of massive gravitons $G^{(\text{k})}$ with mass slitting of order $1/R$. Here we describe in the framework of compact large EDs, the graviton s–channel production at LHC. For the jet$+\not{E}_\text{T}$ and $\gamma+\not{E}_\text{T}$ signatures at ATLAS we remind to [10].

Graviton s–channel processes

Possible signals of compact large EDs come from processes involving virtual graviton G exchange (s-channel) with γ or lepton pair production. In [11] the possible detection of these processes is studied for the ATLAS collaboration. The main di-photon processes are $q\bar{q} \to G \to \gamma\gamma$ and $gg \to G \to \gamma\gamma$, while in the di-lepton case we have $q\bar{q} \to G \to l^+l^-$ and $gg \to G \to l^+l^-$. The main irreducible background, from the SM, has a different rapidity distribution compared to the signal. Furthermore, the ratio s/b can be enhanced using a lower cutoff M_min on the invariant masses $M_{\gamma\gamma}$ and M_{ll}. In Fig. 1 the distribution of the signal and SM background for different values of M_D for the di-lepton channel after $10\,\text{fb}^{-1}$ of integrated luminosity are shown. For a given number n of EDs, the discovery $S/\sqrt{S + B} > 5$ gives the maximal value M_D^max that can be accessed by ATLAS. In Fig. 2, the 5σ contours are drawn in the plane $(M_\text{min}, M_\text{D}^\text{max})$, showing the best M_min for which the discovery is ensured together with a maximal reach value M_D^max.

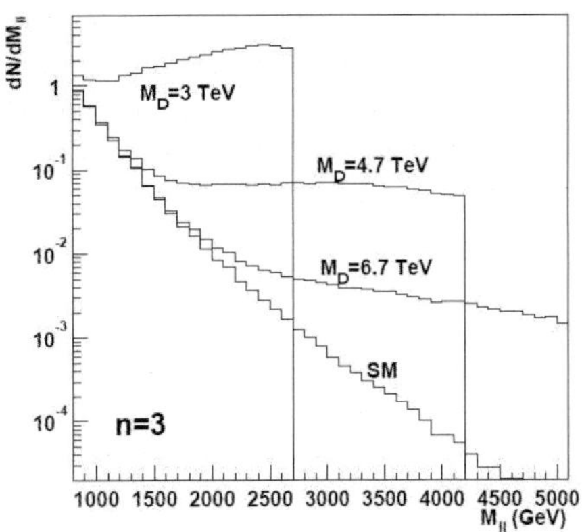

Fig. 1. Di-lepton production: total signal for $n = 3$ and various values of M_D after $10\,\mathrm{fb}^{-1}$ of integrated luminosity

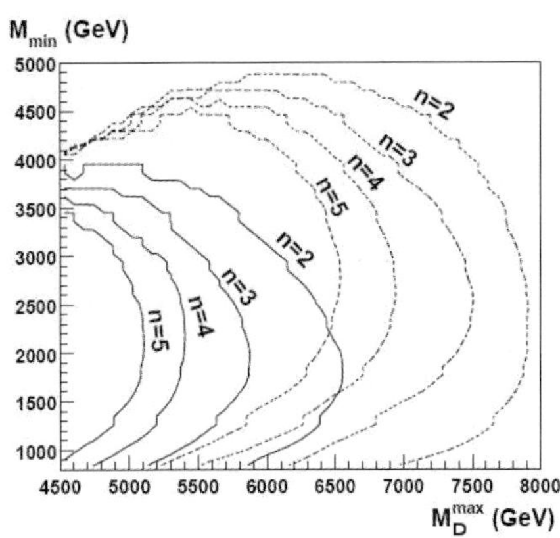

Fig. 2. Di-lepton production: maximal reach at 5σ level in the scale M_D as a function of the lower cut $M_{\min}$. *Solid lines* are for $10\,\mathrm{fb}^{-1}$ of integrated luminosity, *dashed* for $100\,\mathrm{fb}^{-1}$. All lines are labeled by the number of extra dimensions

2.2 Randall–Sundrum model

The model proposed by Randall and Sundrum (RS) [12] assumes a 5D bulk with two 4D branes in at $y = 0$ and $y = \pi r_c$[1]. The metric assumed is non-factorisable

$$\mathrm{d}s^2 = \mathrm{e}^{-2\sigma(y)}\eta_{\mu\nu}\,\mathrm{d}x^\mu\,\mathrm{d}x^\nu - \mathrm{d}y^2 \,, \qquad (1)$$

leading to a gravity scale $\Lambda_\pi = M_P\,\mathrm{e}^{-\pi k r_c}$, where k is the five-dimensional space-time curvature ($\sim M_P$). The scale Λ_π can be of the order $1\,\mathrm{TeV}/c^2$ if $r_c \sim 10^{-32}\,\mathrm{m}$, addressing the hierarchy problem through the warp factor $\mathrm{e}^{-\pi k r_c}$. This leads to the presence of massive well separated Kaluza–Klein (KK) excitations of the gravitons:

$$m_n = k x_n\,\mathrm{e}^{-\pi k r_c}, \qquad (2)$$

---

[1] There exists also a second model by Randall and Sundrum [13], similar to [12] but with an *infinite* warped extra dimension (i. e. $r_c \to \infty$). It was originally of interest because it represented an infinite 5D model which, in many respects, behaved as a 4D model.

x_n being the roots of the Bessel function $J1$. Bounds on the five-dimensional curvature scalar R_5 and on Λ to avoid hierarchy between M_{EW} and $\Lambda\pi$ provide theoretical constraints which limits the region of interest of the parameter space. Experimental constraints [14] on the RS model come from Tevatron Run II data, which exclude M_G up to $200\,\mathrm{GeV}/c^2$ for $c = 0.01$ and $690\,\mathrm{GeV}/c^2$ for $c = 0.1$. At the LHC the main discovery channels are the decay of the first KK excitation into electron or muon pairs.

Graviton detection

A study on the CMS reach in the search for Randall–Sundrum excitations of gravitons decaying into e^+e^- has been done in [15]. At parton level, single gravitons can be produced at LHC via $q\bar{q} \to G$ or $gg \to G$. The SM background consists mostly of the Drell–Yan process: $q\bar{q} \to \gamma, Z^0 \to e^+e^-$.

The reconstruction of the electrons is performed using the CMS ECAL Super Clusters obtained with the island algorithm [16] and accounting for the leakage in the hadronic calorimeter and the saturation of the front-end electronics which occurs when more than $1.7\,\mathrm{TeV}$ is released in one crystal. To select the signal, isolation of the supercomputer, limit on the ratio of the hadronic energy fraction $H/E < 0.1$ and at least two hits for each of the electron tracks are imposed, rejecting very efficiently the QCD background and leaving as a relevant background only the Drell–Yan e^+e^- pairs.

References

1. Review of Particle Properties, K. Hagiwara et al., Phys. Rev. D 66 (2002) 328.
2. M. Shaefer et al, $Z' \to ee$ studies in full simulation, ATL-PHYS-PUB-2005-010(2005).
3. R. Cousins ET AL., CMS Note 2005/002 (2005).
4. T. Sjostrand ET AL, hep-ph/0108264, (2001).
5. J.L. Rosner, Phys. Rev. D 35 2244 (1987) and Phys. Rev. D 54 (1996).
6. R.N. Mohapatra, Unification and Supersymmetry, Springer–Verlag, (2003).
7. For the English version of these works see T. Appelquist, A. Chodos and P.G.O. Freund, Modern Kaluza–Klein theories, Addison–Wensley (1987).
8. N. Arkani–Hamed, S.Dimopoulos and G.R. Dvali, Phys. Rev. D59, 086004 (1998).
9. N. Arkani–Hamed, S.Dimopoulos and G.R. Dvali, Phys. Lett. B 429, 263 (1998).
10. I. Hinchliffe and L. Vacavant, Model independent extra dimension signatures with ATLAS, ATL-PHYS-2000-016 (2000).
11. V. Kabachenko, A. Miagkov and A. Zenin, Sentitivity of the ATLAS detector to extra dimensions in di-photon and di-lepton production processes, ATL-PHYS-2001-002 (2001).
12. L. Randall and R. Sundrum, A Large Mass Hierarchy from a Small Extra Dimension, Phys. Rev. Lett. 83 (1999) 3370.
13. L. R. and R. S., An alternative to compactification, Phys. Rev. Lett. 83 (1999).

14. E. Kajfasz, Search for Extra–Dimensions at pp and ep Colliders, ICHEP 2004, Beijing (2004).
15. C. Collard and M.-C. Lemaire, Search with the CMS detector for Randall–Sundrum excitations of gravitons decaying into electron pairs, CMS Note 2004/024 (2004).
16. E. Meschi et al., Electron Reconstruction in the CMS Electromagnetic Calorimeter, CMS Note 2001-034 (2001).

Search for New Physics at the Tevatron

Simona Rolli

Tufts University, Department of Physics and Astronomy, Medford, MA, 02155, USA
rolli@fnal.gov

1 Introduction

Over the past three decades the Standard Model (SM) of particle physics has been surprisingly successful. Although the precision of experimental tests improved by orders of magnitude no significant deviation from the SM predictions has been observed so far. Still, there are many questions that the Standard Model does not answer and problems it can not solve. Among the most important ones are the origin of the electro–weak symmetry breaking, hierarchy of scales, unification of fundamental forces and the nature of gravity. Recent cosmological observations indicates that the SM particles only account for 4% of the matter of the Universe. Many extensions of the SM (Beyond the Standard Model, BSM) have been proposed to make the theory more complete and solve some of the above puzzles. Some of these extension includes SuperSymmetry (SUSY), Grand Unification Theory (GUT) and Extra Dimensions. At CDF [1] and D0 [2] we search for evidence of such processes in proton–antiproton collisions at $\sqrt{(s)} = 1980$ GeV. The phenomenology of these models is very rich, although the cross sections for most of these exotic processes is often very small compared to those of SM processes at hadron colliders. It is then necessary to devise analysis strategies that would allow to disentagle the small interesting signals, often buried under heavy instrumental and SM background. Two main approaches to search for physics beyond the Standard Model are used in a complementary fashion: model-based analysis and signature based studies. In the more traditional model-driven approach, one picks a favorite theoretical model and/or a process, and the best signature is chosen. The selection cuts for acceptances are optimized based on signal MC. The expected background is calculated from data or Monte Carlo and, based on the number of events observed in the data, a discovery is made or the best limit on the new signal is set. In a signature-based approach a specific signature is picked (i. e. dileptons $+ X$) and the data sample is defined in terms of known SM processes. A signal region (blind box) might be defined with cuts which are kept as loose as possible and the background predictions

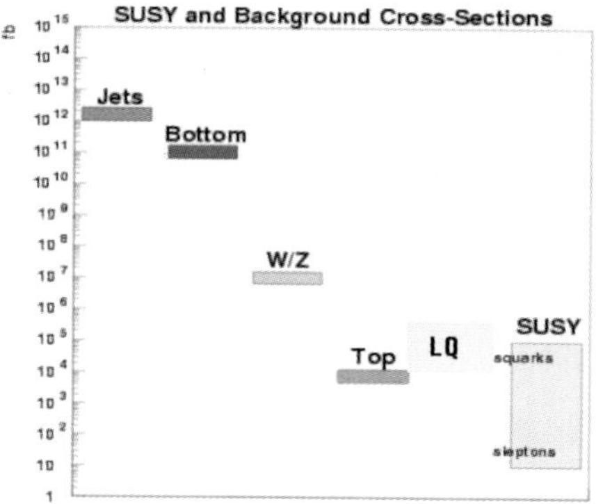

Fig. 1. Cross sections for typical SM processes at the TeVatron and exotic physics

in the signal region are often extrapolated from control regions. Inconsistencies with the SM predictions will provide indication of possible new physics. As the cuts and acceptances are often calculated independently from a model, different models can be tested against the data sample. It should be stressed here, though, that the comparison with a specific model, implies calculating optimized acceptances for a specific BSM signal. In signature-based searches, there is no such an optimization. Since it has been argued that a person external to the collaboration could simulate a specific model and compare it to experimental data, it is necessary to remind that most of the time, optimizing the cuts implies a detailed knowledge of the detector (when using simulated data on which detector simulation, reproducing the running conditions, has been applied) which can only be carried on by the experiments. There are several generic detector simulation packages, which are often used by theorists outside of an experimental collaboration: while these packages give a rough estimate of detector effects, nevertheless they will never be able to reproduce the exact run-time detector conditions. Sometime the run time conditions are so peculiar that very large systematic effects have to be considered or strong correction to the energy scale of the physics objects used (e. g. jet energy corrections).

Given the limited space available for this proceeding, we will focus here on a couple of results based on the aforementioned approaches. Further results are described in [3].

2 Searches for new physics in dilepton final states at CDF

Final states consisting of dileptons are a straightforward signature where to look for new physics, as several resonant states can appear as enhancement of the Drell–Yan cross section. The analysis strategy is very simple: the invariant mass distribution of the dilepton system is compared to the SM expectations,

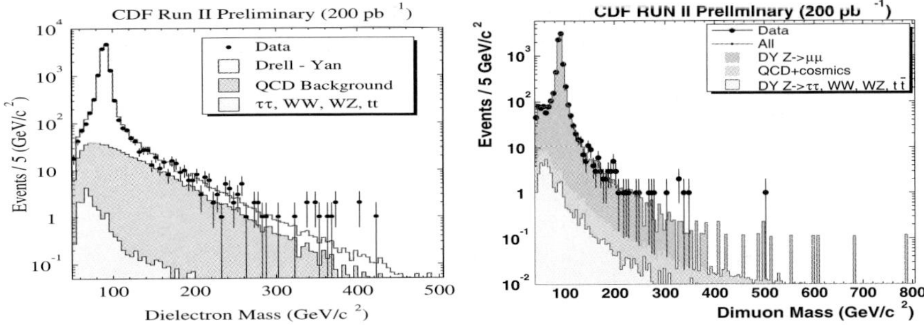

Fig. 2. Dielectron and dimuon mass spectrum. Data are compared to the SM expectations

as shown in Fig. 2 (CDF data). Only identification cuts to select a pair of high P_{T} leptons are placed. To compare this data with specific BSM models, acceptances for different type of resonances, based on their spin are calculated, using MC simulated signal events. Again, only identificatio cuts on the leptons are applied and no optimization is perfomed, in order to reduce the (irreducible) Drell–Yan background. The number of expected events for various scenarios are derived. Since no excess is observed, a 95% C.L. upper limit on the cross section as a function of the mass of the resonance is set [4].

3 Searches for LeptoQuarks at CDF

This is a typical model-based search. Leptoquarks (LQ) are hypothetical color-triplet particles carrying both baryon and lepton quantum numbers and are predicted by many extension of the SM as new bosons coupling to a lepton–quark pair. Their masses are not predicted. They can be scalar particles (spin 0) or vector (spin 1) and at high energy hadron colliders they would be produced directly in pairs, mainly through gluon fusion or quark antiquarks annihilation. Because of theory constraints a LQ would decay into a lepton and quark of the same generation. Traditionally the branching ratio describing the decay of the leptoquark into a charge lepton and and quark is called β. Several analysis have been carried out at the TeVatron: we will report on the CDF search for first-generation and second generation scalar LQ pairs produced in $p\bar{p}$ collisions at $\sqrt{s} = 1.96\,\mathrm{TeV}$ [5, 6]. CDF searched for first and second generation LQ in all three decay channels ($l = e,\mu$): $LQ\overline{LQ} \rightarrow llq\bar{q}$ ($\beta = 1.0$), $LQ\overline{LQ} \rightarrow l\nu q\bar{q}$ ($\beta = 0.5$) and $LQ\overline{LQ} \rightarrow \nu\nu q\bar{q}$($\beta = 0.0$ or $\beta' = \mathrm{Br}(LQ \rightarrow \nu q) = 1.0$). The decay of a pair of leptoquarks would give rise to a signature consisting of one or two high P_{T} leptons, two jets and possibly missing transverse energy. The main SM processes giving rise to the same signature are production of a vector boson (Z or W) accompanied by jets and top quark production, where the W's coming from the decay of the top

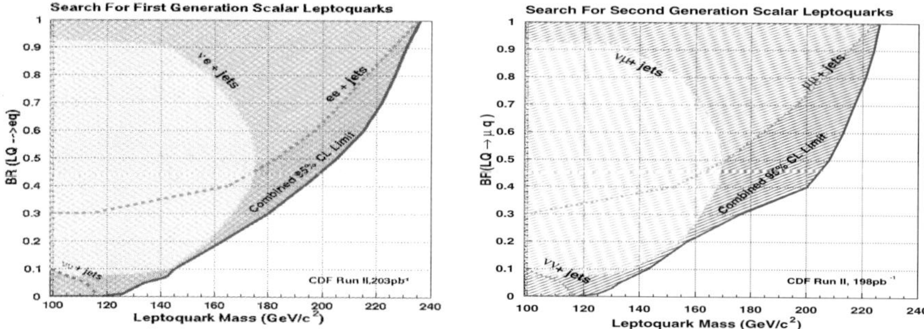

Fig. 3. LQ mass exclusion regions at 95% C.L. as a function of Br(LQ → eq)

quark decay into leptons. The analysis consists of a series of sequential cuts, which are optimized to reduce this type of SM processes, while enhancing the signal retention. Simulated MC events are used to obtain signal acceptances and background estimates. The same cuts are applied on MC events and data, and the number of events observed in the data, after all cuts, are compared to the expectation from SM processes (background) and LQ signal. In all searches no evidence for the existence of leptoquarks is observed. Therefore, upper limits on the production rate × branching ratio of leptoquarks and lower limits on their masses are derived as a function of β (Fig. 3).

4 Conclusions

The CDF and D0 experiments are actively taking physics quality data at the upgraded TeVatron collider. New results on search for physics beyond the Standard Model are released almost daily. So far there is no evidence for new physics and numerous limits on new particle masses and cross sections production are set. Most of the published results contain a statistical sample doubled in respect to Run I (200 to 400 pb^{-1}) but by the end of 2008 there will be numerous updates using a statistical sample corresponding to 1 fb^{-1}. If by that there will still be no evidence for new physics, the baton will be passed to the LHC where the experience of many Tevatron physicists will be nevertheless invaluable.

References

1. CDF Collaboration, F. Abe et al., *Nucl Instrum, Methods,* A271, 387 (1988).
2. D0 Collaboration, S. Abachi et al.,*Nucl Instrum, Methods,,* A**338**, 185 (1994)
3. S. Rolli, `http://ncdf70.fnal.gov:8001/ifae2006/IFAE_Searches4.pdf`.
4. CDF Collaboration, Phys.Rev. Lett**95**, 252001 (2005);
5. D.Acosta et al. (CDF Collaboration), Phys Rev D**72**, 051107 (2005)
6. A.Abulencia et al.,(CDF Collaboration), Phys Rev D**73**, 051102 (2006)

New physics searches in B meson decays

S. Vecchi for the LHCb collaboration

INFN Sezione di Bologna, via Irnerio 46, 40126 Bologna (Italy)
vecchi@bo.infn.it

The B system represents an ideal framework for a phenomenological study of the CKM matrix, the validation of its unitarity and independent tests of the CP violation description in the Standard Model (SM). In particular rare B decays are among the cleanest probes of the flavour sector of the SM available to present experiments and can reveal the existence of new physics (NP).

Radiative decays $b \to (s/d)\gamma$ are flavour changing neutral current (FCNC) processes, which proceed in the SM only at the loop level, so they can be affected by NP interactions through the exchange of super symmetric particles in the loop. This fact makes radiative B decays particularly sensitive to NP, since additional contributions to the SM decay rate can modify Branching Fractions (BF), CP asymmetry or final state angular distribution. The inclusive $\mathrm{BF}(B \to X_s\gamma)$ has been calculated precisely in the SM (at the NLO approximation) [1]. The BF directly probes the Wilson coefficient $|C_7|$, so its measurement can limit some NP model parameters. Experimentally, only B-factories can perform inclusive measurements, since they can use additional kinematic constraints to suppress significantly the huge background present. The values measured by the CLEO, BaBar and Belle experiments, reported in Table 1, are in perfect agreement with the SM prediction. Nevertheless

Table 1. Experimental and theoretical values on the Branching fraction (column 2 and 3) and direct CP asymmetry A_{CP} of the $B \to X_s\gamma$ inclusive transitions

	Inclusive BF $(\times 10^6)$	Semi-inclusive BF $(\times 10^6)$	$A_{\mathrm{CP}}^{(\mathrm{dir})}$ (%)
CLEO [3]	$321 \pm 43 \pm 27^{+18}_{-10}$	$-7.9 \pm 10.8 \pm 2.2$	
Belle [4] [5]	$355 \pm 32^{+30+11}_{-31-7}$	$366 \pm 53 \pm 42^{+50}_{-54}$	$0.2 \pm 5.0 \pm 3.0$
BaBar [6] [7]	$367 \pm 29 \pm 34 \pm 24$	$335 \pm 19^{+56+4}_{-41-9}$	$2.5 \pm 5.0 \pm 1.5$
HFAG^{a} [8]	$355 \pm 24^{+9}_{-10} \pm 3$		0.4 ± 3.6
SM (NLO)^{a} [1]	357 ± 30		0.44 ± 0.20

^{a} Corresponding to an energy cutoff $E > 1.6$ GeV

NP contributions could still be present and some limits on NP parameters can be set. In particular Minimum Flavour Violation (MFV) SUSY model would be favoured. A model independent analysis of data, which accounts for NP contributions, limits the Wilson coefficient ratios $R_7 = C_7^{\text{NP}}/C_7^{\text{SM}}$ and $R_8 = C_8^{\text{NP}}/C_8^{\text{SM}}$ [2]. The results of CP asymmetries A_{CP} in this channel (Table 1) agree with the SM prediction close to zero. At present the experimental values are affected by large errors. In case of NP contributions, depending on the symmetry breaking mechanism, the CP asymmetry could raise up to 10% (in case of non-MFV) so more precise data will be very useful to distinguish between models.

Exclusive $b \to s\gamma$ transitions have recently been measured in several channels. The selection of exclusive channels results in a lower background contaminations compared to inclusive ones, even if low statistics becomes an issue in some cases. The theoretical estimates suffer from large ($\simeq 20\%$) uncertainties, so it is more convenient to deal with BF ratios or A_{CP} asymmetries. The experimental results on BF, omitted here for lack of space (see transparencies or reference [9]), are in total agreement with the SM prediction. Also the direct CP asymmetry in $B \to K^{0*}\gamma$ (Belle: $(-1.5 \pm 4.4 \pm 1.2)\%$; BaBar: $(-1.3 \pm 3.6 \pm 1.0)\%$) agree with the SM ($A_{\text{CP}}^{(\text{dir})} \simeq 1\text{--}2\%$).

The polarization of the photon emitted in the $b \to s\gamma$ transition, mostly left-handed ($> 90\%$) in the SM, provides an additional test of the theory [10]. To access to this polarization, Belle and BaBar measured the time dependent CP asymmetry in $B \to K^{0*}\gamma$ and $B \to K_s^0\pi^0\gamma$, which are expected to be suppressed for a fully-polarized photon. The mean [8] values found ($K^*\gamma$ channel: -0.13 ± 0.32; $K_s\pi^0\gamma$ channel: 0.00 ± 0.28) agree with the SM. The γ polarization can also be retrieved by studying the resonance interferences in $B \to (K\pi\pi)\gamma$ decays. At present only preliminary analyses of the BF are available [11] since the accumulated statistics is too low to perform an angular analysis. The transitions $b \to d\gamma$ are suppressed by a factor $|V_{td}/V_{ts}|^2 = 0.04$ respect to the $b \to s\gamma$ in the SM, so that NP contributions should be more evident. Unfortunately the experimental measurement of the inclusive process is complicated by the huge background contaminantion and only exclusive BF measurements $B^{0/+} \to (\rho^{0/+}/\omega)\gamma$ have been performed so far. If combined with the $B^0 \to K^{0*}\gamma$ value they allow the calculation of the ratio $|V_{td}/V_{ts}| = 0.199^{+0.026\ +0.018}_{-0.025\ -0.015}$ which is in good agreement with the value given by the CKM fit results [12].

Semileptonic penguin transitions. The radiative FCNC electro-weak transition $b \to s\ell^+\ell^-$ is particularly useful to make high precision tests of the theory. The process is ruled by the Wilson coefficients C_7, C_9 and C_{10}. Contributions from NP appear in the experiment as deviations from the SM values, which have been calculated precisely (NNLO). Moreover the additional kinematic observables, like the forward-backward asymmetry (A_{FB}) and the invariant mass distribution of the emitted leptons ($m_{\ell^+\ell^-}$), provide extra sensitivity to NP contributions.

BaBar and Belle performed an inclusive measurement of the $B \to X_s \ell^+ \ell^-$ differential decay rate [13] which was combined with the $BF(B \to X_s \gamma)$ in a model independent analysis to limit the values of the Wilson coefficients involved. SUSY models which forsee a reversed sign of C_7 respect to the SM, while C_9 and C_{10} are unchanged (MSSM MFV with big $\tan \beta$ and light stop), are ruled out [14].

The measurements of BF and A_{CP} of the exclusive decays $B \to K^* (\mu^+ \mu^- / e^+ e^-)$ and $B \to K(\mu^+ \mu^- / e^+ e^-)$ are available. The precise SM estimates of the BF ratios $R_{K^* \mu^+ \mu^- / K^* e^+ e^-} = 0.991 \pm 0.002$ and $R_{K \mu^+ \mu^- / K e^+ e^-} = 1.000 \pm 0.001$ are in perfect agreement with the experimental values by Belle ($0.98^{+0.30+0.06}_{-0.31-0.08}$ and $1.38^{+0.39+0.06}_{-0.41-0.07}$) and BaBar ($0.93 \pm 0.46 \pm 0.12$ and $1.06 \pm 0.48 \pm 0.05$). More data will be needed to test the SM with higher accuracy.

Particularly interesting to validate the SM (or to discover NP) is the study of the forward-backward asymmetry A_{FB} as a function of q^2, whose shape is modulated by the interference between the three operators (O_7, O_9 and O_{10}) involved. The analyses of Belle and BaBar data are in agreement with the SM and allow to exclude some NP scenarios [13].

Pure leptonic decays $B_{d/s} \to \ell^+ \ell^-$ are very rare FCNC processes. NP contribution may raise by a factor $(\tan \beta)^6 / m_A^4 \simeq 100$ the SM BF predictions (non-MFV models). At present only CDF and D0 approach the SM value in the $B_s \to \mu^+ \mu^-$ channel (D0: $< 19 \times 10^{-8}$ CDF: $< 10 \times 10^{-8}$ and SM: 3.4×10^{-9} [15]), while experimental limits of the other channels are far above the predictions. The BF value can be related to the FCNC processes described so far ($b \to s\gamma$, $b \to s\ell^+ \ell^-$, A_{FB}, R_K and R_{K^*}) to put constraints on different SUSY models and limit Wilson coefficients [16]. Improved experimental accuracy is crucial to constraint NP models.

Measurement perspectives at LHCb. The LHCb experiment, that will start collecting data at LHC in 2008, fulfill all the requirements to be successful also in rare B decay measurements. It was optimized to study B physics: the acceptance coverage and the dedicated trigger to select B decays allows the collection of huge statistics of B_d, B_s, Λ_b decays ($\simeq 10^{12}$ $b\bar{b}$ events per year). A high background rejection is guaranteed by a precise particle measurement (which allows invariant mass resolution $\simeq 10-18\,\mathrm{MeV}$, and secondary vertex resolution $\simeq 100-200\,\mu\mathrm{m}$) and an excellent particle identification. As a drawback of working at the proton collider, LHCb will work in a hadron polluted environment, where the seeked signal is overlapped with the underlying event, and some of the background rejection methods or systematic studies done by the B-factory experiments cannot be applied. As a result LHCb will be able to measure only exclusive or semi-inclusive rare B decays. In particular $B_d^0 \to (\omega/\rho/K^{*0})\gamma$ and $B_s \to \phi\gamma$ radiative decays, the electro-weak penguin transitions $B_d \to K^{*0}\mu^+\mu^-$ [17], $B^\pm \to K^\pm(\mu^+\mu^-/e^+e^-)$ and the pure leptonic decays $B_{d/s} \to \mu^+\mu^-$ [18] will be studied with unprecendented statistics and good signal to background ratio [19]. The FB asymmetry in $B_d \to K^{*0}\mu^+\mu^-$ will be studied with high precision already in the first year of data taking ($\delta\mathrm{BF} \simeq 2\%$, $\delta A_{\mathrm{CP}} \simeq 1.5\%$, $\delta\mathrm{BF}(q^2) \simeq [14\% - 6\%]$ depending on q^2 and $\delta A_{\mathrm{FB}} = [0.026, 0.09]$), allowing a precise discrimination between NP models.

Rare B decays are ideal channels to test precisely the SM or discover indirectly NP contributions. Many channels and observables can be used and related to each other by means of model-independent analyses to limit some NP parameters. Thanks to the B-factory experiments, the experimental panorama is rather rich and shows a total agreement with the SM expectations. Nevertheless the experimental precisions obtained so far are still worse than the theoretical ones, due to the limited statistics and systematics. Starting from 2008, the LHCb experiment will give a big improvement in the study of these precesses, at least in the exclusive decay modes.

References

1. T. Hurth, hep-ph/0212304 and Rev. of Mod. Phys. **75** 1159 (2003), A.J. Buras, A. Czarnecki, M. Misiak, J. Urban, Nucl. Phys. B **631**, 219 (2002).
2. Lunghi, hep-ph/0210379;
3. CLEO Collaboration, S. Chen *et al.*, Phys. Rev. Lett. **87**, 251807 (2001).
4. Belle Collaboration, P. Koppenburg *et al.*, Phys. Rev. Lett. **93**, 061803 (2004), Belle Collaboration, K. Abe *et al.*, Phys. Lett. B **511**, 151 (2001).
5. Belle Collaboration, S. Nishida *et al.*, Phys. Rev. Lett. **93**, 031803 (2004).
6. BABAR Collaboration, B. Aubert *et al.*, hep-ex/0507001, BABAR Collaboration, B. Aubert *et al.*, Phys. Rev. D **72**, 052004 (2005).
7. BABAR Collaboration, B. Aubert *et al.*, Phys. Rev. Lett. **93**, 021804 (2004).
8. Heavy Flavor Averaging Group, E. Barberio *et al.*, hep-ex/0603003.
9. CLEO Collaboration, Phys. Rev. Lett. **84**, 5283 (2000), Belle Collaboration, Phys. Rev. D **69**, 112001 (2004), Phys. Rev. Lett. **94**, 111802 (2005), and Phys. Rev. Lett. **89**, 231801 (2002), BABAR Collaboration, Phys. Rev. D **70**, 112006 (2004) and Phys. Rev. D **70**, 091105 (2004).
10. D. Atwood *et al.*, Phys. Rev. Lett. **79**, 185 (1997), B. Grinstein *et al.*, Phys. Rev. D **71**, 011504 (2005), M. Gronau and D. Pirjol, Phys. Rev. D 66, 054008(2002) [hep-ph/0205065].
11. BABAR Collaboration, hep-ex/0507031.
12. UTfit Collaboration http://utfit.roma1.infn.it/ and CKMfitter Collaboration: http://ckmfitter.in2p3.fr/.
13. BABAR Collaboration, Phys. Rev. D 73, 092001 (2006), Belle Collaboration, hep-ex/0603018.
14. Gambino *et al.* Phys. Rev. Lett. **94** (2005) 061803.
15. D0 Collaboration, public note 5009, (2006), CDF Public note 8176 (2006), Hou, Phys. Rev. D **48**, 2342 (1993), Gambino and Misiak, Nucl. Phys. B **611**, 338 (2001), Neubert, Eur. Phys. Jour. C **40**, 165 (2005).
16. G. Hiller and F. Kruger, hep-ph/0310219.
17. J. H. Lopes, public note LHCb 2005-010.
18. B. de Paula, F. Marinho, S. Amato, public note LHCb 2003-165.
19. I. Belyaev, public note LHCb-2005-001.

Fermion mass in E_6 GUT
with discrete family permutation symmetry S_3

Francesco Caravaglios[1] and Stefano Morisi[2]

[1] Dipartimento di Fisica, Università di Milano and INFN sezione di Milano
   `francesco.caravaglios@mi.infn.it`
[2] Dipartimento di Chimica Fisica di Milano and INFN sezione di Milano
   `stefano.morisi@mi.infn.it`

A discrete symmetry S_3 easily explains all neutrino data. However it is not obvious the embedding of S_3 in GUT where all fermions live in the same representation. We show that embedding S_3 in E_6 it is possible to make distinction between neutrinos and the rest of matter fermions[1].

1 Introduction

So far it is not clear how to extend the standard model to explain fermion hierarchies. The mass matrices are not univocally fixed by experimental data. To reduce the remaning arbitrarity flavour and gauge symmetry are used in the model building. We assume a gauge symmetry G_g acting vertically within each generation and a flavour symmetry G_f acting horizontally on the family generations and we study the group $G_\mathrm{g} \times G_\mathrm{f}$. We can get information about G_f flavour symmetry from the observed mass and mixing fermion hierarchies. First we consider the *lepton sector*. The three neutino analysis [1] is well compatible with the following mixing matrix [2]

$$
U_\mathrm{HPS} = \begin{pmatrix} \sqrt{\frac{2}{3}} & \frac{1}{\sqrt{3}} & 0 \\ -\frac{1}{\sqrt{6}} & \frac{1}{\sqrt{3}} & -\frac{1}{\sqrt{2}} \\ -\frac{1}{\sqrt{6}} & \frac{1}{\sqrt{3}} & \frac{1}{\sqrt{2}} \end{pmatrix}
\tag{1}
$$

called tri-bimaximal, where the heaviest third neutrino is maximally mixed between μ and τ flavours and $\theta_{13} = 0$ while the second neutrino is equally mixed between e, μ and τ. Recently discrete symmetries are studied to explain neutrino mixing (1), in particular S_3 [2–4]. In *quark sector* the three mixing angles are small, the only relevant angle is the 1–2 Cabibbo angle

---

[1] Talk given by Stefano Morisi.

which is smaller than the 1–2 and 2–3 leptonic angles and the mass hierarchy is large. We believe that neutrino mixings have a peculiar mixing pattern that give us more interestings hints about flavour symmetry than quarks.

Quark and lepton mixings and mass hierarchies are very different. The neutrino mixing is large (1), while the CKM mixing matrix in quark sector is close to the identity. Very different hierarchy in quark and lepton sectors, could be a problem in GUT models where in general quark and lepton Yukawa couplings are related. We consider the problem to reconcile different mass and mixing hierarchies with lepton–quark unification in unified models. One possibility is that scalar couplings to charged fermion Yukawa and neutrino Yukawa are different, but there are other interesting possibilities like the *screening* mechanism [5]. We have studied [6] the possibility to make a difference between neutrino and charged fermions selecting a particular gauge group G_{g} and the scalar sector.

2 Leptonic flavour symmetry

Neutrino mass matrices M_ν is $\mu \leftrightarrow \tau$ invariant (S_2 symmetric) only if it commutes with the P matrix and it gives maximal atmospheric angle and $\theta_{13} = 0$.

$$P = \begin{pmatrix} 1 & 0 & 0 \\ 0 & 0 & 1 \\ 0 & 1 & 0 \end{pmatrix}, \quad P^{-1} M_\nu P = M_\nu \Rightarrow M_\nu = \begin{pmatrix} a & d & d \\ d & b & c \\ d & c & b \end{pmatrix}.$$

Neutrino mass matrix M_ν and P are diagonalized by the same unitary matrix O which is

$$O(\theta) = \begin{pmatrix} -\cos\theta & \sin\theta & 0 \\ \frac{1}{\sqrt{2}}\sin\theta & \frac{1}{\sqrt{2}}\cos\theta & -\frac{1}{\sqrt{2}} \\ \frac{1}{\sqrt{2}}\sin\theta & \frac{1}{\sqrt{2}}\cos\theta & \frac{1}{\sqrt{2}} \end{pmatrix} \Leftrightarrow \begin{pmatrix} \sqrt{2/3} & \frac{1}{\sqrt{3}} & 0 \\ -\frac{1}{\sqrt{6}} & \frac{1}{\sqrt{3}} & -\frac{1}{\sqrt{2}} \\ -\frac{1}{\sqrt{6}} & \frac{1}{\sqrt{3}} & \frac{1}{\sqrt{2}} \end{pmatrix}.$$

Assuming the charged leptons mass matrix approximatively diagonal, O is the PMNS leptonic mixing matrix where the angle θ is the solar angle and it is not fixed by the $\mu \leftrightarrow \tau$ symmetry while the atmospheric and θ_{13} angles are the same of the tri-bimaximal (1). To obtain the solar angle we need that the singlet eigenstate $(1, 1, 1)$ is and eigenvector of the mass matrix M_ν and there are two possible solutions: i) M_ν is S_3 invariant (S_3 is the permutation group of three objects) or ii) the parameters in M_ν are constrained by $a = b + c - d$ which follows direcly from A_4 symmetry. We are interested in the first case. The S_2 permutation group is contained into S_3 and in general S_3 breaks spontaneously into S_2. We have shown [4] that in case $S_3 \supset S_2$ breaking is *soft*, the solar angle is $\sin^2\theta_{\text{sol}} = 1/3$ that agrees very well with the experimental value.

3 A grand unification model for fermion masses

In previous section we have said that in case S_3 symmetry is softly broken into S_2 ($\mu \leftrightarrow \tau$), we can explain very well neutrino mass and mixing hierarchies. Differently S_2 symmetry is strongly broken in charged fermion sector. The issue is how to embed a neutrino mass matrix in the same unified gauge group where leptons and quarks Yukawa are equal (lepton–quark unification). In $SU(5)$ neutrinos and charged leptons Yukawa couplings are distinct since $SU(5)$ does not contain right-handed neutrinos and we must introduce three additional singlets S

$$L = g_u T^{\alpha\beta} T^{\gamma\delta} H^{\sigma} \varepsilon_{\alpha\beta\gamma\delta\sigma} + g_d T^{\alpha\beta} F_\alpha \bar{H}_\beta + g_v F_\alpha S H^\alpha + M_{ij} S_i \, S_j \qquad (2)$$

However we are interested in models beyond $SU(5)$ for the motivation below. We are interested in non supersymmetric extention of the Standard Model, but in such case gauge couplings do not unify in $SU(5)$. In Lagrangian (2) Yukawa couplings are free parameters and it is not clear the hierarchies among them. One possibility to explain Yukawa hierarchies is assuming that Yukawa couplings are proportional to different scalars embeding $SU(5)$ in bigger groups like $SO(10)$ and E_6. An interesting future of the above groups that contain $SU(5)$, is that their foundamental representation **16** and **27** can accomodate all fermions spectrum including right-handed neutrino.

If we embed $SU(5)$ into $SO(10)$ we have an additional $U(1)$ that commutes with the full $SU(5)$. The $U(1)$ charges for the representation above are $H(+q)$, $\bar{H}(-q)$, $T(-1)$ $F(+3)$ and $\nu_R(-5)$. From these charges we derive that each mass operators have $U_{\rm r}(1)$ charges

$$\begin{array}{|lc|}
\hline
SU(5) \text{ mass operator} & U_{\rm r}(1) \\
\hline
T^{\alpha\beta} F_\alpha & +2 \\
F_\alpha \nu_{\rm R} & -2 \\
\nu_{\rm R}^t \nu_{\rm R} & -10 \\
T^{\alpha\beta} T^{\gamma\delta} & -2 \\
\hline
\end{array} \qquad (3)$$

We observe that the mass operators $T^{\alpha\beta} T^{\gamma\delta}$ and $F_\alpha \nu_{\rm R}$ have the same charges, thus we expect that the same $SU(5)$ singlet is at the origin of their Yukawa interaction. While neutrinos have an approximate S_3 symmetry [4], the same symmetry is not observed in the up sector. If we embed $SU(5)$ into E_6 we have an additional $U_{\rm t}$ (1) and the table above becomes

$$\begin{array}{|lcc|}
\hline
SU(5) \text{ mass operator} & U_{\rm r}(1) & U_{\rm t}(1) \\
\hline
T^{\alpha\beta} F_\alpha & +2 & +2 \\
F_\alpha \nu_{\rm R} & -2 & +2 \\
T^{\alpha\beta} T^{\gamma\delta} & -2 & +2 \\
F_\alpha x_{\rm L} & +3 & +5 \\
\hline
\end{array} \qquad (4)$$

The advantage here is that we have two right-handed neutrinos ν_R and x_L, and the Dirac mass operator $F_\alpha x_L$ has different quantum numbers from all the others and in particular is different from $T^{\alpha\beta}T^{\gamma\delta}$ giving mass to the up sector. Thus we explore the possibility that the fundamental lagrangian has a E_6 unifying gauge symmetry times a S_3 permutation symmetry of the three fermion families that belong to the 27 of E_6. As already said the 27 contains two standard model singlets that will play the role of right-handed neutrinos. Now we have to choose the representation for the Higgs $SU(2)_W$ doublet. We prefer to keep just one Higgs doublet, that will give mass both for the up and the down sector. This is because we want to have the Standard Model with just one higgs at the weak scale such that the FCNC are strongly suppressed due to the GIM mechanism. So we choose the Higgs to be also a S_3 singlet. Now we have to decide to which E_6 representation we have to assign the Higgs doublet. The Yukawa interaction for fermions at the tree level can be

$$27_i^\alpha 27_i^\beta 351'_{\alpha\beta} \tag{5}$$

where $i = 1, 2, 3$ family index, the $351'$ is symmetric under the exchange of α and β the gauge symmetry indices. The reason why we put the Higgs doublet in the $351'$ is that it contains a $SU(2)_W$ doublet with the $(-3, -5)$ charges with respect the $U_r(1) \times U_t(1)$. Also the 351 contains a Higgs doublet with same charges, but this would give zero in (5) since it is antisymmetric. At the tree level of the fundamental high energy lagrangian, we have just one Yukawa interaction that comes from (5), since this is the unique $U_r(1) \times U_t(1)$ gauge invariant operator $g x_{iL}^t \nu_{iL} h_0$. While the quark Yukawa interactions appear as higher order operators only after the S_3 symmetry breaking. The v_L, v_R^c and x_L neutrinos mix through the double seesaw mechanism that will give the light neutrino mass matrix $M_{ij} \sim \lambda \mathbf{27^i 27^j 351^k 78} + \mathbf{27^i 27^i 351^i 78}$

$$M_\nu \approx \lambda \begin{pmatrix} 1 & 1 & 1 \\ 1 & 1 & 1 \\ 1 & 1 & 1 \end{pmatrix} + \begin{pmatrix} \langle 351^1 \rangle & 0 & 0 \\ 0 & \langle 351^2 \rangle & 0 \\ 0 & 0 & \langle 351^3 \rangle \end{pmatrix}$$

where $\langle 351^1 \rangle \neq \langle 351^2 \rangle = \langle 351^3 \rangle$, S_3 is spontaneously broken in S_2. In the limit $\lambda \to \infty$ the above matrix is diagonalized by the tri-bimaximal (1).

References

1. Maltoni *et al.* hep-ph/0309130.
2. P. F. Harrison and W. G. Scott, Phys. Lett. B **557** (2003) 76.
3. W. Grimus and L. Lavoura, JHEP **0508**, 013 (2005); Y. Kajiyama, E. Itou and J. Kubo, arXiv:hep-ph/0511268; N. Haba and K. Yoshioka, arXiv:hep-ph/0511108; Y. Koide, arXiv:hep-ph/0509214; E. Ma, New J. Phys. **6**, 104 (2004); J. Kubo, A. Mondragon, M. Mondragon, E. Rodriguez-Jauregui, O. Felix-Beltran and E. Peinado, J. Phys. Conf. Ser. **18** (2005) 380.
4. F. Caravaglios and S. Morisi, arXiv:hep-ph/0503234.
5. M. Lindner, M. A. Schmidt and A. Y. Smirnov, JHEP **0507**, 048 (2005)
6. F. Caravaglios and S. Morisi, arXiv:hep-ph/0510321.

Search for Supersymmetry
with early ATLAS data

T. Lari, for the ATLAS Collaboration

Universitàdi Milano, Dipartimento di Fisica and INFN
Sezione di Milano, Via Celoria 16, 20126 Milano (Italy)

Summary. The potential of the ATLAS experiment to discover and study the
signals expected from the production of Supersymmetric particles with the first
few fb^{-1} of data delivered by the LHC is illustrated.

1 Introduction

One of the main purposes of the LHC collider is the search for Supersymmetry [1]. If Supersymmetry exists at the electroweak scale, it could hardly
escape detection at the LHC. The center-of-mass energy of 14 TeV extends
the search for SUSY particles up to masses of 2.5 to 3 TeV/c^2 [2].

In the context of the mSUGRA framework with R-parity conservation,
scalar quarks and gluinos are expected to dominate the production cross section. Their decay into the Lightest Supersymmetric Particle (LSP), which
escapes detection, produce an excess of events with multijets, missing energy, and isolated leptons final states compared to the SM expectations. If
squarks and gluinos are lighter than about 1 TeV, as implied by naturalness
arguments, this signature would be observed with high statistical significance
already during the first year of running at the initial LHC luminosity of
10^{33} cm^{-2} s^{-1}.

In the next session, we present the ATLAS strategy to observe the production of SUSY particle through an excess of events with hard jets and large
missing energy, a method to measure the top background using data, and the
reconstruction of specific decays of Supersymmetry particles.

2 Search for Supersymmetry with ATLAS

The distribution of effective mass, defined as the sum of the missing energy
and the transverse momentum of 4 leading jets, is reported in Fig. 1 [3] for
events which pass the following selections: one isolated lepton (electron or

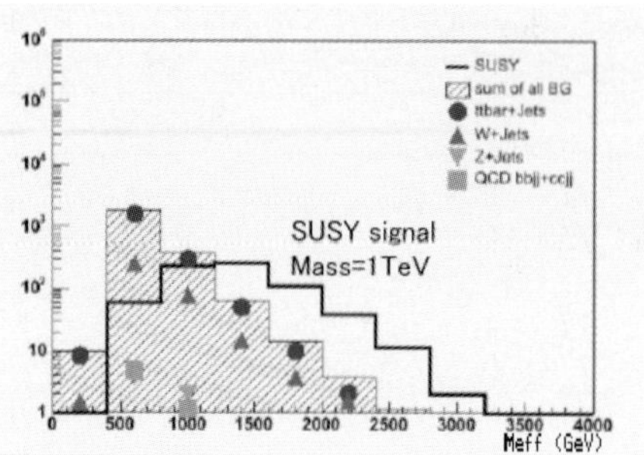

Fig. 1. Distribution of effective mass for the events passing the analysis cuts described in the text. The contributions from mSUGRA Point 2 and Standard Model processes are reported. The statistic corresponds to $10\,\mathrm{fb}^{-1}$ of data and the fast simulation of the ATLAS detector was used. [3]

muon) with $p_\mathrm{T} > 10\,\mathrm{GeV}$, $E^\mathrm{T}_\mathrm{MISS} > 100\,\mathrm{GeV}$, at least one jet with $p_\mathrm{T} > 100\,\mathrm{GeV}$, at least four jets with $p_\mathrm{T} > 50\,\mathrm{GeV}$, transverse sphericity larger than 0.2, transverse mass between lepton and missing energy vector larger than $100\,\mathrm{GeV}$. The most relevant Standard Model processes which contribute to this signature ($t\bar{t}$, W + jets, Z + jets, and multi-jet QCD production) were simulated using ALPGEN [4] (version 1.33) for the hard process, PYTHIA [5, 6] (version 6.2) for emission of soft collinear gluons and hadronization, and ATLFAST [7] to parametrize the detector response. The distribution from Supersymmetric events is reported for LHC Point 2 [2] which has a squark and gluino masses of about $1\,\mathrm{TeV}$. The number of events in the plot correspond to $10\,\mathrm{fb}^{-1}$ of integrated luminosity.

At large effective masses the contribution from Supersymmetry is much larger than the statistical error on the Standard Model background event rate. The SUSY/SM ratio is better than in the 0-lepton channel, and the dominant background is from top pair production, which is expected to be known with a smaller systematic error than the other SM backgrounds.

While an excess of events with large missing energy and jets with high transverse energy will be an hint of the presence of non-SM physics, the existence of Supersymmetry will be more firmly estabilished when the masses and decays of the new particles will be measured. The two body decay chain $\chi^0_2 \to \tilde{l}^\pm l^\pm \to l^\pm l^\pm \chi^0_1$ is particularly promising, as it leads to a very sharp edge in the distribution of the invariant mass of the two leptons, which measures

$$M(\text{edge}) = m(\chi^0_2)\sqrt{1 - m^2(\tilde{l})/m^2(\chi^0_2)}\sqrt{1 - m^2(\chi^0_1)/m^2(\tilde{l})} \qquad (1)$$

In Fig. 2 the invariant mass distribution of opposite sign lepton (electron or muon) pairs is shown, for the LHC benchmark SPS1A [8]. After the analysis cuts the Standard Model beckground is negligible and the dominant background is due to leptons from independent chains in the SUSY events. This background can be estimated using $e^+\mu^-$ and $e^-\mu^+$ pairs. After the subtraction of the background estimate, the expected triangular distribution with a clear edge is observed.

Four other kinematical edges, involving combinations of leptons and jets, can also be observed [8]. The five endpoints are a function of four unknown

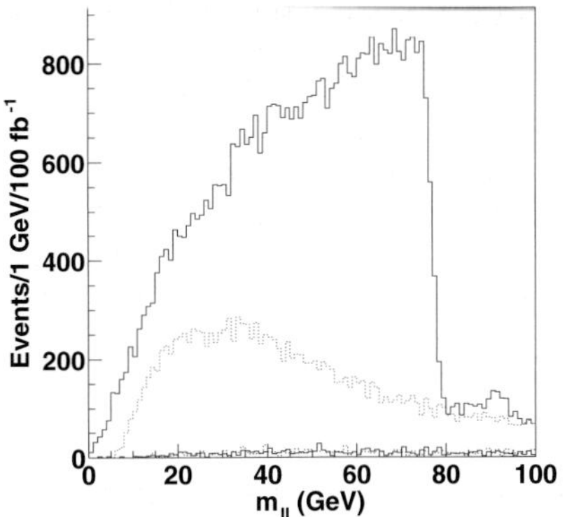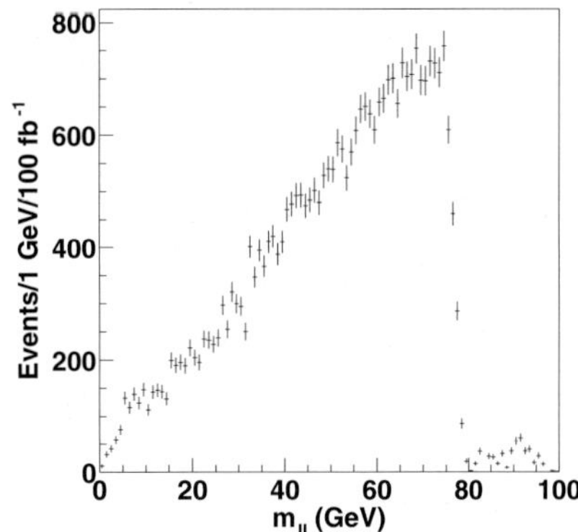

Fig. 2. Distribution of 2-lepton invariant mass, for an integrated luminosity of $100\,\mathrm{fb}^{-1}$ [8]. *Left plot*: the *solid curves* are the opposite sign, same flavour (OSSF) distribution while the *dotted curves* are the opposite sign, opposite flavour (OSOF) distribution. The two *upper curves* are the contribution for the SUSY benchmark SPS1A, the two lower curves the Standard Model contribution. *Right plot*: flavour subtracted OSSF-OSOF distribution. The two plots were obtained using the fast simulation of the ATLAS detector

masses (the two lightest neutralinos, the slepton, and the scalar quark), so these can be extracted from the data.

These results, obtained with the parametrized simulation of the ATLAS detector, were confirmed by studies based on the detailed simulation [9], which demonstrated that several edges will already be visible with a statistics of $5\,\mathrm{fb}^{-1}$ [9].

3 Conclusions

The potential of the ATLAS experiment to discover and study Supersymmetry with the first few fb^{-1} of data was discussed. In most model, a clear excess of events over the Standard Model contribution would be observed, and the first kinematical edges, measuring mass relations between Supersymmetric particles, would be reconstructed.

Acknowledgement. The work presented here was produced by several individuals of the ATLAS collaboration making use of the detector simulation and software tools which are the result of collaboration-wide efforts.

References

1. H.P. Nilles, Phys. Rev. **110**, 1 (1984).
2. The ATLAS Collaboration, "Atlas Detector and Physics Performance TDR", CERN/LHCC-99-15 (1999).

3. S. Asai (for the ATLAS SUSY WG), "Standard Model Backgrounds of SUSY search", presentation at the 4th ATLAS Physics workshop, Rome, 6–11 June 2005.
4. M.L. Mangano *et al.*, JHEP **0307**, 001 (2003).
5. T. Sjöstrand *et al.*, Comp. Phys. Comm. **135**, 238 (2001).
6. T. Sjöstrand *et al.*, "PYTHIA 6.2: physics and manual", hep-ph/0108264 (2001).
7. E. Richter-Was, D. Froidevaux and L. Poggioli, "ATLFAST 2.0 a fast simulation package for ATLAS", ATLAS note, ATL-PHYS-98-131.
8. C.K. Gjelsten *et al.*, "A detailed analysis of the measurement of SUSY masses with the ATLAS Detector at LHC", ATLAS Internal note, ATL-PHYS-2004-007 (2004).
   C.K. Gjelsten *et al.*, JHEP **12** (2004) 003.
9. M. Biglietti *et al.*, "Full Supersymmetry Simulation for ATLAS in DC1, ATLAS note", ATL-PHYS-2004-011.

Detection methods for long lived particles at the LHC

Sara Viganò[1] and Alberto De Min[2]

[1] Università di Milano Bicocca and INFN Milano `sara.vigano@mib.infn.it`
[2] Politecnico di Milano and INFN Milano `alberto.de.min@cern.ch`

1 Introduction

Almost all the extensions of the Standard Model predict the existence of new charged particles; these particles should be very heavy since they have excaped detection so far. In general, such heavy particles decay as soon as they are produced, but under certain circumstances they can be long-lived or even stable. There are two classes of models which predicts the existence of long-lived particles [3]: models with a weakly broken symmetry, where the particle would be stable if the symmetry were exact; and models with an exact symmetry which forbids the decay of heavy exotics into ordinary particles, where the decay of the charged particle into a neutral particle is suppressed either by small couplings or by phase space. Examples of the second kind are supersymmetric models with exact R-parity, where the lightest supersymmetric particle is the gravitino. In this talk we will focus on GMSB (Gauge Mediated Supersymmetry Breaking) which belongs to this kind of models. In GMSB supersymmetry is broken in a so called hidden sector at a scale $\sqrt{F}$ and transmitted to SM particles via a messenger sector. The transmission of the supersymmetry breaking is mediated by gauge fields, in particular by N singlet representations of $SU(5)$.

In this scenario, the gravitino is the LSP (*Lightest Super-Partner*), with a mass given by:

$$m_{\widetilde{G}} = \frac{F}{\sqrt{3}M_{\mathrm{P}}} \propto \frac{M\Lambda}{M_{\mathrm{P}}} \tag{1}$$

where M_{P} is the reduced Planck scale, M is the messenger mass scale and Λ is the effective scale of the SUSY breaking.

Since the gravitino is light and practically non-interacting, properties which make it not detectable. The phenomenology is thus determined by the the Next to Lightest Super-Partner (NLSP) and its decay into the gravitino. The NLSP is not uniquely determined, the two most favorable candidates are the lightest stau $\widetilde{\tau}$ and the lightest neutralino χ_0, depending on the values of the five GMSB parameters.

2 GMSB phenomenology

The decay time of the NSLP $\tilde{X}$ into its Standard Model partner and the gravitino depends on the gravitino mass: $\tau_{\tilde{X}} \propto m_{\tilde{G}}^2$ for fixed NLSP mass. Since the gravitino mass is not tightly constrained at present, the NLSP lifetime can range from microns to hundred of meters:

$$c\tau \approx (1.3\,\mathrm{m}) \left(\frac{100\,\mathrm{GeV}}{m_{\tilde{X}}} \right)^5 \left(\frac{\sqrt{F}}{1000\,\mathrm{TeV}} \right)^4 . \tag{2}$$

When the gravitino mass is above $1\,\mathrm{eV/c^2}$, the NSLP is long-lived and its $c\tau$ is of the order of (or very large with respect to) the detector dimensions.

3 Long-lived stau detection

We first focus on the case of long-lived staus. Such particles will look like heavy muons, differing only by a considerably lower β. The two main observables which are useful in order to distinguish between a stau and a muon are the time of flight and the speficic ionization.

Detection from time of flight measurement

The measurement of the time of flight can be performed with the drift tubes of the muon detector system either in ATLAS [8] or in CMS [4]. If the particle velocity is significantly lower than the speed of light, then the points reconstructed in successive drift tubes with the assumption $\beta \sim 1$ will not align. The position of the particle is determined from the drift time of electrons to anode wires. A starting time t_0 is needed. For the stau t_0 is a function of β. The parameter β is measured in order to minimize the χ^2 of the reconstructed track. The separation between staus and muons can be clearly seen in Fig. 1 (left). The measurement of the lifetime can be performed by selecting the N_1 events in which a stau has been detected and then counting the N_2 times where a second stau is observed, see Fig. 1 (right). Indeed it holds $P(L) = e^{-mL/pc\tau}$, where the P is probability for a particle with mass m, momentum p and proper lifetime τ to travel for a distance L before decaying.

Specific ionization measurement

A complementary approach is given by the measurement of the specific ionization. The tracker is the obvious device, but a precise determination of $\mathrm{d}E/\mathrm{d}x$ could be difficult because of the high density of tracks, the limited resolution and the limited dynamic range of the readout. The electromagnetic calorimeter (either in CMS or in ATLAS) can add useful information thanks to the cleaner environment. More over it is sensitive to particles which don't

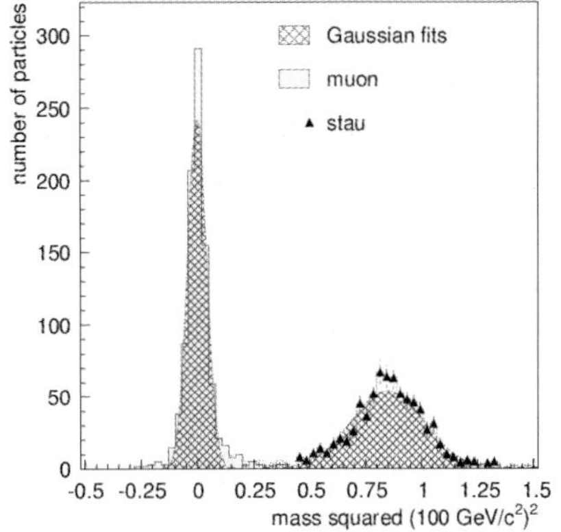
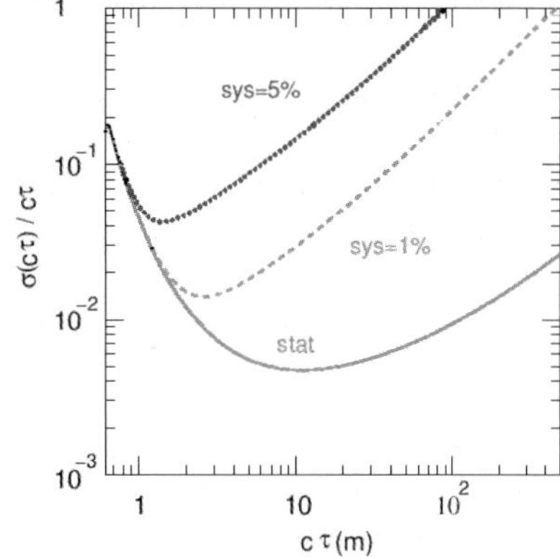

Fig. 1. *Left*: Squared mass distributions obtained from the β measurement with a statistics corresponding to one week at low luminosity [4]. *Right*: Fractional error on the measurement of the slepton lifetime $c\tau$. [8]

reach the muon system for which the measurement of the time of flight is not possible. The electromagnetic calorimeter of CMS, for instance, has shown to have good performance on particles that don't shower into the crystals in the context of the calibration with cosmic rays [7].

4 Neutralino decays and non-pointing photons

When neutralino decays take place in the detector volume, the experimental signature consists of missing energy and two non-pointing photons.

The ATLAS electromagnetic calorimeter can provide directional information since it is longitudinally segmented. In its first compartment it has narrow strips that have good resolution in θ. The resolution on the particle direction can be parametrised as:

$$\Delta\Theta = \frac{60\,\mathrm{mrad}}{\sqrt{E}} \, .$$ (3)

Simulations demonstrate that the efficiency to detect an isolated photon as non-pointing is about 80% requiring $\delta\theta$ (the difference between the photon direction and the nominal vertex direction) to be non-zero by 5 sigma [2].

The electromagnetic calorimeter of CMS is instead not longitudinally segmented; it is though possible to infer the impact direction of the photon from the asymmetry in the shape of the energy deposition among the crystals. Rejection of pointing photons can be performed at a level of 99% keeping an acceptance for the signal photons bigger than 0.8 [5].

An independent way to detect photons coming from neutralino decays is based on the measurement of the time of flight in the calorimeters. If the mass and the momentum distributions of the neutralinos are determined from other measurements it is possible to convert the rate into a lifetime measurement. Photons from neutralino decaying into gravitino are delayed with respect to prompt photons due to the geometry of their path and to the velocity of the neutralino. The mean delay of non-pointing photons of the order of 2 ns which is detectable by both ATLAS and CMS since ATLAS e. m. calorimeter has a time resolution of about 100 ps [2] and CMS ECAL better than 1 ns for signal amplitudes greater than 2 GeV [6].

When neutralino lifetime is in the range 1–100 m a large fraction of decays take place inside the muon detector. A photon created inside the iron yoke develops an electromagnetic shower which can leak to the muon station. The muon chambers should then register from a few tens to few hundreds of hits. Measured flight path (distance between interaction point and the entrance point to the muon station which is supposed to see the accumulation of hits) depends on neutralino lifetime [4].

5 Conclusions

We have shown that both ATLAS and CMS have the sensitivity to massive long-lived particles (such as staus or neutralinos in GMSB models) through direct detection or through their decay products. At the LHC a large amount of SUSY particles of this kind is generated, while the SM background can be reduced at a very low level. Most techniques require some "improper" use of detectors, for instance the use of calorimeters and muon chambers to extract timing information, the detection of showers in muon stations and the determination of impact direction from the asymmetry of shower shapes.

In some cases the decay lifetime can be measured with a good accuracy. Within the GMSB scenario, this will allow the computation of the SUSY breaking scale $\sqrt{F}$.

References

1. CMS Collaboration, *CMS Technical Design Report*, CERN/LHCC 2006-001
2. ATLAS Collaboration, *ATLAS Technical Design Report*, CERN/LHCC 99-15
3. M. Drees, X. Tata, *Signals for heavy exotics at hadron colliders and supercolliders*, Phys. Lett. **B** 252 (December 27th 1990)
4. M. Kazana, G. Wrochna, P. Zalewski, *Study of the NLSP from the GMSB models in the CMS detector at tha LHC*, CMS CR 1999/019
5. G. Franzoni, *The electromagnetic calorimeter of CMS and its sensitivity to non-pointing photons*, PhD Thesis, Università di Milano, 2004
6. R. Brunelière, A. Zabi, *Reconstruction of the signal amplitude of the CMS electromagnetic calorimeter*, CMS NOTE 2006/037
7. M. Bonesini, T. Camporesi *et al.*, *Inter-calibration of the CMS electromagnetic calorimeter with cosmic rays before installation*, CMS NOTE 2005/023
8. S. Ambrosiano, B. Mele *et al.*, *SUSY Long-Lived Massive Particles: Detection and Physics at the LHC*, hep-ph/0012192

A holographic composite Higgs model

Roberto Contino

Dipartimento di Fisica, Università di Roma "La Sapienza" and
INFN Sezione di Roma, Piazzale Aldo Moro 2, I-00185 Roma, Italy
roberto.contino@roma1.infn.it

Composite Higgs models [1] represent an attractive variation of the Technicolor paradigm [2]. In these theories the Standard Model (SM) Higgs doublet is the bound state of a strongly interacting sector with flavor symmetry G. It forms at a scale f_π, the analog of the QCD pion decay constant, as the Goldstone boson associated with the dynamical breaking of the global symmetry G. The couplings of the SM matter and gauge fields to the strong sector break G explicitly, and induce a one-loop Higgs potential that triggers the electroweak symmetry breaking (EWSB) at the scale $v \leq f_\pi$. In the limit of a large separation $\epsilon = v/f_\pi \ll 1$, all the massive bound states of the strong sector decouple, and one is left with the low-energy spectrum of the Standard Model. This means that all corrections to the electroweak precision observables constrained by LEP and SLD experiments will be suppressed by powers of ϵ.

In this talk I will discuss an interesting realization of this old idea which has been recently found in the framework of extra-dimensional warped models [3,4] defined on AdS_5 spacetime with two boundaries [5]. The main virtue of these models is their calculability as effective field theories: various physical quantities, like for example the Higgs potential and the electroweak precision observables, can be computed in 5D in a perturbative expansion. A minimal model was introduced in [4], where a bulk $SO(5) \times U(1)_X \times SU(3)_c$ gauge symmetry is reduced to the SM group $G_{\mathrm{SM}} = SU(2)_L \times U(1)_Y \times SU(3)_c$ on the UV boundary and to $SO(4) \times U(1)_X \times SU(3)_c$ on the IR boundary. Hypercharge is defined as $Y = T_R^3 + X$, where $SO(4) \sim SU(2)_L \times SU(2)_R$. According to the AdS/CFT correspondence [6], such 5-dimensional scenario is equivalent to a 4D composite Higgs theory where the $SO(5)$ flavor symmetry of the strong sector is spontaneously broken to $SO(4)$ in the infrared. This delivers 4 Goldstone bosons that transform as a **4** of $SO(4)$ (a real bidoublet of $SU(2)_L \times SU(2)_R$) and are identified with the Higgs doublet. They correspond, in the 5D theory, to the $SO(5)/SO(4)$ degrees of freedom of the fifth component A_5 of the bulk gauge field.

In [4] the SM fermions were embedded in spinorial representations of the bulk SO(5) gauge symmetry. This choice leads to a generically large modification of the $Zb\bar{b}$ coupling, which in turn rules out a large portion of the parameter space [7]. A more natural model, however, can be simply obtained if the bulk SO(5) symmetry is reduced to O(4) on the IR brane, instead of SO(4), and by embedding the SM fields in fundamental ($\mathbf{5}$) or antisymmetric ($\mathbf{10}$) representations of SO(5). In this case, indeed, a subgroup of the custodial symmetry O(3) that protects the electroweak ϱ parameters from large corrections, also protects the $Zb\bar{b}$ coupling [8]. A possible choice of bulk fields and boundary conditions is the following [9]:

$$\xi_{q_1}(\mathbf{5}_{+2/3}) = \begin{bmatrix} q'_{1L}(-+) \; q'_{1R}(+-) \\ q_{1L}(++) \; q_{1R}(--) \\ s^u_L(--) \; s^u_R(++) \end{bmatrix} \;,\quad \xi_u(\mathbf{5}_{+2/3}) = \begin{bmatrix} q'^u_L(+-) \; q'^u_R(-+) \\ q^u_L(+-) \; q^u_R(-+) \\ u_L(-+) \; u_R(+-) \end{bmatrix} \;,$$

$$\xi_{q_2}(\mathbf{5}_{-1/3}) = \begin{bmatrix} q_{2L}(++) \; q_{2R}(--) \\ q'_{2L}(-+) \; q'_{2R}(+-) \\ s^d_L(--) \; s^d_R(++) \end{bmatrix} \;,\quad \xi_d(\mathbf{5}_{-1/3}) = \begin{bmatrix} q^d_L(+-) \; q^d_R(-+) \\ q'^d_L(+-) \; q'^d_R(-+) \\ d_L(-+) \; d_R(+-) \end{bmatrix} \;.$$

$$(1)$$

Chiralities under the 4D Lorentz group have been denoted with L, R, and $(\pm,\pm)$ is a shorthand notation to denote Neumann $(+)$ or Dirichlet $(-)$ boundary conditions. The fields ξ_{q_1}, ξ_u (ξ_{q_2}, ξ_d) transform as $\mathbf{5}_{2/3}$ ($\mathbf{5}_{-1/3}$) of SO(5)$\times$U(1)$_X$, and their zero modes are identified with a full generation of SM quarks. In particular, only a linear combination of q_{1L} and q_{2L} has Neumann boundary condition on the UV brane – its zero mode is identified with the SM quark doublet q_L – the other combination being Dirichlet.[1] A fundamental of SO(5) decomposes as $\mathbf{5} = \mathbf{4} + \mathbf{1} = (\mathbf{2},\mathbf{2}) + (\mathbf{1},\mathbf{1})$ under SO(4)$\sim$SU(2)$_L\times$SU(2)$_R$, and we have defined the (q,q') fields to transform as $(\mathbf{2},\mathbf{2})$'s, while s^u, s^d, u, d transform as singlets. A similar 5D embedding also works for the SM leptons, although with different U(1)$_X$ charges. Localized on the IR brane, we consider the most general O(4)-invariant set of mass terms:

$$\widetilde{m}_u \left(\bar{q}_{1L} q^u_R + \bar{q}'_{1L} q'^u_R\right) + \widetilde{M}_u \, \bar{s}^u_R u_L + \widetilde{m}_d \left(\bar{q}_{2L} q^d_R + \bar{q}'_{2L} q'^d_R\right) + \widetilde{M}_d \, \bar{s}^d_R d_L + h.c.$$

The most general low-energy Lagrangian for the SM fermions that follows from the embedding (1) is, in momentum-space and at the quadratic order [9]:

$$\mathcal{L}_{\text{eff}} = \bar{q}_L \, \not{p} \left[\Pi^q_0 + \frac{s^2}{2} \left(\Pi^{q1}_1 \, \widehat{H}^c \widehat{H}^{c\dagger} + \Pi^{q2}_1 \, \widehat{H}\widehat{H}^\dagger \right) \right] q_L + \bar{u}_R \, \not{p} \left(\Pi^u_0 + c^2 \Pi^u_1 \right) u_R$$

$$+ \bar{d}_R \, \not{p} \left(\Pi^d_0 + c^2 \Pi^d_1 \right) d_R + \frac{sc}{\sqrt{2}} M^u_1 \, \bar{q}_L \widehat{H}^c u_R + \frac{sc}{\sqrt{2}} M^d_1 \, \bar{q}_L \widehat{H} d_R + h.c.$$

$$(2)$$

---

[1] The same can be obtained by starting with both q_{1L} and q_{2L} having Neumann UV boundary conditions and by adding a mass mixing term on the UV brane.

Here $c = \cos(h/f_\pi)$, $s = \sin(h/f_\pi)$, and $h = \sqrt{(h^a)^2}$, where h^a denote the four real components of the Higgs field:

$$\widehat{H} = \frac{1}{h}\begin{bmatrix} \hat{h}_1 - i\hat{h}_2 \\ \hat{h}_3 - i\hat{h}_4 \end{bmatrix}, \qquad \widehat{H}^c = \frac{1}{h}\begin{bmatrix} -(\hat{h}_3 + i\hat{h}_4) \\ \hat{h}_1 + i\hat{h}_2 \end{bmatrix}. \tag{3}$$

The form factors $\Pi(p^2)$, $M(p^2)$ can be computed in terms of 5D propagators using the holographic approach of [4]. The analog low-energy Lagrangian for the SM gauge fields can be found in [4]; its form implies the following relations:

$$v \equiv \epsilon f_\pi = f_\pi \sin\frac{\langle h\rangle}{f_\pi} = 246\,\mathrm{GeV}, \qquad f_\pi = \frac{2}{\sqrt{g_5^2 k}}\frac{1}{L_1}, \tag{4}$$

where g_5 is the SO(5) gauge coupling in the bulk, k is the AdS$_5$ curvature, and L_1 is the position of the IR brane which sets the mass scale of the new particles $(1/L_1 \sim \mathrm{TeV})$.

The Higgs potential is generated at one loop from the virtual exchange of SM fields. The largest contribution comes from those fields that couple more strongly to the Higgs, namely t_R, t_L and b_L:

$$V(h) =(-6)\int\frac{\mathrm{d}^4 p}{(2\pi)^4}\left\{\log\left(\Pi_0^q + \frac{s^2}{2}\Pi_1^{q2}\right)\right.$$
$$\left. + \log\left[p^2\left(\Pi_0^u + c^2\Pi_1^u\right)\left(\Pi_0^q + \frac{s^2}{2}\Pi_1^{q1}\right) - \frac{s^2 c^2}{2}(M_1^u)^2\right]\right\}, \tag{5}$$

where the form factors refer to $q_L = (t_L, b_L)$ and t_R. Since Π_1 and M_1 drop exponentially for $pL_1 \gg 1$, the logarithm in (5) can be expanded and the potential is well approximated by:

$$V(h) \simeq \alpha\,\sin^2\frac{h}{f_\pi} + \beta\sin^4\frac{h}{f_\pi}, \tag{6}$$

where α and β are integral functions of the form factors. For $\alpha < 0$ and $2\beta > |\alpha|$ the electroweak symmetry is broken, and the minimum is at

$$\sin\frac{\langle h\rangle}{f_\pi} = \epsilon = \sqrt{\frac{-\alpha}{2\beta}}. \tag{7}$$

If $\alpha < 0$ and $2\beta \le |\alpha|$, on the other hand, $\cos\langle h\rangle/f_\pi = 0$ and the EWSB is maximal: $\epsilon = 1$. In this limit an O(4) chiral symmetry is restored, and all the Yukawa couplings (hence the fermion masses) vanish, as one can explicitly see from the effective Lagrangian (2). The model is thus realistic only for $0 < \epsilon < 1$, which can be obtained for natural values of the 5D input parameters [9]. Determining how much of this region is excluded by the electroweak precision constraints gives a measure of how natural is the model in reproducing the EWSB. The strongest bound comes from the Peskin-Takeuchi S parameter:

$$S = \frac{6\pi}{g_5^2 k}\, \epsilon^2 \,. \tag{8}$$

A rough estimate using (8) suggests that imposing $S \leq 0.3$ excludes $\sim 50\%$ ($\sim 75\%$) of the $0 < \epsilon < 1$ region for $1/N = 1/5$ ($1/N = 1/10$), where $1/N \equiv (g_5^2 k/16\pi^2)$ is the 5D expansion parameter. A detailed numerical analysis gives similar results [9]. This shows that a sizable portion of the parameter space is still allowed, and that no large fine-tuning is required in this model to pass all electroweak precision tests. Moreover, the model predicts a light physical Higgs: $100\,\mathrm{GeV} \lesssim m_{\mathrm{Higgs}} \lesssim 150\,\mathrm{GeV}$. This is possible, and the "little hierarchy" puzzle is resolved, since the spectrum of the new vectorial states (which enter the oblique precision observables) is predicted to be heavier than the fermionic resonances (which are responsible for cutting off the SM top loop). In fact, the most important prediction of the model is that, due to the heaviness of the top quark, at least one among its Kaluza–Klein partners is relatively light, with a mass of order 500–$1500\,\mathrm{GeV}$. These new fermions transform as $SU(2)_L$ doublets with hypercharge $Y = 1/6$ or $7/6$, and singlets with $Y = 2/3$. They will be both singly and pair produced at the LHC, decaying to final states populated mostly by tops, bottoms and SM gauge and Higgs bosons. Studying their phenomenology at the LHC will be certainly exciting and challenging at the same time.

Acknowledgement. I would like to thank K. Agashe, A. Pomarol and R. Sundrum for the numerous inspiring discussions and an always stimulating collaboration. I also thank the organizers of IFAE 2006 for inviting me to this interesting conference.

References

1. D. B. Kaplan and H. Georgi, Phys. Lett. B **136**, 183 (1984); B **136**, 187 (1984); H. Georgi, D. B. Kaplan and P. Galison, Phys. Lett. B **143**, 152 (1984); H. Georgi and D. B. Kaplan, Phys. Lett. B **145**, 216 (1984); M. J. Dugan, H. Georgi and D. B. Kaplan, Nucl. Phys. B **254**, 299 (1985).
2. S. Weinberg, Phys. Rev. D **13**, 974 (1976); Phys. Rev. D **19**, 1277 (1979); L. Susskind, Phys. Rev. D **20**, 2619 (1979).
3. R. Contino, Y. Nomura and A. Pomarol, Nucl. Phys. B **671**, 148 (2003).
4. K. Agashe, R. Contino and A. Pomarol, Nucl. Phys. B **719**, 165 (2005).
5. L. Randall and R. Sundrum, Phys. Rev. Lett. **83**, 3370 (1999).
6. J. M. Maldacena, Adv. Theor. Math. Phys. **2**, 231 (1998) [Int. J. Theor. Phys. **38**, 1113 (1999)]; S. S. Gubser, I. R. Klebanov and A. M. Polyakov, Phys. Lett. B **428**, 105 (1998); E. Witten, Adv. Theor. Math. Phys. **2**, 253 (1998). For a review see: O. Aharony, S. S. Gubser, J. M. Maldacena, H. Ooguri and Y. Oz, Phys. Rept. **323** (2000) 183.
7. K. Agashe and R. Contino, Nucl. Phys. B **742** (2006) 59.
8. K. Agashe, R. Contino, L. Da Rold and A. Pomarol, arXiv:hep-ph/0605341.
9. R. Contino, L. Da Rold and A. Pomarol, to appear.

New Physics in Top Events at the LHC

Marina Cobal-Grassmann

University of Udine, Via delle Scienze 208, Udine
`marina.cobal@fisica.uniud.it`

1 Introduction

It is widely believed now that new physics beyond the Standard Model (SM) will appear at the electroweak scale. In almost all models of electroweak symmetry breaking, top either couples strongly to new particles or its properties are modified in some way, and can be therefore considered as a powerful tool for new discoveries findings. The top quark, discovered about 10 years ago [1], is currently being studied in detail by the CDF and D0 experiments at the Tevatron. Its production cross section has been measured in a variety of channels; its mass has been determined to better than 2%, and can be used to constrain the mass of the Higgs. Top quark decays have been tested and non-standard production mechanisms searched for. So far, all of the top quark's properties are consistent with the SM, but these studies have to deal with a quite low statistics. LHC instead will be a top factory, with a NLO $\sigma(t\bar{t})$ equal to about $830\,\mathrm{pb}$ for a luminosity of $2 \times 10^{33}\,\mathrm{cm}^{-2}\,\mathrm{s}^{-1}$: this implies the production of two $t\bar{t}$ events per second. This enormous sample available will allow to increase the precision of the previously mentioned measurements and will also open up new possibilities such as observation of spin correlations, FCNC in top production and decay, and perhaps even CP violation in the top sector. Many different studies are being actively pursued by tha ATLAS and CMS experiments. In this report we only briefly describe some of the the ones performed in ATLAS; for the remaining studied, like charged Higgs decay in top events, we refer the reader [2]. At the moment this presentation was given, the CMS results were not yet public, but now they can be found in [3].

2 Resonances decaying in $t\bar{t}$

Due to the short lifetime of the top quark, the SM predicts that $t\bar{t}$ bound states do not appear. However many other models (like SM Higgs, MSSM

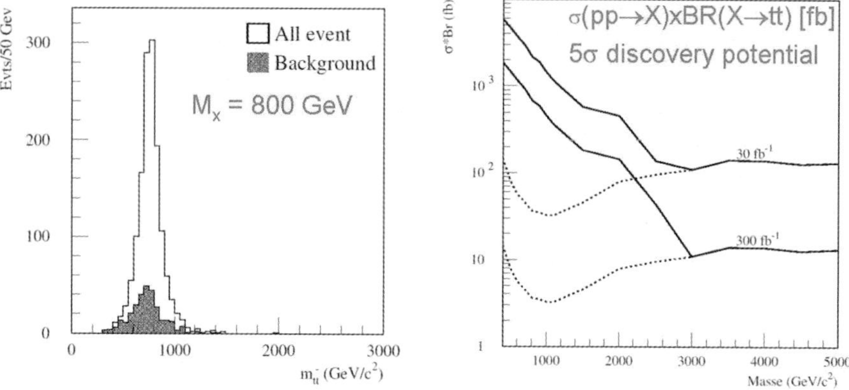

Fig. 1. *Left*: Reconstructed mass of a 800 TeV resonance. *Right*: 5σ discovery potential in ATLAS as a function of the resonance mass

Higgs, Technicolor Models, Topcolor etc.) include the existence of resonances decaying to $t\bar{t}$. ATLAS performed a full detector simulation study of a resonance X, looking at the decay: $X \to t\bar{t} \to WbWb \to l\nu bjjb$ and assuming a given σ_X, Γ_X and $\mathrm{BR}(X \to t\bar{t})$. Figure 1 shows on the left the reconstructed mass peak for a 800 TeV resonance, and on the right the 5σ discovery potential in ATLAS as a function of the resonance mass.

3 New Physics in $t \to bW$

The presence of asymmetries in kinematical distributions which are measured to be different from what predicted by the SM, can give hints on new physics. The ATLAS experiment has studied the LHC sensitivity in the measurement of several angular distributions and asymmetries in the decays of top quark pairs [4], for example the asymmetry in the emission direction of the charged lepton from the W decay with respect to the b-quark from the same top decay. The forward-backward asymmetry A_{FB} as a function of the cosinus of the angle between the charged lepton and the b-quark direction, as calculated in the W cms, is defined as: $A_{\mathrm{FB}} = \frac{N(x \geq 0) - N(x \leq 0)}{N(x \geq 0) + N(x \leq 0)}$ where $x = \cos\theta_{lb}$. Together with this classic definition, two new asymmetries have been defined: $A_{\pm} = \frac{N(x \geq x_{\pm}) - N(x \leq x_{\pm})}{N(x \geq x_{\pm}) + N(x \leq x_{\pm})}$ with $x_{\pm} = \pm(2^{2/3} - 1)$. These asymmetries can be related to the polarization of the W bosons emitted in the decay, for which we take as spin quantisation axis their momentum direction. The SM values of these observables, can be measured with the following precision (systematic and statistical errors are indicated): $A_{\mathrm{FB}} = -0.2237 \pm 0.0035(\mathrm{stat}) \pm 0.0144(\mathrm{sys})$ $(\sigma/A_{\mathrm{FB}} = 7.0\%)$, $A_{+} = 0.5472 \pm 0.0032(\mathrm{stat}) \pm 0.0099(\mathrm{sys})$ $(\sigma/A_{+} = 2.0\%)$, $A_{-} = -0.8387 \pm 0.0018(\mathrm{stat}) \pm 0.0028(\mathrm{sys})$ $(\sigma/A_{-} = 0.4\%)$. The new angular asymmetries $A_{\pm}$ are therefore more precise than A_{FB}. The most general Wtb vertex containing terms up to dimension five can be written – following the convention of [5] – as: $L = -\frac{g}{\sqrt{2}}\bar{b}\gamma^\mu(V_L P_L + V_R P_R)tW_\mu^- - \frac{g}{\sqrt{2}}\bar{b}\frac{i\sigma^{\mu\nu}q^\nu}{M_W}(g_L P_L + g_R P_R)tW_\mu^- + h.c.$ and within the SM the only term different from zero is V_L. Again, if a value for V_R or g is measured different from zero, this can be due to the presence of new phyisics. The measurement of the previously mentioned asymmetries can be used to set limits on the anomalous couplings. The achievable 1σ limits on anomalous couplings are: $-0.05 \leq V_R \leq 0.12$, $-0.05 \leq g_L \leq 0.02$, $-0.021 \leq g_R \leq 0.017$.

4 Flavour Changing Neutral Currents (FCNC)

The FCNC top decays are strongly suppressed within the SM by the GIM mechanism (typical BR are between 10^{-14} and 10^{-12}). However, in some SM extensions (like 2-Higgs doublets, SUSY, exotic fermions), higher BR appears even at the level of 10^{-4}–10^{-6} (like for $t \to qg$ in SUSY). Three channels have been studied in ATLAS [6]: $t \to qZ$, $t \to q\gamma$, $t \to qg$. A common events pre-selection has been applied and, after the preselection, the three channels are isolated by applying additional cuts, specific for each one of the three event topologies, and a likelihood function is defined for the signal and the background, whose ratio (L_R) is used as discriminant variable. The expected BR sensitivities for a 5σ discovery are shown in Table 1 for the three channels, for an integrated luminosity of 10 and $100\,\mathrm{fb}^{-1}$. Even if the SM predicts a much lower BR for the FCNC decays, the expected BR obtained in these analyses are several orders of magnitude better than the present experimental limits.

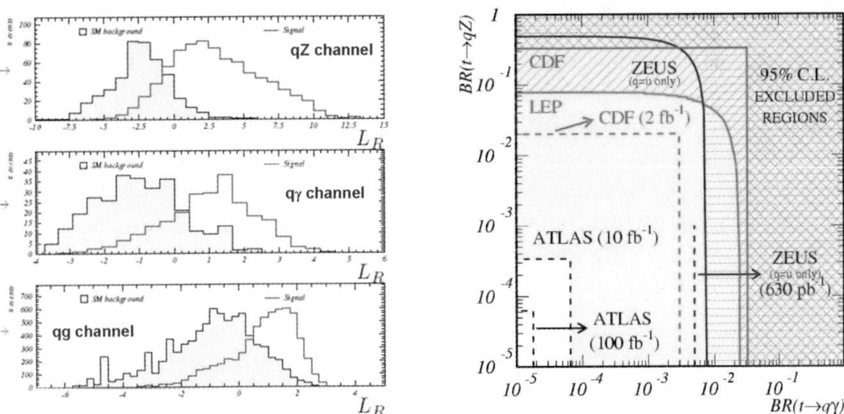

Fig. 2. *Left*: SM background and signal discriminant variable distributions for the three FCNC channels. *Right*: Present 95% CL limits on the BR($t \to q\gamma$) vs. BR($t \to qZ$) plane (*solid lines*)

Table 1. BR sensitivity for each channel in the 5σ discovery hypothesis, for a luminosity of $L = 10$ and $100\,\text{fb}^{-1}$

	$t \to qZ$	$t \to q\gamma$	$t \to qg$ (3 jets analysis)
$L = 10\,\text{fb}^{-1}$	4.4×10^{-4}	1.2×10^{-4}	4.3×10^{-3}
$L = 100\,\text{fb}^{-1}$	1.4×10^{-4}	3.6×10^{-5}	1.4×10^{-3}

5 SUSY virtual effects in top production

Recently, ATLAS looked [7] at the $t\bar{t}$ invariant mass distributions for production of unpolarized and polarized top pairs at LHC, in the theoretical framework of the MSSM. Assuming a light SUSY scenario ($M_{\text{SUSY}} \leq 400\,\text{GeV}$) the leading logarithmic electroweak contributions at one loop in a region of large invariant mass, $M_{t\bar{t}} \simeq 1\,\text{TeV}$, has been evaluated for the unpolarized differential cross section $\mathrm{d}\sigma/\mathrm{d}M_{t\bar{t}}$. The discussion of virtual supersymmetric effects has led to corrections to the lowest level of the partonic $\sigma(t\bar{t})$ production as a function of the partonic centre-of-mass energy. The corrections can be as large as 10–15% and depend on the exact value of the supersymmetric parameters. From a rather conservative evaluation of the most relevant theoretical and experimental uncertainties (like NLO QCD effects, effects of initial and final state radiation, uncertainty on the jet energy scale and on the luminosity), an overall error of approximately 20–25% in the relevant energy range appears realistically achievable. This study has been performed on Monte Carlo data corresponding to $5\,\text{fb}^{-1}$ and passed through a fast detector simulation. Further theoretical and experimental efforts might reduce the uncertainties to a final limit of, say, 15% or 10% size, opening the possibility of obtaining, from accurate measurements of the mass distribution, stringent consistency tests of the model, in particular identifications of large $\tan\beta$ effects.

References

1. CDF Coll.: Phys. Rev. Lett. **74**, 2626 (1995), Phys. Rev. D. **51**, 4623 (1995)
2. K. Assamagan, N. Gollub: Eur. Phys. J. C **39**, 25 (2005), A. Lucotte, F. Chevallier: ATL-PHYS-PUB-2006-014, K. Smolek: Czech. J. Phys. **54** suppl., 451 (2004), F. Hubaut *et al.*: SN-ATLAS-2005-052
3. CMS Coll.: TDR, vol. 2, CERN-LHCC-2006-021 (2006)
4. J.A. Aguilar-Saavedra *et al.*: ATL-PHYS-PUB-2006-018
5. F. del Aguila, J.A. Aguilar-Saavedra: Phys. Rev. D **67**, 014009 (2003)
6. J. Carvalho *et al.*: ATL-PHYS-PUB-2005-026, ATL-PHYS-PUB-2005-009
7. M. Beccaria *et al.*: Phys. Rev. D **71**, 7 (2005)

Searching for micro black holes at LHC

G.L. Alberghi[1,2,3], R. Casadio[1,2], D. Galli[1,2], D. Gregori[1,2], A. Tronconi[1,2], and V. Vagnoni[2]

[1] Physics Department, Bologna University, via Irnerio 46, 40126 Bologna, Italy
[2] I.N.F.N., Sezione di Bologna, via Irnerio 46, 40126 Bologna, Italy
[3] Astronomy Department, Bologna University, via Ranzani 1, 40127 Bologna, Italy

An exciting feature of brane-world models with large extra dimensions [1] is that the fundamental scale of gravity M_G could be as low as the electroweak scale ($M_G \simeq 1\,\text{TeV}$). Since the mass of a black hole must be larger than M_G, otherwise the classical concept of event horizon would be lost, micro black holes of a few TeV's may therefore be produced at the LHC in this scenario (for some recent reviews, see [2]).

According to the "hoop conjecture" [3], the cross section for black hole production should be given by the horizon area of the forming black hole and grow as the centre of mass energy of the colliding particles raised to a power which depends on the number of extra-dimensions [4, 5]. For example, since the no-hair theorems guarantee that a black hole is characterized by its mass, charges and angular momentum only, the horizon radius of an uncharged and non-rotating black hole in $4 + d$ dimensions is related to its mass M by [1]

$$R_{\text{H}} = \frac{1}{\sqrt{\pi}\,M_G} \left(\frac{M}{M_G}\right)^{\frac{1}{d+1}} \left(\frac{8\,\Gamma\left(\frac{d+3}{2}\right)}{d+2}\right)^{\frac{1}{d+1}}, \tag{1}$$

and its Hawking temperature [6] is then given by

$$T_{\text{H}} = \frac{d+1}{4\,\pi\,R_{\text{H}}}. \tag{2}$$

Once formed, the black hole begins to decay. In the standard picture, this process can be divided into three characteristic stages [5]:

1. a *balding phase*, during which multipole moments are quickly radiated away and the black hole becomes "hairless";
2. an *evaporation phase* with the emission of Hawking quanta [6]. Angular momentum and charges should be carried away very quickly, but whether

---

[1] We use units with $c = \hbar = k_{\text{B}} = 1$.

most of the emission occurs on the brane or into the bulk, or there is a significant emission of gravitational waves is still being debated;
3. a *Planck phase* when M approaches the effective Planck scale M_{G}.

The standard description of the evaporation is given by the canonical Planckian distribution for the emitted particles (of specie s and energy ω)

$$n^{(s)}_{(4+d)}(\omega) = \frac{1}{e^{\omega/T_{\mathrm{H}}} \pm 1} \tag{3}$$

and the corresponding black hole decay rate

$$\frac{\mathrm{d}M}{\mathrm{d}t} = A_{(4+d)} \sum_s \int n^{(s)}_{(4+d)}(\omega)\, \Gamma^{(s)}_{(4+d)}(\omega)\, \omega^{3+d}\, \mathrm{d}\omega\,, \tag{4}$$

where $A_{(4+d)}$ is the horizon area in $4+d$ dimensions and $\Gamma^{(s)}_{(4+d)}$ the grey-body factor (whose precise evaluation in brane-world models is rather challenging). Micro black hole life-times should then be so short (order of 10^{-18} sec) that their decay can be viewed as sudden [7]. However, energy conservation is not guaranteed a priori in the canonical ensemble (T_{H} in fact diverges for vanishing M) and is enforced by kinematical constraints in Monte Carlo simulations such as CHARYBDIS [8]. From a theoretical point of view, one should instead employ the microcanonical description [9,10], which yields life-times that vary greatly depending on the details of the brane-world model and could be as long as many seconds [11,12].

The end of the evaporation is an open issue theoretically, because we do not have a reliable theory of quantum gravity. In fact, theoretical physicists are looking forward to detecting micro black holes in order to have experimental data about quantum gravity. Their very detection would be evidence that M_{G} is not the 10^{19} GeV Planck energy and extra dimensions do exist, and the late stage of black hole evaporation (when $M \sim M_{\mathrm{G}}$) could tell us more about the details of quantum gravity. It is therefore natural to study deviations from the Hawking law induced by the underlying theory of quantum gravity. In [13] we employed a modified expression for the temperature of the form

$$T = \frac{R_{\mathrm{H}}^n}{R_{\mathrm{H}}^n + \alpha \ell^n} T_{\mathrm{H}}\,. \tag{5}$$

where ℓ, α, and n are adjustable parameters (see Fig. 1 for an example). Such a modified temperature is inspired by several approaches in the literature (for a partial list, see the References in [13]).

CHARYBDIS [8] simulates the evaporation according to Hawking's law until $M \sim M_{\mathrm{G}}$, afterwards the black hole is fragmented into a few particles simply according to phase space. We instead adjusted the code so that the black hole keeps on evaporating below the fundamental scale to negligible black hole masses, and compared the outputs [13]. The modified temperature (5) produces qualitatively similar results to the standard case when

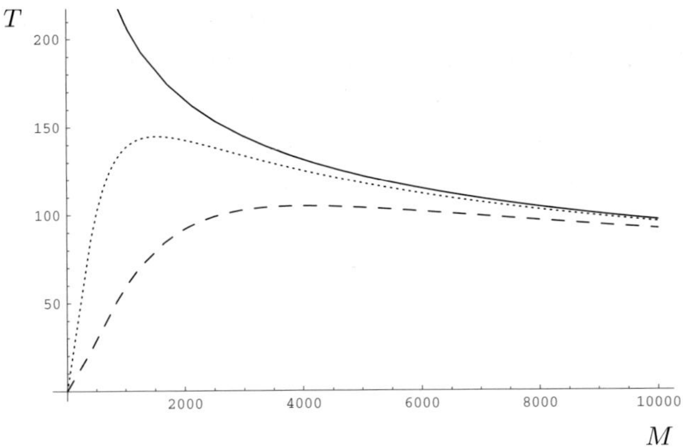

Fig. 1. Black hole temperature (in GeV) as a function of the mass (in GeV). We compare the Hawking law (*solid line*) to the case in (5) for $\ell = 10^{-3}\,\mathrm{GeV}^{-1}, \alpha = 5, n = 5$ (*dashed line*) and $\ell = 10^{-3}\,\mathrm{GeV}^{-1}, \alpha = 5, n = 1$ (*dotted line*)

$R_{\mathrm{H}} \gtrsim \ell$. However, since T tends to zero for $M = 0$, we find a continuous emmission of increasingly softer particles until the black hole evaporates completely. Once the black hole mass has decreased below the threshold for producing a given particle, such a particle can no longer be emitted. Hence,

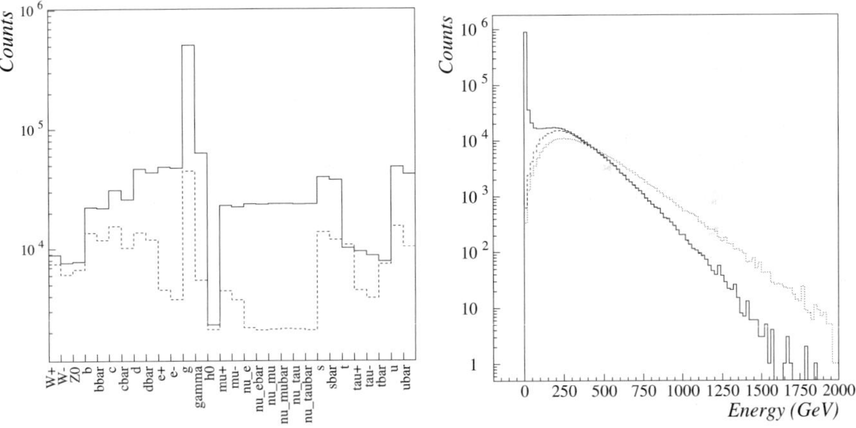

Fig. 2. *Left panel*: abundance of Standard Model particles produced during the evaporation of 10 TeV black holes, for a run of 10 000 events with the modified law (5) (*solid histogram*) and with standard CHARYBDIS (*dashed histogram*). $M_{\mathrm{G}} = 1\,\mathrm{TeV}$, the number of extra dimensions $d = 2$ and temperature parameters are the same as those for the dashed line in Fig. 1. *Right panel*: energy distribution of the emitted particles obtained with the modified law (5) (*solid histogram*), including only particles emitted when the black hole has a mass exceeding 1 TeV (*dashed histogram*), and with standard CHARYBDIS (*dotted histogram*)

the production of the heavier particles (i. e., massive gauge bosons and top quarks) is scarcely affected, while the emission of soft low mass (or massless) particles is largely enhanced. A comparison of the relative abundance of the Standard Model particles produced by the black hole evaporation with and without using the modified law of the form (5) – before parton evolution and hadronization – is shown in the left panel of Fig. 2. In the right panel of Fig. 2 we show the energy distribution of the emitted particles. As expected, the modified law dramatically changes the spectrum at low energy, leaving the spectrum at large energy moderately affected. This translates into a much larger multiplicity of isotropically emitted soft particles which might be a possible phenomenological signature of quantum gravity effects. The detection of the isotropic emission of a large number of soft isolated particles (i. e., not associated to jet-like topologies) could be evidence in favour of a specific decay law or of a specific theory of quantum gravity. In our analysis we have not explicitly considered the case in which black holes leave stable remnants [14]. Let us just mention that very large values of α and/or ℓ in the modified temperature (5) (together with energy conservation) would make it very hard for the black holes to continue the emission for $M \ll M_\mathrm{G}$ and stable remnants would be effectively produced.

References

1. N. Arkani-Hamed, S. Dimopoulos and G. Dvali, Phys. Lett. **B 429**, 263 (1998); Phys. Rev. D **59**, 0806004 (1999); I. Antoniadis, N. Arkani-Hamed, S. Dimopoulos and G. Dvali, Phys. Lett. **B 436**, 257 (1998); L. Randall and R. Sundrum, Phys. Rev. Lett. **83**, 4690 (1999); Phys. Rev. Lett. **83**, 3370 (1999).
2. M. Cavaglia, Int. J. Mod. Phys. A **18**, 1843 (2003); P. Kanti, Int. J. Mod. Phys. A **19**, 4899 (2004).
3. K.S. Thorne, in *Magic without magic*, ed. J. Klauder (Freeman, 1972).
4. T. Banks and W. Fishler, hep-th/9906038; S.B. Giddings and E. Katz, hep-th/0009176;
5. S.B. Giddings and S. Thomas, hep-ph/0106219.
6. S.W. Hawking, Nature **248**, 30 (1974); Comm. Math. Phys. **43**, 199 (1975).
7. S. Dimopoulos and G. Landsberg, Phys. Rev. Lett. **87** (2001) 161602.
8. C. M. Harris, P. Richardson and B.R. Webber, JHEP **0308**, 033 (2003).
9. R. Casadio, B. Harms and Y. Leblanc, Phys. Rev. D **57** 1309 (1998); R. Casadio and B. Harms, Phys. Rev. D **58**, 044014 (1998); Mod. Phys. Lett. **A14**, 1089 (1999).
10. R. Casadio and B. Harms, Phys. Lett. B **487** (2000) 209; Phys. Rev. D **64** (2001) 024016; S. Hossenfelder, S. Hofmann, M. Bleicher and H. Stoecker, Phys. Rev. D **66**, 101502 (2002)
11. R. Casadio and B. Harms, Int. J. Mod. Phys. A **17**, 4635 (2002)
12. T. G. Rizzo, hep-ph/0510420; arXiv: hep-ph/0601029.
13. G. L. Alberghi, R. Casadio, D. Galli, D. Gregori, A. Tronconi and V. Vagnoni, hep-ph/0601243.
14. B. Koch, M. Bleicher and S. Hossenfelder, hep-ph/0507138.

PARALLEL SESSION:
Flavour Physics
(S. Giagu and L. Silvestrini, conveners)

Lepton Flavor Violation and Rare Kaon Decays

Paride Paradisi

Department of Physics, Technion-Israel Institute of Technology,
Technion City, 32000 Haifa, Israel
paride@technion.ac.il

The impact of rare K decays and Lepton-Flavor Violating processes in shedding light on physics beyond the Standard Model is reviewed. Moreover, we show that tests of Lepton Universality in K decays can represent an interesting handle to obtain relevant information on New Physics scenarios.

1 Lepton Flavor Violating τ decays

The observation of neutrino oscillation have established the existence of lepton flavor violation (LFV). As a natural consequence, one would expect flavor mixing to appear also in the charged leptons sector, in rare decay processes such as $\mu \to e\gamma$, $\tau \to \mu\gamma$ etc. However, it is well known that, within a Standard Model framework, LFV processes are highly suppressed, at a level which is well below any optimistic future experimental sensitivity.

On the other hand, within a super symmetric (SUSY) framework the situation can be completely different because of new direct sources of LFV,

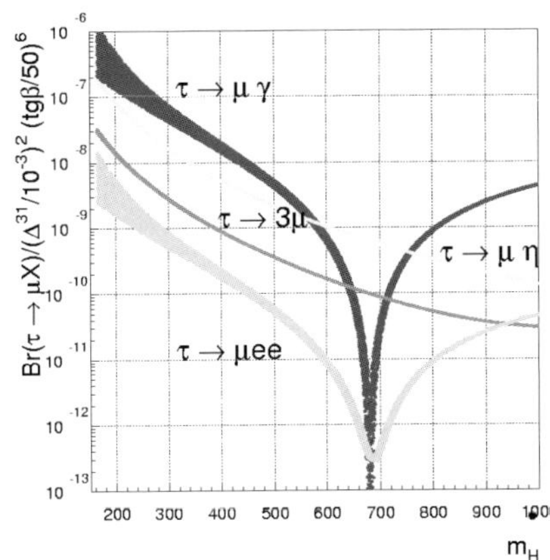

Fig. 1. Branching ratios of various $\tau \to \mu$ and transitions vs. the heaviest Higgs boson mass m_H [3]. In the figures we assume $X = \gamma, \mu\mu, \eta$. The Δ^{ij} terms are the sources of LFV which are induced at one loop level by non-holomorphic interactions

namely the off-diagonal soft terms in the slepton mass matrices and in the trilinear couplings [1]. In practice, LFV would originate from any misalignment between fermion and sfermion mass eigenstates. LFV processes arise at one loop level through the exchange of neutralinos (charginos) and charged sleptons (sneutrinos).

Another potential source of LFV in the minimal supersymmetric standard model (MSSM) is the Higgs sector. Higgs-mediated FCNC are induced at one loop level by non holomorphic corrections through the exchange of gauginos and sleptons, provided LFV mixing among the sleptons [2] (for a comprehensive phenomenological analysis, see [3]). In Fig.1, we report the branching ratios (Br) for different LFV τ decays.

2 LF universality tests in $K_{\ell 2}$ and $\pi_{\ell 2}$

An alternative phenomenological tool to search for new sources of LF violation is provided by precise LF universality tests in charged-current meson decays. In particular, the ratios

$$R_P^{\mu/e} = \frac{\mathcal{B}(P \to \mu\nu)}{\mathcal{B}(P \to e\nu)} \tag{1}$$

can be predicted with excellent accuracies in the SM, both for $P = \pi$ (0.02% accuracy [4]) and $P = K$ (0.04% accuracy [4]), allowing for some of the most significant tests of LF universality.

As recently pointed out in [5], large departures from the SM expectations can be generated in realistic supersymmetric scenarios. The key ingredients which allows visible non-SM contributions in $R_P^{\mu/e}$ within the MSSM are:

i)   large values of $\tan\beta$ (the ratio of the two Higgs vacuum expectation values), such that the overall normalization of Y_E – and correspondingly the $H^\pm$-exchange contribution to $P \to \ell\nu$ – is enhanced;

ii) large mixing angles in the right-slepton sector (Δ_R^{ij}, with $i \neq j$) such that the $P \to \ell_i \nu_j$ rate becomes non negligible.

Denoting by $\Delta r_{\rm NP}^{e-\mu}$ the deviation from μ–e universality in R_K due to new physics, i.e.: $R_K^{\mu/e} = (R_K^{\mu/e})_{\rm SM} \left(1 + \Delta r_K^{e-\mu}\right)$, it turns out that:

$$\Delta r_K^{e-\mu} \simeq \left(\frac{m_K^4}{M_H^4}\right)\left(\frac{m_\tau^2}{m_e^2}\right)|\Delta_R^{31}|^2 \tan^6\beta \,. \tag{2}$$

In the most favorable scenarios, the deviations from the SM could reach $\sim 1\%$ in the $R_K^{\mu/e}$ case [5] (not far from the present experimental resolution [6]) and $\sim$ few $\times\, 10^{-4}$ in the $R_\pi^{\mu/e}$ case. In the pion case the effect is quite below the present experimental resolution [7], but could well be within the reach of the new generation of high-precision $\pi_{\ell 2}$ experiments planned at TRIUMPH and at PSI.

In principle, larger violations of LF universality are expected in $B \to \ell\nu$ decays, with $\mathcal{O}(10\%)$ deviations from the SM in $R_B^{\mu/\tau}$ and even order-of-magnitude enhancements in $R_B^{e/\tau}$ [8]. However, the difficulty of precision measurements of the highly suppressed $B \to e/\mu\ \nu$ modes makes these non-standard effects undetectable (at least at present).

Similarly to the FCNC decays, also for the LF universality tests the low-energy systems ($K_{\ell 2}$ and $\pi_{\ell 2}$) offer a unique opportunity in shedding light on physics beyond the SM: the smallness of NP effects is more than compensated (in terms of NP sensitivity) by the excellent experimental resolution and the good theoretical control.

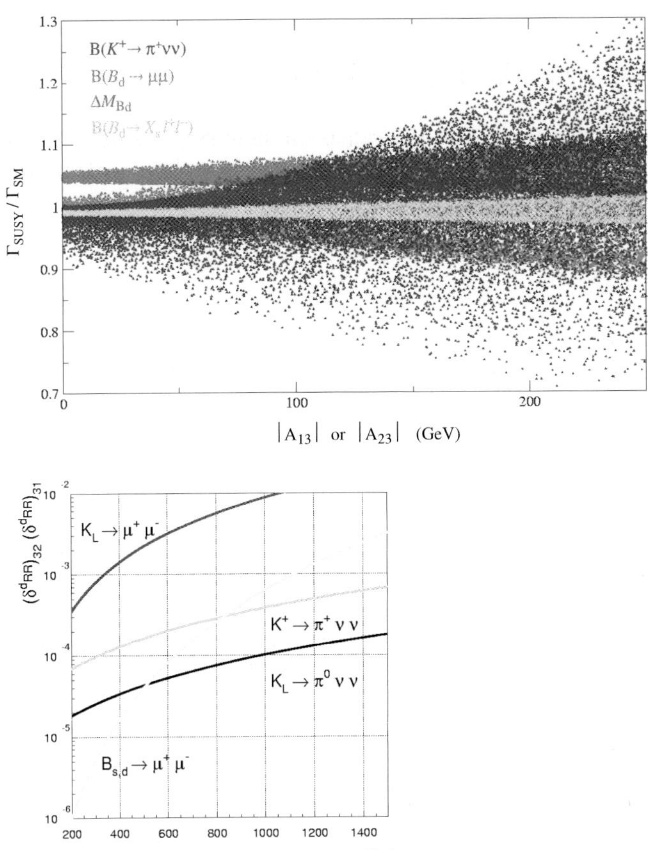

Fig. 2. *Upper plot*: dependence of various FCNC observables (normalized to their SM value) on the up-type trilinear terms (A_{13}) and on squark masses in the general MSSM [15]. *Lower plot*: sensitivity to the off-diagonal down-type squark terms $(\delta^d_{RR})_{23}(\delta^d_{RR})_{31}$ of various rare K and B decays as a function of the charged-Higgs boson mass setting $\tan\beta = 50$ assuming a 10% measurements of $K \to \pi\nu\bar{\nu}$ branching ratios [14]. The $B_{s,d} \to \mu^+\mu^-$ bounds refer to the latest experimental limits [16]

3 Rare K decays: $K \to \pi\nu\bar{\nu}$

The strong suppression within the Standard Model (SM) and the high sensitivity to physics beyond the SM of the two $K \to \pi\nu\bar{\nu}$ rates, signal the unique possibilities offered by these rare processes in probing the underlying mechanism of flavor mixing. This statement has been reinforced by the recent theoretical progress in the evaluation of NNLO [9] and power-suppressed [10] contributions to $K^+ \to \pi^+\nu\bar{\nu}$. These allow to obtain predictions for the corresponding branching ratio of high accuracy ($\sim 6\%$), not far from exceptional level of precision ($\sim 2\%$) already reached in the $K_L \to \pi^0\nu\bar{\nu}$ case [11, 12].

Within the MSSM with R-parity, sizable non-standard contributions to $K \to \pi\nu\bar{\nu}$ decays can be generated only going beyond the so-called Minimal Flavor Violation (MFV) hypothesis and are mediated or by the exchange of chargino/up-squark loops [13] or by charged Higgs/top-quark loops [14]. In the first case, large effects are generated only if the left–right mixing of the up squarks has a flavor structure substantially different from the MFV hypothesis [13]. In the second case, deviations from the SM expectations are induced by the presence of off-diagonal mixing terms in the right–right down squark sector (with sizable effects even for tiny, CKM-type, mixing) but only for large values of $\tan\beta$.

The effects of the above contributions are examined separately in Fig. (2) where the constraints of B physics observables (that are sensitive to same susy parameters) are also reported.

References

1. F. Borzumati and A. Masiero, Phys. Rev. Lett. **57**, 961 (1986).
2. K.S. Babu and C. Kolda, Phys.Rev.Lett.**89**, 241802 (2002).
3. P. Paradisi, JHEP **0602**, 050 (2006); JHEP **0608** (2006) 047.
4. W. J. Marciano and A. Sirlin, Phys. Rev. Lett. **71** (1993) 3629; M. Finkemeier, Phys. Lett. B **387** (1996) 391.
5. A. Masiero, P. Paradisi and R. Petronzio, Phys. Rev. D **74** (2006) 011701.
6. A. Ceccucci, eConf **C060409**, 033 (2006), [arXiv:hep-ex/0605120]; T. Mori, eConf **C060409** (2006) 034 [arXiv:hep-ex/0605116]; L. Fiorini [NA48/2 Collaboration], talk presented at EPS 2005 July 21st-27th 2005 (Lisboa, Portugal).
7. G. Czapek *et al.*, Phys. Rev. Lett. **70** (1993) 17; D. I. Britton *et al.*, Phys. Rev. Lett. **68** (1992) 3000.
8. G. Isidori and P. Paradisi, Phys. Lett. B **639** (2006) 499.
9. A. J. Buras *et al.*, hep-ph/0508165; hep-ph/0603079.
10. G. Isidori, F. Mescia and C. Smith, Nucl. Phys. **B718** (2005) 319.
11. A. J. Buras, F. Schwab and S. Uhlig, hep-ph/0405132; D. Bryman, A. J. Buras, G. Isidori, L. Littenberg, Int. J. Mod. Phys. A **21**, 487 (2006).
12. G. Buchalla and G. Isidori, Phys. Lett. **B440** (1998) 170.
13. G. Colangelo and G. Isidori, JHEP **09** (1998) 009.
14. G. Isidori and P. Paradisi, Phys. Rev. D **73**, 055017 (2006).
15. G. Isidori *et al.*, JHEP **0608**, 064 (2006).
16. R. Bernhard *et al.* [CDF Collaboration], hep-ex/0508058.

Neutral Kaon Physics at KLOE

Marco Dreucci

INFN dei Laboratori Nazionali di Frascati
Marco.Dreucci@lnf.infn.it

1 The DAΦNE collider and the KLOE detector

KLOE data taking at DAΦNE (the Frascati e^+e^- collider) has been terminated on March 2006, after collecting more than $2\,\mathrm{fm}^{-1}$. The results presented in this article are based on the $450\,\mathrm{pb}^{-1}$ of data taken in 2001 and 2002. The KLOE detector consists of a large cylindrical drift chamber (DC), surrounded by a lead scintillating-fiber electromagnetic calorimeter (EMC) with 4880 PMT's. A superconducting coil around the calorimeter provides a $0.52\,\mathrm{T}$ field. The drift chamber [1] is $4\,\mathrm{m}$ in diameter and $3.3\,\mathrm{m}$ long, filled with a 90%He+10%IsoB gas. The momentum resolution is $\sigma_{p_\perp}/p_\perp \approx 0.4\%$. The vertex resolution is $\sim 2\,\mathrm{mm}$. The calorimeter [2] is divided into a barrel and two endcaps. It covers 98% of the solid angle. Cells close in time and space are grouped into calorimeter clusters. The energy and time resolutions are rispectively $\sigma_E/E = 5.7\%/\sqrt{E\,(\mathrm{GeV})}$ and $\sigma_T = 54\,\mathrm{ps}/\sqrt{E\,(\mathrm{GeV})} \oplus 100\,\mathrm{ps}$. The two-level KLOE trigger [3] uses calorimeter and chamber informations. Recognition and rejection of cosmic-ray events is also performed at the trigger level. The KLOE Monte Carlo (MC) program, GEANFI [4], includes a full description of the KLOE detector. Machine background is extracted for each run and overlaid with the event generator. Radiative contributions are implemented in the kaon decay generators [5]. An example of Data-Monte Carlo agreement is shown in Fig. 1, for the main K_L charged decays.

2 Neutral kaon at KLOE

KLOE provides monochromatic and pure beam of kaons in a $J^{PC} = 1^{--}$ antisymmetric quantum state, with a momentum of $\sim 110\,\mathrm{MeV}/c$. Observation of K_S (K_L) signals the presence of a K_L (K_S) on the other side. This fact is the basis for many measurements of $K_S\,K_L$ system. In particular the tagging technic allows precise measurement of absolute BR's. Since the K_L and the K_S are initially in a pure quantum state, it is also possible to observe effect

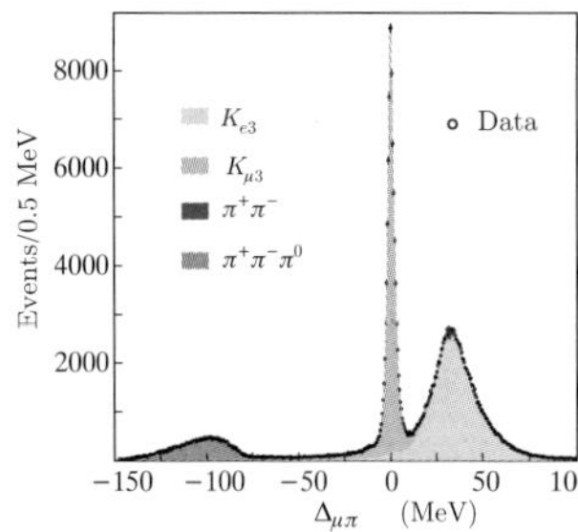
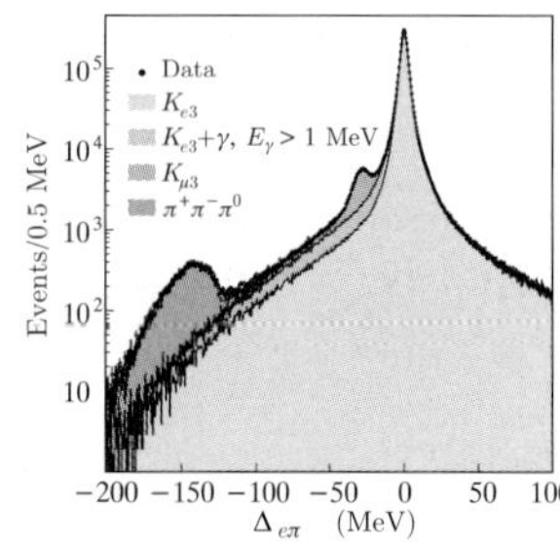

Fig. 1. $E_{\mathrm{miss}}-p_{\mathrm{miss}}$ distribution in different mass hypotheses

of quantum-mechanical interference. All K_S's decay in the beam pipe, while K_L acceptance is about 40%, and a fraction of about 50% of K_L's enter the calorimeter without decaying. A typical K_L tag is the $K_S \to \pi^+\pi^-$ decay (efficiency $\sim 70\%$). The K_L's entering the calorimeter ('K_L crash') are used to tag K_S (efficiency $\sim 30\%$).

3 Semileptonic decays and V_{us}

The most precise test of CKM unitarity comes from its first row, in which the smallness of V_{ub} matrix element allows to test unitarity up to 10^{-3} level. The best determination of $|V_{us}|$ is obtained from semileptonic kaon decays. At the present PDG04 quotes a 2σ deviation from unitarity. From the decay rate (inclusive of radiation) KLOE, by measuring all experimental input, can extract $|V_{us}|$.

3.1 BR($K_S \to \pi e \nu(\gamma)$) and first observation of $K_S \to \pi\mu\nu$

The relative BR for $K_S \to \pi^\pm e^\mp \nu(\gamma)$ events, normalized to $K_S \to \pi^+\pi^-(\gamma)$ in the same dataset, has been measured. The excellent timing performance of EMC allows to use time of flight (TOF) to identify signal events. The counting is performed separately for each charge. The combined result is $\mathrm{BR}(K_S \to \pi^\pm e^\mp \nu) = (7.046 \pm 0.077 \pm 0.049) \times 10^{-4}$. Using this result we can measure for the first time the charge asymmetry A_S, giving $(1.5 \pm 9.6 \pm 2.9) \times 10^{-3}$. The analysis of $K_S \to \pi\mu\nu$ (still in progress) is very similar to that of $K_S \to \pi^\pm e^\mp \nu$ events, but measurement of $\mathrm{BR}(K_S \to \pi\mu\nu)$ is more difficult due to the similarity of pion and muon mass.

3.2 Dominant K_L BR's and K_L lifetime

The main BR's of K_L have been measured with a precision below 1% [6]. The efficiency of the $K_S \to \pi^+\pi^-$ decay tag depends slightly on the fate of the K_L. The main analysis problem has been to reduce as much as possible this bias. In Table 1 is shown the result. Free parameters of the fit are the BR's and the K_L lifetime, with the constraint of the unitarity.

Table 1. Result of main K_L BR's and lifetime with the unitary constraint

	KLOE	PDG 2004
$\mathrm{BR}(K_L \to \pi^\pm e^\mp \nu)$	$40.07 \pm 0.06_{\mathrm{stat}} \pm 0.14_{\mathrm{syst}}$	38.81 ± 0.27
$\mathrm{BR}(K_L \to \pi \mu \nu)$	$26.98 \pm 0.06_{\mathrm{stat}} \pm 0.14_{\mathrm{syst}}$	27.19 ± 0.25
$\mathrm{BR}(K_L \to \pi^0 \pi^0 \pi^0)$	$19.97 \pm 0.05_{\mathrm{stat}} \pm 0.19_{\mathrm{syst}}$	21.05 ± 0.23
$\mathrm{BR}(K_L \to \pi^+ \pi^- \pi^0)$	$12.63 \pm 0.05_{\mathrm{stat}} \pm 0.11_{\mathrm{syst}}$	12.59 ± 0.19
τ_{KL} (ns)	$(50.72 \pm 0.17_{\mathrm{stat}} \pm 0.33_{\mathrm{syst}})$	51.8 ± 0.4

Another independent measurement of K_L lifetime has been performed [7] from a fit to the $K_L \to \pi^0 \pi^0 \pi^0$ proper-time decay distribution. The result is $\tau_L = (50.92 \pm 0.17_{\mathrm{stat}} \pm 0.25_{\mathrm{syst}})$ ns.

3.3 Semileptonic form factors in $K_L \to \pi^\pm e^\mp \nu(\gamma)$

In this analysis we use the quadratic parametrization: $f_+(t) = f_+(0) \times (1 + \lambda'_+ \, t/m^2 + 1/2\lambda''_+ t^2/m^4)$. Using the t distribution, both a linear fit and a quadratic fit have been considered. Free parameters of the fit are the slope parameters. A fit in the natural form of pole approximation has also been made. The result is: $\lambda_+ = (28.6 \pm 0.5_{\mathrm{stat}} \pm 0.4_{\mathrm{syst}}) \times 10^{-3}$ (linear fit), $\lambda'_+ = (25.5 \pm 1.5_{\mathrm{stat}} \pm 1.0_{\mathrm{syst}}) \times 10^{-3}$ and $\lambda''_+ = (1.4 \pm 0.7_{\mathrm{stat}} \pm 0.4_{\mathrm{syst}}) \times 10^{-3}$ (quadratic fit), $M_V = (870 \pm 6_{\mathrm{stat.}} \pm 7_{\mathrm{syst.}})$ MeV (polar fit).

3.4 V_{us} extraction

We use CKM-working group prescription to extract $|V_{us}| \, f_+^{K^0 \pi}(0)$ from the decay rate. The quadratic parametrization is used for form factors, $f_+^{K^0 \pi}(0)$ is taken from Leutwyler-Roos $(0.961(8))$ [9] and λ_0, λ'_+ and λ''_+ slopes are obtained from a combined fit to K_{e3} and $K_{\mu3}$ results from KTeV [11] and ISTRA+ [12, 13]. In Table 2 is shown the result.

4 $K_L \to \pi^+ \pi^-$ decay and CP violation

We count signal events fitting data distribution of $\sqrt{E^2_{\mathrm{miss}}(\pi\pi) + p^2_{\mathrm{miss}}}$ with MC shapes and normalize to the events $K_L \to \pi \mu \nu$ in the same data set. The

Table 2. Result of main K_L BR's and lifetime with the unitary constraint

KLOE result:	$|V_{us}| f_+^{K^0 \pi}(0) \; K_{Lc3} = 0.2164 + 0.0007$
	$|V_{us}| f_+^{K^0 \pi}(0) \; K_{L\mu3} = 0.2174 \pm 0.0009$
	$|V_{us}| f_+^{K^0 \pi}(0) \; K_{Se3} = 0.2169 \pm 0.0017$
unitary band:	$\sqrt{1 - |V_{ud}|^2} f_+^{K^0 \pi}(0) = 0.2187 \pm 0.0022$ [10]

result is $\mathrm{BR}(K_L \to \pi^+\pi^-) = (1.963 \pm 0.012_{\mathrm{stat}} \pm 0.017_{\mathrm{syst}}) \times 10^{-3}$ confirming the 4-σ discrepancy with PDG04. Using the KLOE measurements and τ_S from PDG04 after subtracting the direct emission contribution (from E731), we evaluate the direct CP violation term: $|\varepsilon| = (2.216 \pm 0.013) \times 10^{-3}$ It result to be about 4σ from PDG04 but, in any case, still in agreement ($\sim 1.5\sigma$) with prediction from Unitarity Triangle.

5 CPT test and quantum mechanic test

Measurement of K_S and K_L observables can be used for the CPT test from unitarity using the Bell-Steinberger relation. We can extract $\mathrm{Re}(\varepsilon)$ and $\mathrm{Im}(\delta)$ after inserting input measurements from KLOE and from PDG. The result is: $\mathrm{Re}(\varepsilon) = (1.602 \pm 0.013) \times 10^{-3}$ and $\mathrm{Im}(\delta) = (1.2 \pm 1.9) \times 10^{-5}$.

In the quantum mechanic sector the decay time difference distribution for the $K_S \to \pi^+\pi^-$, $K_L \to \pi^+\pi^-$ final state allows to measure the mass difference Δm and the decoherence term ζ_{D} (QM prevides $\zeta_{\mathrm{D}} = 0$). The preliminary result is : $\Delta m = (5.34 \pm 0.34) \times 10^9 \mathrm{s}^{-1}$ if $\zeta_{\mathrm{D}} = 0$; $\zeta_{0,0} = 0.24^{+0.21}_{-0.19} \pm 0.01_{\mathrm{syst}} \times 10^{-5}$ and $\zeta_{S,L} = 0.043^{+0.038}_{-0.035} \pm 0.008_{\mathrm{syst}}$ if $\Delta m = \Delta m_{\mathrm{PDG}}$.

6 Conclusion

KLOE results on neutral kaon physics made a considerable step during the 2005 with the following measurement: $\mathrm{BR}(K_S \to \pi^0\pi^0\pi^0)$, τ_L, main BR's(K_L), $R^S_{\pi\pi}$, $\mathrm{BR}(K_{Se3})$, $\mathrm{BR}(K_L \to \pi^+\pi^-)$, $K_{\ell e3}$ form factors. The following analyses are in progress and will be ended in 2006: $\Gamma(K_S \to \pi^+\pi^-(\gamma))/\Gamma(K_S \to \pi^0\pi^0)$, $\mathrm{BR}(K_{S\mu3})$ and the sector of rare K_S decays: $\pi^+\pi^-\pi^0$, $\pi^+\pi^-e^+e^-$, $\gamma\gamma$. All KLOE measurement will be increased by a factor of five when the full KLOE data sample will be used.

References

1. M. Adinolfi, *et al.*, Nucl. Instrum. Meth. A 488 (2002) 51.
2. M. Adinolfi, *et al.*, Nucl. Instrum. Meth. A 482 (2002) 364.
3. M. Adinolfi, *et al.*, Nucl. Instrum. Meth. A 492 (2002) 134.
4. F. Ambrosino, *et al.*, Nucl. Instrum. Meth. A 534 (2004) 403.
5. C. Gatti, Eur. Phys. J. C45 (2006) 417.
6. F. Ambrosino, *et al.*, Phys. Lett. B 632 (2006) 43.
7. F. Ambrosino, *et al.*, Phys. Lett. B 626 (2005) 15.
8. F. Ambrosino, *et al.*, Phys. Lett. B 636 (2006) 166.
9. H. Leutwyler, M. Roos, Z. Phys. C 25 (1984) 91.
10. J. Marciano, A. Stirling, Phys. Rev. Lett. 96 (2006).
11. T. Alexopolous, *et al.*, Phys. Rcv. D 70 (2004) 092007.
12. O.P. Yushchenko, *et al.*, Phys. Lett. B 581 (2004) 31.
13. O.P. Yushchenko, *et al.*, Phys. Lett. B 589 (2004) 111.

Charged kaons and V_{us} at KLOE

KLOE collaboration

presented by P. Massarotti
Università degli Studi "Federico II" e Sezione INFN, Napoli
http://www.lnf.infn.it/kloe/

At the Frascati phi-factory Dafne e^+ and e^- beams collide in the center of mass energy of the the ϕ-meson which decays into anti-collinear $K\bar{K}$ pairs. In the laboratory this remains approximately true because of the small crossing angle of the e^+e^- beams. Therefore the detection of a $K(\bar{K})$ tags the presence of a $\bar{K}(K)$ of given momentum and direction. The decay products of the $K^{\pm}$ pair define two spatially well separated regions called the tag and the signal hemispheres. Identified $K^{\mp}$ decays tag a $K^{\pm}$ beam and provide an absolute count, using the total number of tags as normalization. This procedure is a unique feature of a ϕ-factory and provides the means for measuring absolute branching ratios. Charged kaons are tagged using the two body decays $K^{\pm} \to \mu^{\pm}\nu_{\mu}$ and $K^{\pm} \to \pi^{\pm}\pi^0$. Since the two body decays correspond to about 85% of the charged kaon decays [4] and since $\mathrm{BR}(\phi \to K^+K^-) \simeq 49\%$ [4], there are about $1.5 \times 10^6 K^+K^-$ events/pb^{-1}. The two body decays are identified as peaks in the momentum spectrum of the secondary tracks in the kaon rest frame and in the pion mass hypothesis $p^*(m_\pi)$ In order to minimize the impact of the trigger efficiency, the tagging kaon by itself must provide the event trigger (self-trigger).

1 Measurement of the absolute branching ratio $K^+ \to \mu^+\nu_\mu(\gamma)$

The measurement has been performed using $175\,\mathrm{pb}^{-1}$ of data collected in 2002 [5]. The data sample has been split in two uncorrelated subsamples, $60\,\mathrm{pb}^{-1}$ have been used for the BR measurement, the remaining $115\,\mathrm{pb}^{-1}$ have been used to evaluate the efficiencies and the background. The $K^- \to \mu^-\nu_\mu$ tag has been required in order to minimize the effect of the nuclear interactions on the signal side. The signal is given by a K^+, moving outwards in the DC with momentum $70 < p_K < 130\,\mathrm{MeV}/c$ and having point of closest approach to beams pipe with $0 < \sqrt{x_{\mathrm{PCA}}^2 + y_{\mathrm{PCA}}^2} < 10\,\mathrm{cm}$ and $|z_{\mathrm{PCA}}| < 20\,\mathrm{cm}$.

The kaon decay in the DC fiducial volume ($40 < \sqrt{x_\mathrm{V}^2 + y_\mathrm{V}^2} < 150\,\mathrm{cm}$) is required. The number of $K^+ \to \mu^+ \nu_\mu(\gamma)$ decays is obtained counting the events with $225 < p^* < 400\,\mathrm{MeV}/c$. The background is given by $K^+ \to \pi^+ \pi^0$ and $K^+ \to \pi^{0+} \nu_l$. The detection of the photons form π^0 decay allow the identification of background events and then to obtain directly from data the p^* distribution of the background. The number of signal events we found is $N_{K^+ \to \mu^+ \nu_\mu(\gamma)} = 865\,283$. The efficiency has been evaluated on a control sample of pure $K\mu\nu$ events selected using only calorimeter information. The efficiency is $\epsilon = 0.3153 \pm 0.0002$. Many possible sources of systematic effects have been taken into account; the most imporant being related to trigger, energy cuts and fiducial volume acceptances. The branching is given by

$$\mathrm{BR}(K^+ \to \mu^+ \nu_\mu(\gamma)) = \frac{N_{K^+ \to \mu^+ \nu_\mu(\gamma)}}{N_\mathrm{Tag}} \times \frac{1}{\epsilon} \text{ and is:}$$

$$\mathrm{BR}(K^+ \to \mu^+ \nu_\mu(\gamma)) = 0.6366 \pm 0.0009_\mathrm{stat.} \pm 0.0015_\mathrm{syst.}$$

corresponding to a total accuracy of 0.27%.

2 Measurement of the $K^\pm$ semileptonic decays absolute branching ratios

The measurement of the semileptonic branching ratios has been performed on $410\,\mathrm{pb}^{-1}$ of self-triggering tags collected in 2001 and 2002. The whole sample has been split in four subsamples defined by different decay modes for the tagging kaon: two charged and 2 possible tags ($K_{\mu 2}$ and $K_{\pi 2}$; for the latter the identification of the π^0 is also required). This redundancy allows us the study of systematics. The kaon tracks and decay vertexes are defined as in the $K_{\mu 2}$ analysis. Moreover it is required that the secondary track, extrapolated to the calorimeter, points to an energy deposit. In order to reject the two body decay events with $p^* > 195\,\mathrm{MeV}$ have been discarded; p^* is the momentum of the secondary particle on the signal side, evaluted in the pion mass hypothesis. The π^0 reconstruction is then required looking for two neutral clusters in the EMC having time of flight consistent with the one expected for photons emitted at the kaon decay vertex. Other sources of background are $K^\pm \to \pi^\pm \pi^0 \pi^0$ with a π^0 undergoing a Dalitz decay and $K^\pm \to \pi^\pm \pi^0$ with an early decay $\pi^\pm \to \mu^\pm \nu$. The former are discarded requiring ($E_\mathrm{miss} - P_\mathrm{miss} < 90\,\mathrm{MeV}$, the latter using the missing momentum of the secondary track in the pion rest frame $P^*_\mathrm{sec} < 90\,\mathrm{MeV}$. Then the spectrum of the lepton squared massa (m^2_lept) is obtained from the speed of the lepton computed from its time of flight. The number of K_{e3} and $K_{\mu 3}$ decays is then obtained by fitting the m^2_lept distribution with the MC distributions for the signals and background sources. The branching ratios have been evaluated separately for each tag sample. Corrections have been applied in order to account for data-MC differences in tracking and calorimeter clustering. About $190\,000\,K^\pm_{e3}$ and $100\,000\,K^\pm_{\mu 3}$ decays have been selected. The preliminary results for the branching ratios obtained are:

$$\mathrm{BR}(K_{e3}^{\pm}) = (5.047 \pm 0.046_{\mathrm{Stat+Tag}})\% \tag{1}$$

$$\mathrm{BR}(K_{\mu3}^{\pm}) = (3.310 \pm 0.040_{\mathrm{Stat+Tag\,ns}})\% \tag{2}$$

The values are averages over the four different tag samples for each channel. Correlations have been taken into account. The error is dominated by the uncertainty on data/MC efficiency corrections and the systematic error evaluation from the signal selection efficiency still has to be completed.

3 Measurement of the charged kaon lifetime

The measurement is performed using $230\,\mathrm{pb}^{-1}$. The data sample has been split in two uncorrelated subsamples, $150\,\mathrm{pb}^{-1}$ have been used for the measurement, the remaining $80\,\mathrm{pb}^{-1}$ have been used to evaluate the efficiencies. Both charges $K_{\mu2}$ tags have been used. There are two methods available for the measurement: the kaon decay length and the kaon decay time. The two methods allow us to do cross checks and studies of systematics; their resolutions are comparable. The method relying on the measurement of the charged kaon decay length requires as first the reconstruction of the kaon decay vertex in the fiducial volume (as previously defined) using only DC information. Once the decay vertex has been identified, the kaon track is extrapolated backward to the interaction point with 2-mm steps, taking into account the energy loss to evaluate its velocity βc. Then the proper time is given by $\tau^* = \sum_i \Delta T_i = \sum_i \frac{\sqrt{1-\beta_i^2}}{\beta_i} \Delta l_i/c \; [c = c_{\mathrm{light}}]$ The efficiency has been evaluated directly on data. The control sample has been selected using calorimetric information only, looking for a neutral vertex: two clusters in time fired by the photons coming from the π^0 decay. The proper time is fitted between 16 and 30 ns correcting for the efficiency. Resolution effects have been taken into account. The preliminary result we have obtained for the K^+ is:

$$\tau^+ = (12.377 \pm 0.044 \pm 0.065)\,\mathrm{ns} \tag{3}$$

with $\chi^2 = 17.7/15$, corresponding to a χ^2 probability $P(\chi^2) = 28.4\%$. The second method relies on the measurement of the kaon decay time. It requires the backward extrapolation to the interaction point of the tagging kaon track and the forward extrapolation of the virtual helix of the kaon on the signal side. Stepping along the helix we look for the π^0 decay vertex without looking at the real kaon track. For each photon it is possible to measure the proper time as $\tau^* = (t_\gamma - \frac{r_\gamma}{c} - t_\phi) \cdot \sqrt{1 - \beta_K^2}$ and then, fitting the obtained distribution, the charged kaon lifetime. The work concerning this second method is in progress.

4 V_{us} extraction

The three measurements presented here allow the extraction V_{us} since the branching ratio of the semileptonic decays and the charged kaon lifetime are connected to the value of V_{us} [6]: $\frac{\mathrm{BR}(K\to\pi l\nu_l)}{\tau_K} \propto |V_{us}\, f_+(0)|^2$.

Using as input the KLOE measurements of the semileptonic branching ratios for neutral and charged kaons (K_{Le3}, $K_{L\mu3}$, K_{Se3}, $K^{\pm}_{e3}$, $K^{\pm}_{\mu3}$), and of the K_L lifetime and using a quadratic parametrization of the form factors ($\lambda'_+ = 0.02496(80)$, $\lambda''_+ = 0.00162(35)$ and $\lambda_0 = 0.01587(95)$) we have obtained:

$$V_{us} \times f_+(0) = 0.2160 \pm 0.005 \tag{4}$$

with $\chi^2/dof = 1.9/4$ (Fig. 1) The branching ratio of the purely leptonic decay, $K_{\mu2}$, can be used, together with the lattice calculation of the ratio f_K/f_π to evaluate V_{us} as pointed out in [7]: $\frac{\mathrm{BR}(K\to\mu\nu_\mu(\gamma))}{\mathrm{BR}(\pi\to\mu\nu_\mu(\gamma))} \propto \left|\frac{V_{us}}{V_{ud}}\right|^2 \times \left(\frac{f_K}{f_\pi}\right)^2$.

Taking $f_K/f_\pi = 1.198 \pm 0.003^{+0.016}_{-0.005}$ from [8], we obtain:

$$\left|\frac{V_{us}}{V_{ud}}\right| = 0.2294 \pm 0.0026 \,. \tag{5}$$

It is then possible to perform a fit in the V_{us}–V_{ud} plane (Fig. 1) taking $V_{ud} = 0.97377 \pm 0.00027$ from [9] and the KLOE estimates for V_{us}/V_{ud} from $K_{\mu2}$ and V_{us} from K_{l3}. The results we obtain are:

$$V_{us} = 0.2243 \pm 0.0016 \tag{6}$$
$$V_{ud} = 0.97377 \pm 0.00027 \tag{7}$$

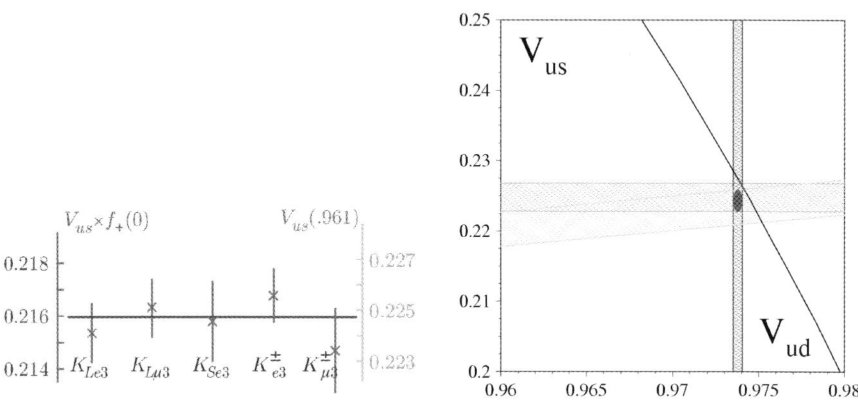

Fig. 1. *On the left* the fit of $|V_{us}f_+(0)|$ using KLOE measurements of BRs and of K_L lifetime and a quadratic parametrization of the form factors. *On the right* view of the V_{us}–V_{ud} plane. The bounds from measurements and the unitarity line are shown

with $P(\chi^2) = 0.43$. Assuming also the unitarity:

$$V_{us} = 0.2264 \pm 0.0009 \tag{8}$$

which is in agreement within the errors.

References

1. M. Adinolfi *et al.*, [KLOE Collaboration], *The tracking detector of the KLOE experiment, Nucl. Instrum. Meth* **A 488** 2002 51
2. M. Adinolfi *et al.*, [KLOE Collaboration], *The KLOE electomagnetic calorimeter, Nucl. Instrum. Meth* **A 482** 2002 364
3. M. Adinolfi *et al.*, [KLOE Collaboration], *The trigger system of the KLOE experiment, Nucl. Instrum. Meth* **A 492** 2002 134
4. S. Eidelman *et al.*, Particle Data Group, *Phys. Lett.* **B 592** 2004 1
5. F. Ambrosino *et al.* [KLOE Collaboration], Phys. Lett. B **632** (2006) 76 [arXiv:hep-ex/0509045].
6. H. Leutwyler and M. Roos, Z. Phys. C **25**, 91 (1984).
7. W. J. Marciano, Phys. Rev. Lett. **93** (2004) 231803 [arXiv:hep-ph/0402299].
8. C. Bernard *et al.* [MILC Collaboration], PoS **LAT2005** (2006) 025 [arXiv:hep-lat/0509137].
9. W. J. Marciano and A. Sirlin, Phys. Rev. Lett. **96** (2006) 032002 [arXiv:hep-ph/0510099].

Rare decays at the B-Factories

Concetta Cartaro

University and INFN Trieste – Via Valerio, 2 – 34127 Trieste – Italia
cartaro@ts.infn.it

The present contribution summarizes some results from the BaBar and Belle collaborations on the following categories of rare decays:

- Radiative decays
- Leptonic decays

Hadronic rare decays will not be treated in this contribution.

1 The B-Factories

The BaBar and Belle detectors collect data respectively at the PEPII and KEK e^+e^- high luminosity asymmetric storage rings operating at the $\Upsilon(4S)$ resonance at $10.58\,\text{GeV}$. The $\Upsilon(4S)$ decays in couples of B mesons charged and neutral with the same probability. The two detectors are described in detail elsewhere [1,2]. The total dataset accumulated by the two experiments consists of $350\,\text{fb}^{-1}$ for BaBar and $550\,\text{fb}^{-1}$ for Belle. The analyses here presented are based on samples of about $(88\text{–}232) \times 10^6$ of $B\bar{B}$ couples for BaBar and $(111\text{–}450) \times 10^6$ of $B\bar{B}$ for Belle.

2 Radiative rare decays

In the Standard Model (SM) the radiative decay of a B meson to a strange hadronic state occurs via the electroweak penguin transition $b \rightarrow s\gamma$. This process is interesting because of the possibility of non-Standard-Model high-mass virtual particles in the loop. The SM prediction at the next-to-leading order is $\text{BR}(B \rightarrow X_s\gamma)[E_\gamma > 1.6\,\text{GeV}] = (3.61^{+0.37}_{-0.49}) \times 10^{-4}$. The next-to-next-to-leading-order (NNLO) calculation will decrease the uncertainty to about 5%.

In the b quark rest frame the two decay products, s and γ, would be monochromatic, each with momentum $m_b/2$. But the strong force results in motion, the Fermi momentum p_f, of the b quark inside the B meson, and in a smearing of the E_γ spectrum such that $E_\gamma \approx m_b/2$ and the width (second moment) is related to p_f. The experimental techniques to determine the E_γ spectrum and the $\mathrm{BR}(B \to X_s\gamma)$ are two.

2.1 Semi-inclusive analysis

In the semi-inclusive analysis [3] the X_s final states are reconstructed in 38 decay modes with kaons and pions (both neutral and charged) and η. These modes represent 55% of the total inclusive rate in the region $M(X_s) = 1.1\text{--}2.8\,\mathrm{GeV}$. The missing part of the spectrum, given by the X_s modes that we do not reconstruct, are modelled with a Monte Carlo simulation.

The measured branching fraction for $E_\gamma > 1.9\,\mathrm{GeV}$ is $\mathrm{BR}(b \to s\gamma) = (3.27 \pm 0.18(\mathrm{stat.})^{+0.55}_{-0.40}(\mathrm{syst.})^{+0.04}_{-0.09}(\mathrm{theory})) \times 10^{-4}$. The hadronic mass and photon energy spectra are shown in Fig. 1 where the K^* peak is clearly seen in both.

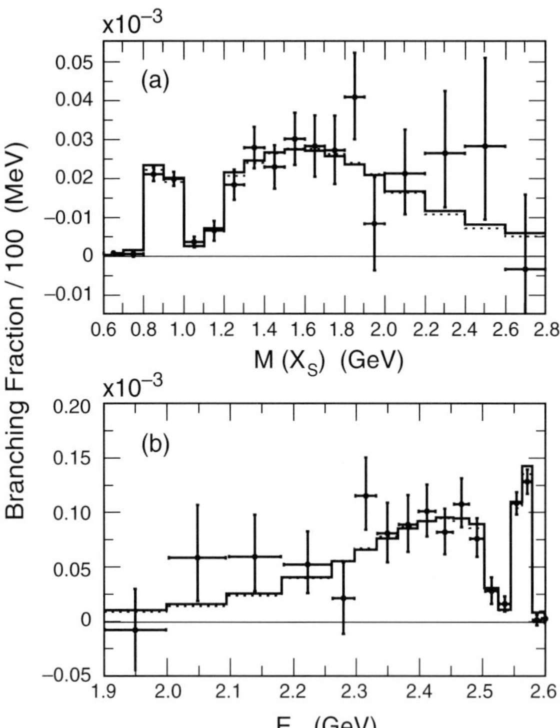

Fig. 1. The hadronic mass spectrum (**a**), and the photon energy spectrum (**b**). The data points are compared to theoretical predictions (*histograms*) obtained using the shape function (*solid line*) and kinetic (*dashed line*) schemes

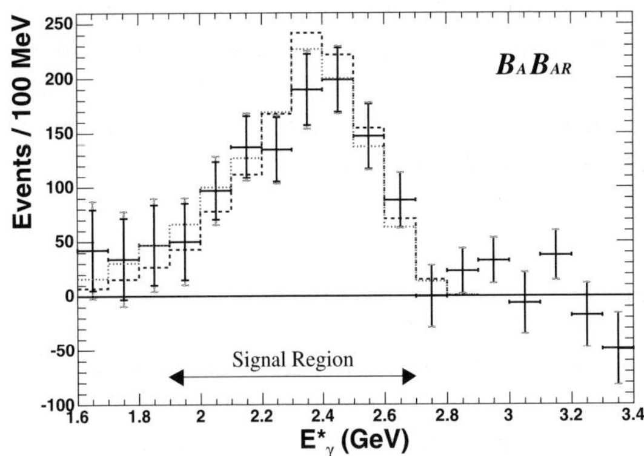

Fig. 2. The photon energy spectrum after background subtraction. The histograms show the spectra for the best fits to the moments in the kinetic scheme (*dashed*) and shape function scheme (*dotted*), normalized to the data in the signal region

2.2 Inclusive analysis

In the inclusive analysis the photon from the decay of one B meson is measured, but the X_s is not reconstructed. This allow the integration of the uncertainties associated with the X_s fragmentation. The identification of a high-momentum lepton from the non-signal B decay strongly reduces the continuum background (including a contribution from initial state radiation). The residual continuum background is modelled using off-resonance data (CM energy 40 MeV below the $\Upsilon(4S)$ resonance) and a Monte Carlo simulation. The measured branching fraction is $\mathrm{BR}(b \to s\gamma) = 3.67 \pm 0.29(\text{stat.}) \pm 0.34(\text{syst.}) \pm 0.29(\text{theory})) \times 10^{-4}$. Figure 2 shows the measured spectrum for signal and control regions after the $B\bar{B}$ and continuum backgrounds have been subtracted.

The Belle collaboration has also performend an inclusive measurement [4]:
$\mathrm{BR}(b \to s\gamma) = (3.55 \pm 0.32(\text{stat.})^{+0.30}_{-0.31}(\text{syst.})^{+0.11}_{-0.07}(\text{theory})) \times 10^{-4}$.

Asymmetries

New physics can also significantly enhance the direct CP asymmetry for $b \to (s,d)\gamma$ decays: $A_{\mathrm{CP}} = \frac{\Gamma(b \to s\gamma + b \to d\gamma) - \Gamma(\bar{b} \to \bar{s}\gamma + \bar{b} \to \bar{d}\gamma)}{\Gamma(b \to s\gamma + b \to d\gamma) + \Gamma(\bar{b} \to \bar{s}\gamma + \bar{b} \to \bar{d}\gamma)}$ where $A_{\mathrm{CP}} \approx 10^{-9}$ in the SM. The direct CP asymmetry $A_{\mathrm{CP}}(B \to X_{s+d}\gamma)$ is measured to be $-0.110 \pm 0.115(\text{stat.}) \pm 0.017(\text{sys.})$. Belle reports $A_{\mathrm{CP}} = (2 \pm 50 \pm 30) \times 10^{-3}$.

2.3 Other rare raditive measurements

The $b \to d\gamma$ process proceeds via a loop diagram in the SM and is suppressed with respect to $b \to s\gamma$ by a factor $|V_{td}/V_{ts}|^2$ where V_{td} and V_{ts} are elements of the Cabibbo–Kobayashi–Maskawa (CKM) matrix. The first observation has been reported by the Belle collaboaration in [5] at 5.1σ. The BaBar collaboration reports a 2σ measurment in [6].

3 Rare Leptonic decays

3.1 $B \to \tau\nu$

In the SM the amplitude of the process is due to the annihilation of the b and $\bar{u}$ quarks into a virtual W boson and is proportional to the product of the CKM element V_{ub} and the B meson decay constant f_B . The resulting expression of the branching fraction is then: $\mathrm{BR}(B \to \tau\nu) = \frac{G_F^2 m_B}{8\pi} m_\tau^2 \left(1 - \frac{m_\tau^2}{m_B^2}\right)^2 f_B^2 V_{ub}^2 \tau_B$ where G_F is the Fermi coupling constant, m_τ and m_B are the τ lepton and the B meson masses, τ_B is the B lifetime. The SM expectation is $\approx 1 \times 10^{-4}$. The $B \to \tau\nu$ decay can also proceed through a Higgs boson providing a constraint on the Higgs mass. The most recent result is by the Belle collaboration that reports an observation [7] at 3.5σ quoting $\mathrm{BR}(B \to \tau\nu) = (1.79^{+56}_{49}(\mathrm{stat})^{+39}_{0.46}(\mathrm{syst})) \times 10^{-4}$. The BaBar colaboration has published a 90% CL upper limit [8]: $\mathrm{BR}(B \to \tau\nu) < 2.8 \times 10^{-4}$ with a central value of $\mathrm{BR}(B \to \tau\nu) = 0.88^{0.68}_{0.67}(\mathrm{stat}) \pm 0.11(\mathrm{syst}) \times 10^{-4}$.

3.2 Other rare leptonic decays

The same experimental techniques that are used to search for $B \to \tau\nu$ decays are useful to search $B \to K\nu\nu$ and $B \to \pi\nu\nu$ as well. The BaBar collaboration reports upper limits on both these decays [9] while the Belle collaboration has reported the most stringent upper limit on $B \to K\nu\nu$ [10]: $\mathrm{BF}_{\mathrm{BaBar}}(B \to K\nu\nu) < 5.2 \times 10^{-5}$, $\mathrm{BF}_{\mathrm{BaBar}}(B \to \pi\nu\nu) < 1.0 \times 10^{-4}$, $\mathrm{BF}_{\mathrm{Belle}}(B \to K\nu\nu) < 3.6 \times 10^{-5}$. All 90% CL.

References

1. B. Aubert *et al.* [BaBar Collaboration], "The BaBar detector," Nucl. Instrum. Meth. A **479** (2002) 1 [arXiv:hep-ex/0105044].
2. A. Abashian *et al.* [Belle Collaboration], "The Belle detector," Nucl. Instrum. Meth. A **479** (2002) 117.
3. B. Aubert *et al.* [BaBar Collaboration], Phys. Rev. D **72**, 052004(2005)
4. P. Koppenburg *et al.* [Belle Collaboration], Phys. Rev. Lett. **93**, 061803(2004).
5. D. Mohapatra *et al.* [Belle Collaboration], Phys. Rev. Lett. **96**, 221601(2006).
6. B. Aubert *et al.* [BaBar Collaboration], Phys. Rev. Lett. **94**, 011801(2005).
7. T. Browder's talk at ICHEP 2006, submitted to Phys. Rev. Lett.
8. B. Aubert *et al.* [BaBar Collaboration], Phys. Rev. D **73**, 057101(2006).
9. B. Aubert *et al.* [BaBar Collaboration], Phys. Rev. Lett. **94**, 101801(2005)
10. HEP2005 Europhysics Conference in Lisbon, Portugal [arXiv:hep-ex/0507034]

CP violation and CKM parameters determination in BaBar

Nicola Neri

INFN Sezione di Pisa, Largo B. Pontecorvo 3, 56127 Pisa, Italy
nicola.neri@pi.infn.it

Introduction

CP violation (CPV), first detected in $K_L \to \pi^+\pi^-$ decays in 1964 [1], is accomodated in the Standard Model (SM) by a CP-violating phase in the CKM matrix, the matrix which describes the mixing of the quarks (under the weak interaction) [2]. From the unitarity constraints of the CKM matrix, we have $V_{ud}V_{ub}^* + V_{cd}V_{cb}^* + V_{td}V_{tb}^* = 0$, the so-called Unitarity Triangle relation, represented in Fig. 1. CPV is proportional to the area of the triangle and requires the angles and sides to be different from zero. The main goal of the BaBar experiment [3] is to overconstrain the Unitarity Triangle parameters, measuring the sides and the angles directly from the data of B_u and B_d meson decays, in order to test the consistency of the SM.

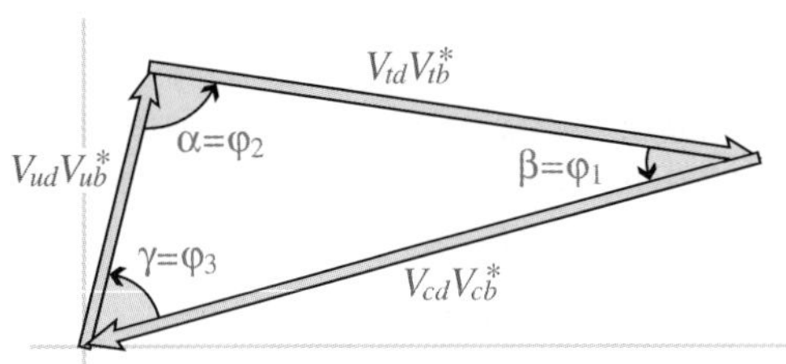

Fig. 1. Graphical representation of the unitarity constraint $V_{ud}V_{ub}^* + V_{cd}V_{cb}^* + V_{td}V_{tb}^* = 0$ as a triangle in the complex plane

1 CP violation in decay and the angle γ

Direct CPV occurs in both charged and neutral B meson decays when the amplitude for a decay and its CP-conjugate process have different magnitudes. BaBar provided the first evidence of direct CPV in $B^0 \to K^-\pi^+$ decays [4], which was then confirmed by Belle [5]. The measured asymmetry is $A_{K\pi} = \frac{n_{K^-\pi^+} - n_{K^+\pi^-}}{n_{K^-\pi^+} + n_{K^+\pi^-}} = -0.133 \pm 0.030 \pm 0.009$. CPV in decay arises from the interference of tree and penguin diagrams with non-vanishing CP-even and CP-odd relative phases. It is not possible to extract any CKM parameter from this measurement due to large theorethical uncertainties in the hadronic parameterization of the decay amplitude.

BaBar was able to measure the CKM angle $\gamma \equiv \arg\left(-\frac{V_{ud}V_{ub}^*}{V_{cd}V_{cb}^*}\right)$ studying the $B^- \to D^{(*)}K^{(*)-}$ decay chain, reconstructing the D in a final state f accessible by both the D^0 and $\bar{D}^0$. The sensitivity to γ comes from the interference between $b \to c\bar{u}s$ and the $b \to u\bar{c}s$ amplitudes. The state f can be a CP eigenstate (GLW method) [6,7], or a Double-Cabibbo-Suppressed state (ADS method) [8], or a three-body final state (Dalitz plot method) [9]. The Dalitz plot method has the highest sensitivity to the angle γ and requires the analysis of the distribution of the $B^- \to D^{(*)}K^{(*)-}$ events in the $D^0 \to K_S^0\pi^+\pi^-$ Dalitz plane. This analysis measures $\gamma = (92\pm41\pm11\pm12)$ degrees, where the first error is statistical, the second is experimental and the third is due to the D^0 decay model [10]. The combination of all the different methods yields $\gamma = (73 \pm 29)$ degrees [11].

2 CP violation in mixing

CPV in mixing occurs when the two neutral mass eigenstates cannot be chosen to be CP eigenstates. We define the mass eigenstates B_H and B_L,

$$|B_H\rangle = p\sqrt{1-z}|B^0\rangle + q\sqrt{1+z}|\bar{B}^0\rangle \quad |B_L\rangle = p\sqrt{1+z}|B^0\rangle - q\sqrt{1-z}|\bar{B}^0\rangle .$$

CPT invariance requires the complex parameter z to be zero. Similarly, T invariance leads to $\frac{|q|}{|p|} = 1$, while CP invariance requires both $\frac{|q|}{|p|} = 1$ and $z = 0$. While CPV in mixing has been observed in the neutral kaon system, there is currently no experimental evidence of CPV in mixing (or T or CPT violation) in the neutral B meson system. Reconstructing the inclusive decays of $B^0 \to Xl\nu$ it is possible to measure $A_{T/CP}(\Delta t) = \frac{N(l^+l^+) - N(l^-l^-)}{N(l^+l^+) + N(l^-l^-)} \simeq \frac{1-|q/p|^4}{1+|q/p|^4}$, where $N(l^\pm l^\pm)$ are the events with same lepton charge from the B decays. The sign of the lepton charge is correlated with the flavor of the neutral B meson and it is sensitive to the mixing probability. Another observable, is sensitive to CP/CPT violation $A_{CP/CPT}(|\Delta t|) = \frac{N(l^+l^-,\Delta t>0) - N(l^+l^-,\Delta t<0)}{N(l^+l^+,\Delta t>0) + N(l^-l^-,\Delta t<0)} \simeq 2\frac{\mathrm{Im}(z)\sin(\Delta m\Delta t) - \mathrm{Re}(z)\sinh(\Delta\Gamma\Delta t/2)}{\cosh(\Delta\Gamma\Delta t/2) + \cos(\Delta m\Delta t)}$. BaBar measures $|q/p| = 1.008 \pm 0.027 \pm 0.019$, $\mathrm{Im}(z) = 0.0139 \pm 0.0073 \pm 0.0032$ and $\Delta\Gamma \times \mathrm{Re}(z) = -0.0071 \pm 0.0039 \pm 0.0020\,\mathrm{ps}^{-1}$ [12]. These measurements help to constrain possible models of New Physics.

3 CP violation in the interference between decays with and without mixing

A third type of CPV occurs in decays into final states that are common to B^0 and $\bar{B}^0$, for example neutral B decays into CP eigenstates, f_{CP}. The physically meaningful quantity of interest, independent of phase conventions is $\lambda \equiv \frac{q}{p}\frac{\bar{A}_{f_{CP}}}{A_{f_{CP}}} \equiv \eta_{f_{CP}}\frac{q}{p}\frac{\bar{A}_{\bar{f}_{CP}}}{A_{f_{CP}}}$, with $\eta_{f_{CP}}$ the CP eigenvalue of the final state f. When CP is conserved, $|q/p| = 1$, $|\bar{A}_{\bar{f}_{CP}}/A_{f_{CP}}| = 1$, and the relative phase between (q/p) and $(\bar{A}_{\bar{f}_{CP}}/A_{f_{CP}})$ vanishes. Therefore, $\lambda \neq \pm 1$ implies CPV. It is possible to define a time-dependent asymmetry, $a_{f_{CP}} = \frac{\Gamma(B^0_{\text{phys}}(t)\to f_{CP})-\Gamma(\bar{B}^0_{\text{phys}}(t)\to f_{CP})}{\Gamma(B^0_{\text{phys}}(t)\to f_{CP})+\Gamma(\bar{B}^0_{\text{phys}}(t)\to f_{CP})}$, where $B^0_{\text{phys}}(t)$ is the state vector of a B^0 meson at time $t = t_{f_{CP}} - t_{\text{tag}}$. The flavor of the B_{CP} meson at time $t = 0$ is determined from the other B of the event (B_{tag}). The BaBar detector is able to measure the time difference t of the two B decays reconstructing their decay vertex positions, separated by the boost of the center of mass system ($\beta\gamma = 0.56$). The silicon vertex tracker (SVT), reconstructs the decay vertices of the B mesons achieving a resolution on the spatial separation ($\sim 100\,\mu$m), which is fundamental for these measurements. The time-dependent asymmetry is given by $a_{f_{CP}} = \frac{(1-|\lambda_{f_{CP}}|^2)\cos(\Delta m_B t)-2\text{Im}(\lambda_{f_{CP}})\sin(\Delta m_B t)}{1+|\lambda_{f_{CP}}|^2}$ and is sensitive to the physics parameter λ.

3.1 The angle β

In the decay modes which proceed via $b \to c\bar{c}s$ transitions, we have $\text{Im}(\lambda) = \eta_{f_{CP}}\sin 2\beta$. BaBar has measured $\sin 2\beta = 0.710 \pm 0.034 \pm 0.019$ and $\lambda = 0.932 \pm 0.026 \pm 0.017$ [13] reconstructing CP eigeinstates containing a charmonium meson. Decay modes which proceed via $b \to d\bar{d}s$ or $b \to s\bar{s}s$ penguin diagrams, could have contributions from particles not predicted in the SM. The precise measurement of time dependent CP asymmetries for those decay modes compared with the results of the charmonium modes could reveal effects of New Physics. BaBar measured CP asymmetries in several penguin modes: $\sin 2\beta_{\text{eff}} = 0.55 \pm 0.11 \pm 0.02$ in $\eta'K^0$ [14], $\sin 2\beta_{\text{eff}} = 0.50 \pm 0.25^{+0.07}_{-0.04}$ in ϕK^0 [15], and $\sin 2\beta_{\text{eff}} = -0.66 \pm 0.26 \pm 0.08$ in $K^0_S K^0_S K^0_S$ [16]. Results are compatible with SM expectations within the errors.

3.2 The angle α

The angle α can be measured in $B^0 \to \rho^-\rho^+$ and $B^0 \to \pi^-\pi^+$ time-dependent CP-asymmetries. The decay amplitude has penguing contributions which make the extraction of the weak phase from the CP-asymmetry measurement difficult. An isospin analysys, which requires also the measurement of the branching ratios of isospin related modes, could in principle extract the value of α [17]. This is currently not possible with the available statistics, instead we can evaluate the penguin pollution measuring the ratio of the branching ratios $r = \frac{\text{BR}(B^0 \to h^0 h^0)}{\text{BR}(B^\pm \to h^\pm h^0)}$ $(h = \pi, \rho)$ and bound effect on the measured value of α [19]. In the case of $B \to \pi^+ \pi^-$ BaBar constrain $\alpha \in [0, 27] \cup [63, 180]$ degrees [18] while the case of $B^0 \to \rho^+ \rho^-$ BaBar measures $\alpha \in [74, 117]$ degrees [20, 21]. A time-dependent Dalitz plot analysis of the $B \to \pi^+ \pi^- \pi^0$ decay, which takes advantage of the interference between the ρ resonances in the Dalitz plot, is also sensitive to the angle α and measures $\alpha \in [75, 152]$ degrees [22].

4 Perspectives and conclusions

CPV has been estabilished in the B meson system. Measurement of the angle β has reached a good precision, while the measurements of the angles α and γ require more statistics. Several approaches have been explored and with $1 \ \text{ab}^{-1}$ data BaBar can reach $\simeq 10{-}15$ degree precision on both α and γ. The present BaBar results are fully compatible with the SM predictions. New Physics searches require much larger statistics, $\mathcal{O}(50) \ \text{ab}^{-1}$, achievable with a SuperB Factory [23].

References

1. J. H. Christenson, J. W. Cronin, V. L. Fitch and R. Turlay,Phys. Rev. Lett. **13**, 138 (1964).
2. N. Cabibbo, *Phys. ReV. Lett.* **10**, 531 (1963);
   M. Kobayashi and T. Maskawa, *Prog. Theor. Phys.* **49**, 652 (1973).
3. B. Aubert *et al.* [BABAR Collaboration], Nucl. Instrum. Meth. A **479**, 1 (2002)
4. B. Aubert *et al.* [BABAR Collaboration],Phys. R ev. Lett. **87**, 091801 (2001)
5. K. Abe *et al.* [Belle Collaboration],Phys. Rev. Lett. **87**, 091802 (2001)
6. M. Gronau and D. London, *Phys. Lett.* **B253**, 483 (1991).
7. M. Gronau and D. Wyler, *Phys. Lett.* **B265**, 172 (1991).
8. D. Atwood, I. Dunietz and A. Soni, *Phys. Rev. Let t.* **78**, 3257 (1997);
9. A. Giri, Yu. Grossman, A. Soffer and J. Zupan, *Ph ys. Rev.* **D68**, 054018 (2003).
10. B. Aubert [BABAR Collaboration], arXiv:hep-ex/0607104.
11. M. Ciuchini *et al.*, JHEP **0107**, 013 (2001)
12. B. Aubert *et al.* [BABAR Collaboration], Phys. Rev. Lett. **96**, 251802 (2006)
13. B. Aubert *et al.* [BABAR Collaboration], arXiv:hep-ex/0607107.
14. B. Aubert *et al.* [BABAR Collaboration], arXiv:hep-ex/0607100.
15. B. Aubert *et al.* [BABAR Collaboration], Phys. Rev. D **71**, 091102 (2005)
16. B. Aubert *et al.* [BABAR Collaboration], arXiv:hep-ex/0607108.
17. M. Gronau and D. London, Phys. Rev. Lett. **65**, 3381 (1990).
18. B. Aubert *et al.* [BABAR Collaboration], arXiv:hep-ex/0607106.
19. Y. Grossman and H. R. Quinn, Phys. Rev. D **58**, 017504 (1998)
20. B. Aubert *et al.* [BABAR Collaboration], arXiv:hep-ex/0607097.
21. B. Aubert *et al.* [BABAR Collaboration], arXiv:hep-ex/0607098.
22. B. Aubert [BABAR Collaboration], arXiv:hep-ex/0608002.
23. J. Albert *et al.*, arXiv:physics/0512235.

Semileptonic and nonleptonic decays of B_c

Mikhail A. Ivanov[1], Jürgen G. Körner[2], and Pietro Santorelli[3]

[1] Bogoliubov Laboratory of Theoretical Physics, Joint Institute for Nuclear Research, 141980 Dubna, Russia
[2] Institut für Physik, Johannes Gutenberg-Universität, D-55099 Mainz, Germany
[3] Dipartimento di Scienze Fisiche, Università di Napoli "Federico II" and INFN, Sezione di Napoli, Via Cintia, I-80126 Napoli, Italy
   `Pietro.Santorelli@na.infn.it`

Summary. Using our relativistic constituent quark model we present results on the exclusive nonleptonic and semileptonic decays of the B_c-meson. The nonleptonic decays are studied in the framework of the factorization approximation. We calculate the branching ratios for a large set of exclusive nonleptonic and semileptonic decays of the B_c meson and compare our results with the results of other models.

1 Introduction

The B_c-meson is the lowest bound state of two heavy quarks (charm and bottom) with open flavor. The B_c-meson therefore decays weakly via (i) b-quark decay, (ii) c-quark decay, and (iii) the annihilation channel. Starting from the pioneering paper [1], the modern state of art in the spectroscopy, production and decays of the B_c-meson can be found in the review [2].

The first observation of the B_c meson was reported by the CDF Collaboration at Fermilab [3] in the semileptonic decay mode $B_c \to J/\psi + l + \nu$ with the J/ψ decaying into muon pairs. Values for the mass and the lifetime of the B_c meson were given as $M(B_c) = 6.40 \pm 0.39 \pm 0.13\,\mathrm{GeV}$ and $\tau(B_c) = 0.46^{+0.18}_{-0.16}(\mathrm{stat})\pm 0.03(\mathrm{syst})\,\mathrm{ps}$. Recently, CDF reported new value for the mass of B_c meson, $6.2857 \pm 0.0053(\mathrm{stat.}) \pm 0.0012(\mathrm{syst.})\,\mathrm{GeV}$ with errors significantly smaller than in the first measurement. Also D0 has observed the B_c in the semileptonic mode $B_c \to J/\psi + \mu + X$ and reported preliminary evidence that $M(B_c) = 5.95^{+0.14}_{-0.13}\pm 0.34\,\mathrm{GeV}$ and $\tau(B_c) = 0.45^{+0.12}_{-0.10} \pm 0.12\,\mathrm{ps}$ [5].

In the following we report on the results of an analysis of almost all accessible low-lying exclusive nonleptonic two-body and semileptonic three-body modes of the B_c-decays [6] within our relativistic constituent quark model [7–10]. In [6] we updated the free parameters of the model by using the latest experimental data on the B_c-mass [4] and the weak decay constant f_D [11]. We give a set of numerical values for the leptonic, semileptonic and nonleptonic partial decay widths of the B_c-meson and compare them with the results of other approaches.

2 Results and discussions

The constituent relativistic quark model we employ to study B_c decays was developed in [7–10] and successfully applied to a very large class of weak decays (see for example [12]). For technical details regarding the model we refer the interested reader to [6]. Here we present our results on the semileptonic

Table 1. Branching ratios (in %) of exclusive semileptonic B_c decays. ψ indicates $\psi(3836)$. $\tau(B_c) = 0.45$ ps

Mode	[6]	[13, 14]	[15]	[16]	[17]	[18]
$B_c^- \to \eta_c e\nu$	0.81	0.75	0.97	0.40	0.76	0.51
$B_c^- \to \eta_c \tau\nu$	0.22	0.23	–	–	–	–
$B_c^- \to J/\psi e\nu$	2.07	1.9	2.35	1.21	2.01	1.44
$B_c^- \to J/\psi \tau\nu$	0.49	0.48	–	–	–	–
$B_c^- \to \overline{D}^0 e\nu$	0.0035	0.004	0.006	0.001	0.003	0.0014
$B_c^- \to \overline{D}^0 \tau\nu$	0.0021	0.002	–	–	–	–
$B_c^- \to \overline{D}^{*0} e\nu$	0.0038	0.018	0.018	0.008	0.013	0.0023
$B_c^- \to \overline{D}^{*0} \tau\nu$	0.0022	0.008	–	–	–	–
$B_c^- \to \overline{B}_s^0 e\nu$	1.10	4.03	1.82	0.82	0.98	0.92
$B_c^- \to \overline{B}_s^{*0} e\nu$	2.37	5.06	3.01	1.71	3.45	1.41
$B_c^- \to \overline{B}^0 e\nu$	0.071	0.34	0.16	0.04	0.078	0.048
$B_c^- \to \overline{B}^{*0} e\nu$	0.063	0.58	0.23	0.12	0.24	0.051

Mode	[6]	[19]
$B_c^- \to \chi_{c0} e\nu$	0.17	0.12
$B_c^- \to \chi_{c0} \tau\nu$	0.013	0.017
$B_c^- \to \chi_{c1} e\nu$	0.092	0.15
$B_c^- \to \chi_{c1} \tau\nu$	0.0089	0.024
$B_c^- \to h_c e\nu$	0.27	0.17
$B_c^- \to h_c \tau\nu$	0.017	0.024
$B_c^- \to \chi_{c2} e\nu$	0.17	0.19
$B_c^- \to \chi_{c2} \tau\nu$	0.0082	0.029
$B_c^- \to \psi e\nu$	0.0066	–
$B_c^- \to \psi \tau\nu$	9.9×10^{-5}	–

decays into charmonia, into $(\overline{B}_s^0, \overline{B}_s^{0*}, \overline{D}^0, \overline{D}^{0*}, \overline{B}^0, \overline{B}^{0*})$(cf. Table 1) and on the two body nonleptonic B_c decays (cf. Tables 2 and 3). Tables 2 and 3 contain numerical results corresponding to processes with branching ratios larger than 0.1%. For the complete list see the tables in [6].

Table 2. Branching ratios (in %) of exclusive nonleptonic B_c decays with the choice of Wilson coefficient: $a_1^c = 1.20$ and $a_2^c = -0.317$ for c-decay, and $a_1^b = 1.14$ and $a_2^b = -0.20$ for b-decay. Modes with branching ratios smaller than 0.1% can be found in Table V of [6]

Mode	This work	[13, 14]	[15]	[16]	[17]
$B_c^- \to \eta_c \pi^-$	0.19	0.20	0.18	0.083	0.14
$B_c^- \to \eta_c \rho^-$	0.45	0.42	0.49	0.20	0.33
$B_c^- \to J/\psi \pi^-$	0.17	0.13	0.18	0.060	0.11
$B_c^- \to J/\psi \rho^-$	0.49	0.40	0.53	0.16	0.31
$B_c^- \to \eta_c D_s^-$	0.44	0.28	0.054	–	0.26
$B_c^- \to \eta_c D_s^{*-}$	0.37	0.27	0.044	–	0.24
$B_c^- \to J/\psi D_s^-$	0.34	0.17	0.041	–	0.15
$B_c^- \to J/\psi D_s^{*-}$	0.97	0.67	–	–	0.55
$B_c^- \to \overline{B}_s^0 \pi^-$	3.9	16.4	5.75	2.46	1.56
$B_c^- \to \overline{B}_s^0 \rho^-$	2.3	7.2	4.41	1.38	3.86
$B_c^- \to \overline{B}_s^{*0} \pi^-$	2.1	6.5	5.08	1.58	1.23
$B_c^- \to \overline{B}_s^{*0} \rho^-$	11	20.2	14.8	10.8	16.8
$B_c^- \to \overline{B}_s^0 K^-$	0.29	1.06	0.41	0.21	0.17
$B_c^- \to \overline{B}_s^{*0} K^-$	0.13	0.37	0.29	0.11	0.13
$B_c^- \to \overline{B}_s^{*0} K^{*-}$	0.50	–	–	–	1.14
$B_c^- \to \overline{B}^0 \pi^-$	0.20	1.06	0.32	0.10	0.10
$B_c^- \to \overline{B}^0 \rho^-$	0.20	0.96	0.59	0.13	0.28
$B_c^- \to \overline{B}^{*0} \rho^-$	0.30	2.57	1.17	0.67	0.89
$B_c^- \to B^- K^0$	0.38	1.98	0.66	0.23	0.27
$B_c^- \to B^- K^{*0}$	0.11	0.43	0.47	0.09	0.32
$B_c^- \to B^{*-} K^{*0}$	0.32	1.67	0.97	0.82	1.70

Table 3. The same as of Table 2

Mode	[6]	[19]	[20]	[21]
$B_c^- \to h_c \pi^-$	0.11	0.05	1.60	–
$B_c^- \to \chi_{c0} \rho^-$	0.13	0.072	3.29	–
$B_c^- \to h_c \rho^-$	0.25	0.12	5.33	–
$B_c^- \to \chi_{c2} \rho^-$	0.12	0.051	3.20	0.023

From the tables we observe that our results are generally close to the QCD sum rule results of [13, 14] and the constituent quark model results of [15–17] for the $b \to c$ induced decays. In exception are the $(b \to c; c \to (s, d))$ results of [15] which are considerably smaller than our results, and smaller than the results of the other model calculations. Summing up the exclusive contributions one obtains a branching fraction of 8.8%. Considering the fact that the $b \to c$ contribution to the total rate is expected to be about 20% [2] this leqaves plenty of room for nonresonant multibody decays

For the $c \to s$ induced decays our branching ratios are considerably smaller than those predicted by QCD sum rules [13, 14] but are generally close to the other constituent quark model results. When we sum up our exclusive branching fractions we obtain a total branching ratio of 27.6% which has to be compared with the 70% expected for the $c \to s$ contribution to the total rate [2]. The sum rule model of [13, 14] gives a summed branching fraction of 73.4% for the $c \to s$ contribution, i. e. the model of [13, 14] predicts that the exclusive channels pretty well saturate the $c \to s$ part of the total rate.

3 Conclusions

In the coming few years one can expect large data samples on exclusive B_c decays at the TEVATRON and at the LHC. We are looking forward to a comparison of our model results with the upcoming experimental data.

References

1. M. Lusignoli and M. Masetti, Z. Phys. C **51**, 549 (1991).
2. V. V. Kiselev in N. Brambilla *et al.*, CERN Yellow Report "Heavy quarkonium physics", CERN-2005-005, Geneva: CERN, 2005. 487 p. arXiv:hep-ph/0412158.
3. F. Abe *et al.* [CDF Collaboration], Phys. Rev. D **58**, (1998) 112004 [arXiv:hep-ex/9804014]; Phys. Rev. Lett. **81**, (1998) 2432 [arXiv:hep-ex/9805034].
4. D. Acosta *et al.* [CDF Collaboration], arXiv:hep-ex/0505076.
5. M. D. Corcoran [CDF Collaboration], arXiv:hep-ex/0506061.
6. M. A. Ivanov, J. G. Körner and P. Santorelli, Phys. Rev. D **73**, 054024 (2006) [arXiv:hep-ph/0602050].
7. M. A. Ivanov, J. G. Körner and P. Santorelli, Phys. Rev. D **63**, 074010 (2001) [arXiv:hep-ph/0007169].
8. A. Faessler, T. Gutsche, M. A. Ivanov, J. G. Körner and V. E. Lyubovitskij, Eur. Phys. J. directC **4**, 18 (2002) [arXiv:hep-ph/0205287].
9. M. A. Ivanov, J. G. Körner and O. N. Pakhomova, Phys. Lett. B **555**, 189 (2003) [arXiv:hep-ph/0212291].
10. M. A. Ivanov, J. G. Körner and P. Santorelli, Phys. Rev. D **71**, 094006 (2005) [arXiv:hep-ph/0501051].
11. M. Artuso *et al.* [CLEO Collaboration], Phys. Rev. Lett. **95** (2005) 251801 [arXiv:hep-ex/0508057].

12. M. A. Ivanov, P. Santorelli, Phys. Lett. B **456**, 248 (1999) [arXiv:hep-ph/9903446].

13. V. V. Kiselev, A. E. Kovalsky and A. K. Likhoded, Nucl. Phys. B **585**, 353 (2000) [arXiv:hep-ph/0002127]; arXiv:hep-ph/0006104.

14. V. V. Kiselev, arXiv:hep-ph/0211021.

15. C. H. Chang and Y. Q. Chen, Phys. Rev. D **49**, 3399 (1994).

16. D. Ebert, R. N. Faustov and V. O. Galkin, Mod. Phys. Lett. A **17**, 803 (2002) [arXiv:hep-ph/0204167]; Eur. Phys. J. C **32**, 29 (2003) [arXiv:hep-ph/0308149]; Phys. Rev. D **68**, 094020 (2003) [arXiv:hep-ph/0306306].

17. A. Abd El-Hady, J. H. Munoz and J. P. Vary, Phys. Rev. D **62**, 014019 (2000) [arXiv:hep-ph/9909406].

18. M. A. Nobes and R. M. Woloshyn, J. Phys. G **26**, 1079 (2000) [arXiv:hep-ph/0005056].

19. C. H. Chang, Y. Q. Chen, G. L. Wang and H. S. Zong, Phys. Rev. D **65**, 014017 (2002) [arXiv:hep-ph/0103036]; Commun. Theor. Phys. **35**, 395 (2001) [arXiv:hep-ph/0102150].

20. V. V. Kiselev, O. N. Pakhomova and V. A. Saleev, J. Phys. G **28**, 595 (2002) [arXiv:hep-ph/0110180].

21. G. Lopez Castro, H. B. Mayorga and J. H. Munoz, J. Phys. G **28**, 2241 (2002) [arXiv:hep-ph/0205273].

Aspects of non leptonic B_s decays

R. Ferrandes

Phys. Dept. University of Bari and INFN, via Orabona 4, I-70126 Bari, Italy
`rossella.ferrandes@ba.infn.it`

Non perturbative strong interaction effects make difficult a theoretical description of non leptonic weak decays of hadrons; so it is relevant the possibility to obtain predictive parametrizations of decay amplitudes. It is noticeable that the widths of a set of two-body B_s transitions can be predicted using the symmetries of QCD and available information on B decays [1].

We consider in particular a class of decay modes induced by the transitions $b \to c\bar{u}q$ ($q = d$ or s) which are collected in Table 1. These quark transitions are described by the following effective hamiltonian, obtained by a renormalization group evolution from the electroweak scale down to $\mu = m_b$: $H = \frac{G_F}{\sqrt{2}} V_{cb} V_{ud}^* [c_1(\mu) Q_1 + c_2(\mu) Q_2]$, where $Q_1 = (\bar{c}_i b_i)_{V-A} (\bar{d}_j u_j)_{V-A}$ and $Q_2 = (\bar{c}_i b_j)_{V-A} (\bar{d}_j u_i)_{V-A}$.

The most popular approach to compute hadronic matrix element of four quark operators is based on naive factorization [2], consisting in factoriz-

Table 1. $SU(3)$ amplitudes for B^-, $\overline{B}^0$ and $\overline{B}_s^0$ decays to $D_{(s)} P$ (P is a light pseudoscalar meson), induced by $b \to c\bar{u}d(s)$ transitions. The predicted $\overline{B}_s^0$ branching fractions are also reported

$B^-, \overline{B}^0$	amplitude	$\mathrm{BR}_{\exp}$ (10^{-4})	$\overline{B}_s^0$	amplitude	$\mathrm{BR}_{\mathrm{th}}$ (10^{-4})
$D^0 \pi^-$	$V_{ud}^* V_{cb} (C + T)$	49.8 ± 2.9	$D_s^+ \pi^-$	$V_{ud}^* V_{cb} T$	29 ± 6
$D^0 \pi^0$	$V_{ud}^* V_{cb} \frac{(C-E)}{\sqrt{2}}$	2.91 ± 0.28	$D^0 \overline{K}^0$	$V_{ud}^* V_{cb} C$	8.1 ± 1.8
$D^+ \pi^-$	$V_{ud}^* V_{cb} (T + E)$	27.6 ± 2.5	$D^0 \eta_8$	$V_{us}^* V_{cb} \frac{(2C-E)}{\sqrt{6}}$	
$D_s^+ K^-$	$V_{ud}^* V_{cb} E$	0.38 ± 0.13	$D^0 \eta_0$	$V_{us}^* V_{cb} D$	
$D^0 \eta_8$	$-V_{ud}^* V_{cb} \frac{(C+E)}{\sqrt{6}}$		$D^0 \eta$		0.21 ± 0.12
$D^0 \eta_0$	$V_{ud}^* V_{cb} D$		$D^0 \eta'$		0.10 ± 0.08
$D^0 \eta$		2.2 ± 0.5	$D^0 \pi^0$	$-V_{us}^* V_{cb} \frac{E}{\sqrt{2}}$	0.010 ± 0.003
$D^0 \eta'$		1.7 ± 0.4	$D^+ \pi^-$	$V_{us}^* V_{cb} E$	0.020 ± 0.006
$D^0 K^-$	$V_{us}^* V_{cb} (C + T)$	3.7 ± 0.6	$D_s^+ K^-$	$V_{us}^* V_{cb} (T + E)$	1.8 ± 0.3
$D^0 \overline{K}^0$	$V_{us}^* V_{cb} C$	0.50 ± 0.14			

ing them in current matrix elements determined in terms of meson decay constants and semileptonic form factors. This approach has some troubles: firstly, the Wilson coefficient $c_i(\mu)$ depend on renormalization scale μ, while form factors, decay constants and physical amplitudes are scale independent. Moreover, the amplitudes are real, so that strong phases are neglected; finally, annihilation amplitudes are predicted to be tiny. Instead, data point to sizeable strong phases and to non negligible annihilation contributions, as we shall see.

An alternative approach is a model independent analysis based on flavor symmetry and experimental data. The key observation is that the various B_s decay modes are governed, in the $SU(3)_F$ limit, by few independent amplitudes [3] that can be constrained, both in moduli and in phase differences, from corresponding B decay processes. Since $\overline{B} \to DP$ decays induced by $b \to c\bar{u}d(s)$ transitions involve a weak Hamiltonian transforming as a flavor octet, using the notation $T_\nu^{(\mu)}$ for the $\nu = (Y, I, I_3)$ component of an irreducible tensor operator of rank (μ) [4], one can write: $H_W = V_{cb}V_{ud}^* T^{(8)}_{0-1-1} + V_{cb}V_{us}^* T^{(8)}_{-1-\frac{1}{2}-\frac{1}{2}}$. When combined with the initial $\overline{B}$ mesons, which form a (3^*)-representation of $SU(3)$, this leads to (3^*), (6) and (15^*) representations, which are also those formed by the combination of the final octet light pseudoscalar meson and triplet D meson. Therefore, using the Wigner–Eckart theorem, the decay amplitudes can be written as linear combinations of three reduced amplitudes $\langle \phi^{(\mu)} | O^{(8)} | B^{(3^*)} \rangle$, with $\mu = 3^*, 6, 15^*$. By appropriate linear combinations of these amplitudes one can obtain a correspondence with the color suppressed, color enhanced and W-exchange diagrams, C, T and E, respectively, as in Table 1. The transition in the $SU(3)$ singlet η_0 involves another amplitude D in principle not related to the previous ones. The $SU(3)$ representation for B decays is reported in Table 1.

We note that $\overline{B} \to D_s K$ only fixes the modulus of E, which is not small at odds with the expectations by factorization, where W-exchange processes are suppressed by ratios of decay constants and form factors and are usually considered to be negligible. What can be done is to use all the information on $\overline{B} \to D\pi, D_s K$ and DK (7 experimental data) to determine T, C and E (5 parameters). A similar strategy has been recently adopted in [6]. Noticeably, the combined experimental information is enough accurate to tightly determine the ranges of variation for all these quantities. In Fig. 1 we have depicted the allowed regions in the C/T and E/T planes, obtained fixing the other variables to their fitted values, with the corresponding confidence levels. It is worth noticing that the phase differences between the various amplitudes are close to be maximal; this signals sizeable deviation from naive (or generalized) factorization, provides contraints to QCD-based approaches proposed to evaluate non leptonic B decay amplitudes [7–9] and points towards large long-distance effects in C and E [10]. We obtain $|\frac{C}{T}| = 0.53 \pm 0.10$, $|\frac{E}{T}| = 0.115 \pm 0.020$, $\delta_C - \delta_T = (76 \pm 12)°$ and $\delta_E - \delta_T = (112 \pm 46)°$.

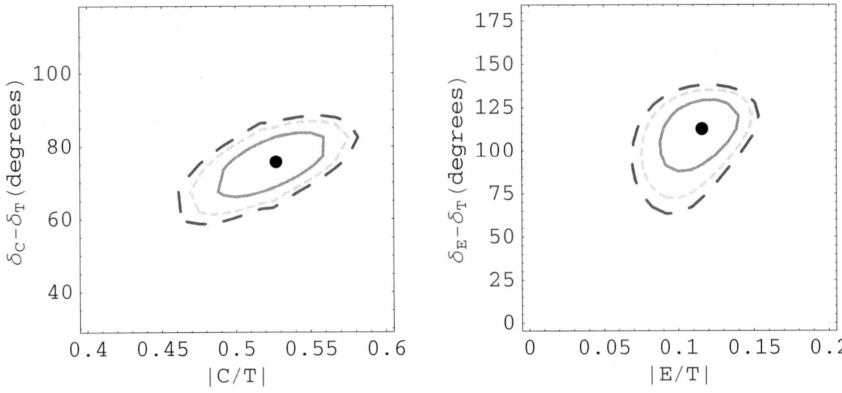

Fig. 1. Ratios of $SU(3)$ amplitudes obtained from B data in Table 1. The contours correspond to the confidence level of 68% (*continuous line*), 90% (*dashed line*) and 95% (*long-dashed line*); the *dots* show the result of the fit

With the results for the amplitudes we can determine a number of B_s decay rates, and the predictions are collected in Table 1. The uncertainties in the predicted rates are small; in particular, the W-exchange induced processes $\overline{B}^0_s \to D^+\pi^-, D^0\pi^0$ are precisely estimated [11]. The predicted ratio $\frac{\Gamma(B_s \to D_s^-\pi^+)}{\Gamma(B^0 \to D^-\pi^+)} = 1.05 \pm 0.24$ can be compared to the experimental value $1.32 \pm 0.18 \pm 0.38$ recently obtained by CDF Collaboration [12].

Table 2. Experimental branching fractions of $\bar{B} \to D_{(s)}V,\ D^*_{(s)}P$ decays and predictions for $\overline{B}^0_s$ decays

$B^-, \overline{B}^0$	BR_{exp} (10^{-4})	$\overline{B}^0_s$	BR_{th} (10^{-4})
$D^0\rho^-$	134 ± 18	$D_s^+\rho^-$	72 ± 35
$D^0\rho^0$	2.9 ± 1.1	$D^0\overline{K}^{*0}$	9.6 ± 2.4
$D^+\rho^-$	77 ± 13		
$D_s K^{*-}$	< 9.9		
$D^0 K^{*-}$	6.1 ± 2.3	$D^0\rho^0$	0.28 ± 1.4
$D^0\overline{K}^{*0}$	0.48 ± 0.12	$D^+\rho^-$	0.6 ± 2.8
$D^+ K^{*-}$	3.7 ± 1.8	$D_s^+ K^{*-}$	4.5 ± 3.1
$D^{*0}\pi^-$	46 ± 4	$D_s^{*+}\pi^-$	32 ± 2
$D^{*0}\pi^0$	2.7 ± 0.5	$D^{*0}\overline{K}^0$	4.7 ± 2.2
$D^{*+}\pi^-$	27.6 ± 2.1		
$D_s^* K^-$	0.20 ± 0.06		
$D^{*0} K^-$	3.6 ± 1.0	$D^{*0}\pi^0$	0.0057 ± 0.0017
		$D^{*+}\pi^-$	0.0115 ± 0.0035
$D^{*+} K^-$	2.0 ± 0.5	$D_s^{*+} K^-$	1.3 ± 0.2

We have performed an analogous analysis for $\bar{B} \to D_{(s)}V$, $D^*_{(s)}P$ decays, for which the same $SU(3)$ decomposition holds, with amplitudes T', C', E'. B decay data are collected in Table 2, including the recently observed W-exchange mode $\bar{B}^0 \to D^*_s K^-$ [14], together with the predictions for B_s decays.

$SU(3)_\mathrm{F}$ breaking terms can modify our predictions. Those effects in general cannot be reduced to well defined and predictable patterns without new assumptions. Their parametrization would introduce additional quantities [13] that at present cannot be sensibly bounded since their effects seem to be smaller than the experimental uncertainties. It will be interesting to investigate their role when the B_s decay rates will be measured and more precise B branching fractions will be available.

Acknowledgement. I thank P. Colangelo for collaboration and F. De Fazio for useful discussions.

References

1. P. Colangelo, R. Ferrandes, Phys. Lett. B **627**, 77 (2005).
2. M. Neubert and B. Stech, Adv. Ser. Direct. High Energy Phys. **15**, 294 (1998).
3. D. Zeppenfeld, Z. Phys. C **8**, 77 (1981);
   B. Grinstein and R. F. Lebed, Phys. Rev. D **53**, 6344 (1996).
4. J. J. de Swart, Rev. Mod. Phys. **35**, 916 (1963).
5. S. Eidelman *et al.* [Particle Data Group], Phys. Lett. B **592**, 1 (2004).
6. C. S. Kim, Sechul Oh, Chaehyun Yu, Phys. Lett. B **621**, 259 (2005).
7. M. Beneke *et al*, Nucl. Phys. B **591**, 313 (2000).
8. C. W. Bauer, D. Pirjol and I. W. Stewart, Phys. Rev. Lett. **87**, 201806 (2001).
9. Y. Y. Keum *et al.*, Phys. Rev. D **69**, 094018 (2004).
10. P. Colangelo *et al.*, Phys. Lett. B **542**, 71 (2002); Phys. Rev. D **69**, 054023 (2004); Phys. Lett. B **597**, 291 (2004); H. Y. Cheng *et al.*, Phys. Rev. D **71**, 014030 (2005); C. K. Chua and W. S. Hou, Phys. Rev. D **72**, 036002 (2005).
11. R. Aleksan, I. Dunietz and B. Kayser, Z. Phys. C **54**, 653 (1992).
12. A. Abulencia *et al* [CDF]: Phys. Rev. Lett. **96**, 191801 (2006).
13. M. Gronau *et al.*, Phys. Rev. D **52**, 6356 (1995).
14. B. Aubert *et al.* [BABAR], hep-ex/0604012.

Measurement of the B_s oscillation frequency at CDF

Giuseppe Salamanna *on behalf of the CDF Collaboration*

Dipartimento di Fisica, Univ. di Roma *La Sapienza* and INFN Sez.di Roma,
P.le A.Moro, 2 00182 Rome, Italy
`giuseppe.salamanna@roma1.infn.it`

Summary. The first precise measurement of the B_s^0–$\bar{B}_s^0$ oscillation frequency Δm_s with the CDFII experiment [1] is summarized in this talk.

Flavour oscillations occur via $\Delta F = 2$ flavour changing weak interactions; in the B meson sector ($b\bar{q}$, with $q = d, s$ for B_d^0, B_s^0) the amplitude of such a process is proportional to the mass difference Δm_q of the Hamiltonian eigenstates $B_{q,\mathrm{H}}^0$ and $B_{q,\mathrm{L}}^0$. The expressions of Δm_q includes the product $|V_{tq}^* V_{tb}|$ of the CKM matrix elements. Thus, provided we can calculate the non-perturbative hadron matrix part of the process (which is done using lattice techniques), the measurement of the Δm_q gives information on the CKM matrix. Furthermore, models beyond the SM predict the existence of new particles, some of which can contribute to give a Δm_q value away from SM predictions ($\Delta m_s(\mathrm{SM}) \sim 20\,\mathrm{ps}^{-1}$).

To extract the value of Δm_s we perform a maximum likelihood-based fit in the B_s proper time domain. We use B_s hadronic and semileptonic decays corresponding to $1\,\mathrm{fb}^{-1}$ of data collected by the CDF II detector in $p\bar{p}$ collisions at $\sqrt{s} = 1.96\,\mathrm{TeV}$ at the Fermilab Tevatron collider.

1 Measurement

1.1 Final state selection

Both hadronic and semileptonic decay modes are selected using a dedicated 3-level trigger based on the Silicon Vertex Trigger: this exploits the kinematics of charm and bottom hadron decays and their feature to be long-lived particles. It requires 2 tracks of opposite electric charge and an impact parameter $120\,\mu\mathrm{m} \leq |d_0| \leq 1000\,\mu\mathrm{m}$; both tracks are also required to have $p_\mathrm{t} \geq 2.0\,\mathrm{GeV}/c$ and $p_{\mathrm{t},1} + p_{\mathrm{t},2} \geq 5.5\,\mathrm{GeV}/c$. We reconstruct several D_s final states: $D_s^+ \to \phi\pi^+, K^*(892)^0 K^+, \pi^+\pi^-\pi^+$, where the resonances decay as $\phi \to K^+ K^-$ and $K^*(892)^0 \to K^+\pi^-$ respectively. These are also required to be compatible with the known mass and width values [2]. The resulting D_s mesons are then associated with other tracks to form $D_s^+\ell^-$, $D_s^+\pi^-$ and $D_s^+\pi^-\pi^+\pi^-$; a spatial constraint is introduced for them plus any other tracks associated to the B_s to originate from the same 3-D decay vertex. In the semileptonic case, muons and electrons are identified via a likelihood using muon chamber and electromagnetic calorimeter information.

A total of 3600 hadronic signal events is used. To remove contributions from partially reconstructed decays we require that the decay candidate have $M > 5.3\,\mathrm{GeV}/c^2$. Candidates with $M > 5.5\,\mathrm{GeV}/c^2$ are used to build Probability Density Functions for combinatorial background.

A total of 37 000 semileptonic B_s events have been selected. Due to the unmeasured momentum from neutral particles, no B_s mass can be reconstructed. Nevertheless, we see that the combined use of the D_s mass and the ℓD_s mass $(m(\ell D_s))$ is effective in rejecting most of the various sources of background present, among which the association of a signal D_s and a fake lepton coming from the primary vertex and decays of the kind $\bar{B}_s^0 \to X D_s D(\to \ell\nu Y)$.

1.2 Proper time determination

The decay time in the B_s rest frame is given by $t = \kappa \times [L_\mathrm{T} m(B_s)/p_\mathrm{T}]$, where L_T is the proper decay length of the B_s meson in the transverse plane; the factor κ accounts for missing momentum in the semileptonic decays. The trigger lower cuts on the impact parameter and other proper decay time cuts introduce a bias on the proper time distribution. To correct for this and obtain an unbiased determination of the proper decay time of our B_s, a detailed simulation of the trigger and detector effects is performed. The parameterization of the trigger efficiency is then checked by performing a lifetime measurement of the various B mesons accordingly. We find the lifetime values well in agreement with the values in [4]. The associated systematic uncertainty is found to be negligible for mixing measurements.

The estimate of the decay-time resolution $\sigma(ct_i)$ is performed for each event starting from the measured track parameters and the relative uncertainties. We also calibrate such estimate on a large D^+ data sample combined with one or three prompt tracks to mimic the B^0-like decay topologies. In the case of the fully reconstructed decays, the uncertainty on the absolute time scale is dominated by the resolution on the position of the primary vertex. The average decay time resolution obtained is $\langle \sigma(ct) \rangle = 25.9\,\mu\mathrm{m}$, which corresponds to $\frac{1}{5}$ an oscillation period at the lower limit on Δm_s ($14.5\,\mathrm{ps}^{-1}$, [2]). For semileptonic decays, the distribution of the κ factor is determined from Monte Carlo simulation. The fraction of B_s momentum carried by undetected particles is a relevant contribution to the final proper-time resolution and cannot be neglected with respect to the one coming from vertex determination. To reduce this effect and increase the sensitivity of semileptonic modes to mixing, we determine the κ factor distribution as a function of $m(\ell D_s)$.

1.3 Flavour tagging

The flavour of the B_s at production is determined using both opposite-side and same-side flavour tags. The Opposite Side Taggers infer the initial b-flavour of the mixing candidates by looking at the decay products of the *other* b produced in the incoherent $b\bar{b}$ pair production at the Tevatron. We use lepton ($\ell = e, \mu$) charge and jet charge, whose sign is correlated with the flavour of the away b. The leptons are selected using a likelihood technique. The Opposite Side Taggers are combined exclusively such that the lepton taggers are preferred to the jet charge. The overall Opposite-Side tag effectiveness Q is measured on a mixed B^- and B_d^0 sample, using a maximum-likelihood fit. We find $Q = 1.47 \pm 0.10\%$ on hadronic decays and $Q = 1.44 \pm 0.04\%$ on semileptonic modes. We also make use of Same-Side tags: these identify the flavour of the candidate B by looking at the charge of the leading product of the fragmentation process that produced also the reconstructed B. In particular, a K^- (K^+) is likely to be produced in association with a B_s^0 ($\bar{B}_s^0$). Particle Identification from the specific ionization $\mathrm{d}E/\mathrm{d}x$ in the CDF central drift chamber and the particle's Time-Of-Flight is used to separate Kaons from other particle species. The track with the largest Kaon likelihood is selected to tag the b flavour. We predict the performances using a simulated sample generated with the PYTHIA Monte Carlo [5], after an extensive data-MC comparison of fragmentation-related quantities. Such a prediction is checked on other species and systematic uncertainties on the agreement are evaluated accordingly. We find $Q = 3.5 \pm 0.5\%$ ($Q = 4.0 \pm 0.6\%$) for hadronic (semileptonic) decays.

2 Results

The search for B_s oscillations is performed using an unbinned maximum likelihood fit, combining mass, decay-time, decay-time resolution and flavour tag for each candidate, separately for signal and the various types of background for each decay mode. We perform an *amplitude scan* [6] by introducing a term $\mathcal{A}$ in the mixing part of the signal likelihood, $\mathcal{L} \sim \frac{1}{\tau}\mathrm{e}^{-t/\tau}(1 \pm \mathcal{A} \cdot D \cdot (\Delta m_s t))$, and fitting for oscillations while fixing Δm_s at a probe value. $\mathcal{A}$ is expected to be consistent with unity at a given Δm_s if mixing is detected and with 0 elsewhere. The sensitivity is defined as the maximum value of Δm_s where $\mathcal{A} = 1$ is excluded at 95% C.L., if the measured value of $\mathcal{A}$ were 0. Figure 1 shows the result of the combined hadronic and semileptonic scan we perform with the $1\,\mathrm{fb}^{-1}$ data sample. CDF has a sensitivity of $25.8\,\mathrm{ps}^{-1}$ and exceeds the combined sensitivity of all previous experiments [2]. A value $\mathcal{A} = 1.03 \pm 0.28$ (stat.) is found at $\Delta m_s = 17.3\,\mathrm{ps}^{-1}$. The value is $3.7\,\sigma$ away from 0. The significance of such a peak is evaluated by calculating the quantity $\Lambda = \log(\mathcal{L}^{\mathcal{A}=0}/\mathcal{L}^{\mathcal{A}=1})$ for each probe frequency Δm_s. The likelihood ratio is shown again in Fig. 1. The minimum at $\Delta m_s = 17.3\,\mathrm{ps}^{-1}$ has

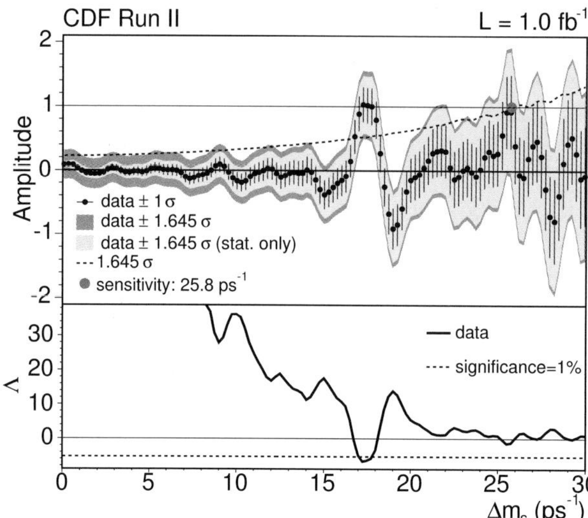

Fig. 1. *Upper*: amplitude scan; the *lighter band* represents the statistical uncertainty, the *darker band* the statistical+systematic one. *Lower*: likelihood ratio profile with the *line* indicating a 1% null hypothesis probability.

a value $\Lambda = -6.75$. Given this profile, we can evaluate the probability of null hypothesis to be $p = 0.2\%$, by randomizing the flavour tags on data. The p value corresponds to a significance $> 3\sigma$. From the likelihood itself we measure $\Delta m_s = 17.31^{+0.33}_{-0.18}(\text{stat.}) \pm 0.07(\text{syst.})\,\text{ps}^{-1}$, in the signal hypothesis. The only non-negligible systematic uncertainty comes from our knowledge of the absolute decay-time scale.

This value agrees with the SM predictions (e.g. [3]) well within 2σ. This can also be translated into an estimate of the contribution of New Physics to the mixing hamiltonian, $C_{B_s} = \frac{\mathcal{H}^{\text{SM+NP}}}{\mathcal{H}^{\text{SM}}} = 0.97 \pm 0.27$ [3], which is compatible with unity (*SM-only* contributions) with an uncertainty already smaller than for the B_d system. Finally, using the input values $m(B^0)/m(B_s^0) = 0.98390$ [7], $\Delta m_d = 0.505 \pm 0.005$ [2] and $\xi = 1.210^{+0.047}_{-0.035}$ [8], where ξ accounts for the QCD non-perturbative part of the mixing amplitude, from the $\Delta m_d/\Delta m_s$ ratio we infer $|V_{td}/V_{ts}| = 0.208^{+0.001}_{-0.002}(\text{exp.})^{+0.008}_{-0.006}(\text{theo.})$.

References

1. A. Abulencia *et al.* (CDF Collaboration), Phys. Rev. Lett. **97**, 062003 (2006)
2. S. Eidelman *et al.*, Phys.Lett. B **592**, 1(2004) and 2005 partial update for the 2006 edition available on the PDG www pages (http://pdg.lbl.gov)
3. M. Bona *et al.* (UTFit Collaboration) http://www.utfit.roma1.infn.it
4. The Heavy Flavor Averaging Group http://www.slac.stanford.edu/xorg/hfag
5. T. Sjöstrand *et al.*, Computer Phys.Commun. **135**, 238 (2001)
6. H.G. Moser and A. Roussarie, Nucl. Instr. Methods Phys. Res., Sect.A **A384**, 491 (1997)
7. D. Acosta *et al.* (CDF Collaboration), Phys. Rev. Lett. **96**, 202001 (2006)
8. M. Okamoto, PoS LAT2005 (2005) 013 (hep-lat/0510113)

Prospects for heavy flavor physics at LHC

G. Passaleva

INFN-Firenze, via G. Sansone 1 – I-50019 – Sesto Fiorentino (FIRENZE) –
ITALY
giovanni.passaleva@fi.infn.it

1 Introduction

In the last decade, the flavor sector of the Standard Model, parametrized
by the Cabibbo–Kobayashi-Maskawa (CKM) matrix [1], has been studied
and constrained with unprecedented precision thanks to the beautiful results obtained on Kaon CP violation experiments E731, KLOE and NA48,
by the B-factories KEKB and PEPII, by the charm factory CESR and by
the experiments at Tevatron. The impact of these results on the experimen-
tal tests of the Standard Model is pictorially represented on the $\bar{\rho}$–$\bar{\eta}$ plane
where the constraints on the Unitarity Triangle (UT) are displayed [2]. The
Large Hadron Collider (LHC) in construction at CERN, is expected to deliver
a large amount of events containing heavy flavors due to the high luminosity
$(10^{32}$–$10^{34}\,\mathrm{cm}^{-2}\mathrm{s}^{-1})$ and high production cross sections $(\sigma(bb) \sim 0.5\,\mathrm{mb})$.

The two LHC general purpose experiments ATLAS and CMS and the
B–physics dedicated experiment LHCb are devising specialized triggers and
analysis strategies to exploit this exceptionally high statistics and work out
a rich and complex B-physics program. In this paper the prospects for B-
physics at LHC will be reviewed along with some benchmark physics channels.

2 B-Physics at LHC

B decays turn out to be particularly sensitive to new physics effects. These
effects can show up either in single measurements, e. g. anomalously large
branching ratios in very rare decays, or through deviations from the Standard
Model predictions in precision measurements of the UT parameters. In both
cases large statistics is required and, therefore, the LHC represents a unique
tool for this kind of studies.

2.1 LHC experiments

The B-physics program at LHC will be pursued by three experiments. Two of them ATLAS [3] and CMS [4], are omni-purpose and optimized for discovery physics. Their B-physics program is mainly concentrated in the first years of running, when the LHC luminosity is expected to be $10^{33}\,\mathrm{cm}^{-2}\,\mathrm{s}^{-1}$. In the subsequent years at high luminosity ($10^{34}\,\mathrm{cm}^{-2}\,\mathrm{s}^{-1}$), when several pp collisions per bunch crossing will pile up, only search for very rare B decays with clear signatures will be performed. The third experiment, LHCb [5], is the LHC experiment dedicated to B-physics. It will locally tune the luminosity, by de–focusing the beams, to $2 \times 10^{32}\,\mathrm{cm}^{-2}\,\mathrm{s}^{-1}$, in order to limit pile up of pp interactions. LHCb is a single–arm forward spectrometer in the polar region 10–300 mrad, with good acceptance for b events thanks to the forward peaked production of b-hadrons at LHC. For B-physics studies at hadron colliders very sophisticated trigger systems are mandatory. ATLAS [6] and CMS [7] have a rather small bandwidth dedicated to B-physics, at the level of 10–15 Hz on tape. Both triggers are organized in two levels. The first level selects high p_T muons or di–muons. CMS also exploits on-line tracking for the selection of exclusive B events at High Level Trigger [8], while ATLAS foresees a flexible trigger strategy with the progressive addition of other triggers [9]. The LHCb [10] trigger is operating in two stages as well. The Level-0 is a hardware trigger that requires the presence of leptons, photons or hadrons with high p_T, while the High Level Trigger is a software trigger using the full information of the event. Its output contains 200 Hz of exclusive B candidates and about 1.8 KHz of inclusive channels to be used also for calibration purposes and systematic studies. Along with a specialized trigger, the experiments must have excellent tracking and particle identification capabilities to reconstruct exclusive B decays and a high flavor tagging level of 10–20 MeV is expected in B-hadron reconstruction and a proper time resolution of 40–100 fs. The tagging efficiency is about 4% for ATLAS and CMS while for LHCb is expected to be at the level of 7–9% [11]. A unique feature of LHCb is its excellent particle identification capability, due mainly to the RICH, that allows a K/π separation of more than 5σ over a momentum range of 2–100 GeV/c [12].

2.2 Benchmark physics channels

$B_s \bar{B}_s$ mixing

Oscillations of the B_s meson are particularly sensitive to new physics contributions through the oscillation frequency Δm_s and the mixing phase ϕ_s. The CDF collaboration has recently reported [13] a measurement of $\Delta m_s = 17.31^{+0.33}_{-0.18}\pm0.07\,\mathrm{ps}^{-1}$ in agreement with the Standard Model expectations. With a sample of $2\,\mathrm{fb}^{-1}$ LHCb expects to collect 120k events $B_s \to D_s\pi$ with a signal–to–noise ratio of 0.4 and a tagging efficiency $\varepsilon D^2 = 9\%$. Given the measured value of Δm_s LHCb will reach the same statistical precision of CDF with an integrated luminosity of $0.5 \times 10^{32}\,\mathrm{cm}^{-2}\,\mathrm{s}^{-1}$. ATLAS will also make a 5σ observation of oscillations in $10\,\mathrm{fb}^{-1}$ while, due to a smaller allocated trigger bandwidth, CMS will need twice the luminosity. Much more interesting is the measurement of the $B_s\bar{B}_s$ mixing phase ϕ_s, according to some recent theoretical works [14, 15]. The phase ϕ_s is expected to be very small in the Standard Model ($\phi_s \equiv 2\chi = 2\lambda^2\eta \simeq 0.04$), resulting in a high sensitivity to possible new physics contributions in b $\to$ s transitions. Hints of new physics could also be found in the measurement of the decay width difference between the two CP eigenstates $\Delta\Gamma_s = \Gamma(B_\mathrm{L}) - \Gamma(B_\mathrm{H})$ which is expected to be of the order of 10% in the Standard Model. Both quantities can be measured using $B_s \to J/\psi\varphi$ decays. For $\Delta m_s = 17.5\,\mathrm{ps}^{-1}$ LHCb expects to collect 1.3×10^5 events $B_s \to J/\psi\varphi$ with $2\,\mathrm{fb}^{-1}$ and to obtain a precision on $\sin\phi_s$ of about 0.023 and a precision on $\Delta\Gamma_s/\Gamma_s$ of ~ 0.011 CMS and ATLAS, with $30\,\mathrm{fb}^{-1}$, expect a sensitivity on $\sin\phi_s$ of 0.03–0.04 and a sensitivity on $\Delta\Gamma_s/\Gamma_s$ of 0.011–0.012, respectively [16, 17]. It is interesting to note that, according to references [15, 18], the actual UT data do not exclude large values of $\sin\phi_s$; with only $0.2\,\mathrm{fb}^{-1}$ LHCb can reach a sensitivity of $\sigma(\phi_s) = \pm 0.1$, enough to set an interesting limit or to measure ϕ_s if large.

Rare decays

ATLAS and LHCb have a very rich measurement program concerning rare B decays which are sensitive to new physics contributions. Sensitivities and event yields can be found in [19, 20]. As a benchmark channel the three LHC experiments have studied in detail the $B_s \to \mu^+\mu^-$ rare decay, whose branching ratio is estimated to be $BR(B_s \to \mu^+\mu^-) = (3.5 \pm 0.1) \times 10^{-9}$ in the Standard Model [21]. In various supersymmetric extensions of the Standard Model the branching ratio can be enhanced by one to three orders of magnitude as it is proportional to $(\tan\beta)^6$. In the Standard Model context, LHCb expects to select 17 events per $2\,\mathrm{fb}^{-1}$ [20]. For this search ATLAS and CMS will also exploit the high luminosity runs. In $100\,\mathrm{fb}^{-1}$ (1 year at 10^{34}) 92 and 26 signal events are expected respectively, while in the first year at low luminosity ($10\,\mathrm{fb}^{-1}$) both experiments expect 7 events each [19, 22]. At the moment no reliable background estimation is available because of lack of Monte Carlo events.

CP violation measurements

The UT angle γ can be measured in several different ways by LHCb [5], both from tree level decays which are not sensitive to new physics effects and from decays with large penguin contributions where new physics effects are potentially large. Among the proposed decay channels, $\mathrm{B}^- \to \mathrm{D}^0(\mathrm{K}^\pm\pi^\mp)\mathrm{K}^-$ [26], $\mathrm{B_d} \to \pi\pi/\mathrm{B_s} \to \mathrm{KK}$ [24] and $\mathrm{B_d} \to \mathrm{D}^0(\mathrm{K}^\pm\pi^\mp, \mathrm{KK})\mathrm{K}^{*0}$ [25] give the best statistical precision on γ, at the level of 5–8° with $2\,\mathrm{fb}^{-1}$. The same precision can be reached by the B-factories with at least $2\,\mathrm{ab}^{-1}$.

3 Conclusions

LHC represents a unique tool for heavy flavor and, in particular, for B-physics studies. High luminosity and production rates and access to all the B meson and baryon species will allow the three experiments ATLAS CMS and LHCb to investigate a vast range of B-physics items with unprecedented precision. The sensitivity on many B decay channels is largely sufficient to detect new physics signals at the TeV scale. The LHC heavy flavor program is complementary to direct new physics searches and is essential to understand the flavor structure of any new physics that LHC will eventually discover.

References

1. N. Cabibbo : Phys. Rev. Lett. **10**, 531 (1963)
   M. Kobayashi and T. Maskawa : Prog. Theor. Phys. **49**, 652 (1973)
2. M. Bona *et al.*, UTfit Collaboration : hep-ph/0606167 (2006)
3. ATLAS Collaboration: CERN-LHCC-99-014 (1999)
4. CMS Collaboration : CERN/LHCC 97-10 (1997)
   CERN/LHCC 97-33, (1997)
   CERN/LHCC 97-32 (1997)
   CERN/LHCC 98-06 (1998)
5. LHCb Collaboration: CERN-LHCC-2003-030 (2003)
6. ATLAS Collaboration: CERN/LHCC/2003-022 (2003)
7. CMS Collaboration: CERN/LHCC 2000-38 (2000)
8. W. Adam *et al.*: Eur. Phys. J. C **46**, 605 (2006)
9. M. Smizanska: Eur. Phys. J. C **34**, S385 (2004).
10. LHCb Collaboration: CERN-LHCC-2003-031 (2003)
11. LHCb Collaboration: LHCB 2003-115 (2003)
12. S. Easo *et al.*: Nucl. Instrum. Meth. A **553**, 333 (2005)
13. A. Abulencia *et al.*, CDF Collaboration: Phys. Rev. Lett. **97**, 062003 (2006)
14. P. Ball and R. Fleischer: hep-ph/0604249 (2006).
15. Z. Ligeti, M. Papucci and G. Perez: hep-ph/0604112 (2006)
16. R. Jones, ATLAS Collaboration: Nucl. Phys. Proc. Suppl. **156**, 147 (2006)
17. CMS Collaboration: CERN/LHCC/2006-021, 113 (2006)
18. M. Bona *et al.*, UTfit Collaboration: hep-ph/0605213 (2006)
19. C. Padilla, ATLAS Collaboration: Nucl. Phys. Proc. Suppl. **156**, 99 (2006)
    N. Nikitin *et al.*, ATLAS Collaboration : Nucl. Phys. Proc. Suppl. **156**, 119 (2006)
20. I. Belyaev, LHCb Collaboration: Czech. J. Phys. **55**, B485 (2005)
21. A. Ali: Nucl. Phys. Proc. Suppl. **59**, 86 (1997)
22. A. Nikitenko, A. Starodumov and N. Stepanov, CMS Collaboration: hep-ph/9907256 (1999)
23. R. Aleksan, I. Dunietz and B. Kayser: Z. Phys. C **54**, 653 (1992)
24. R. Fleischer: Phys. Lett. B **459**, 306 (1999)
25. M. Gronau and D. Wyler: Phys. Lett. B **265**, 172 (1991)
26. D. Atwood, I. Dunietz and A. Soni: Phys. Rev. Lett. **78**, 3257 (1997)

QED corrections and the $B \to K\pi$ puzzle

Elisabetta Baracchini

University of Rome "La Sapienza"and INFN, Sezione di Roma, P.le A. Moro 2, 00185 Rome, Italy

Summary. We discuss the theoretical and experimental treatment of radiative corrections in hadronic B decays. We present analytic expressions to describe the leading effects induced by both real and virtual (soft) photons in the generic process $H \to P_1 P_2(\gamma)$, where H and $P_{1,2}$ are scalar or pseudoscalar particles. After discussing the procedures to be adopted in experimental analysis for a clear definition of the observables, we show how it is possible to explain the "$B \to K\pi$ puzzle" in view of an $SU(2)$ analysis taking into account QED corrections as a source of isospin breaking.

1 Introduction

The $B \to K\pi$ decays have been considered for a long time a rich source of theoretical informations and recently a revitalized interest for these processes has been caused by the so-called "$K\pi$ puzzle". The puzzle arises mainly from two sources: the measured branching ratios seem to be isospin violating and experimental data seem to contradict the relations among A_{CP}, as they are calculated by models [1]. In particular, the isospin breaking is often assessed with the two quantities R_{c} and R_{n} [2], defined as

$$
R_{\text{c}} = 2 \left[\frac{BR(B^+ \to \pi^0 K^+) + BR(B^- \to \pi^0 K^-)}{BR(B^+ \to \pi^+ K^0) + BR(B^- \to \pi^- \overline{K^0})} \right] ,
$$

$$
R_{\text{n}} = \frac{1}{2} \left[\frac{BR(B_d^0 \to \pi^- K^+) + BR(\overline{B_d}^0 \to \pi^+ K^-)}{BR(B_d^0 \to \pi^0 K^0) + BR(\overline{B_d}^0 \to \pi^0 \overline{K^0})} \right] . \tag{1}
$$

As the Standard Model (SM) expectation implies that $R_{\text{n}} \simeq R_{\text{c}}$, up to (small) corrections dependent on the hadronic model, recent works [3] have claimed that the present experimental values ($R_{\text{c}} = 1.01 \pm 0.09$ and $R_{\text{n}} = 0.83 \pm 0.08$ from HFAG averages [4]) can be explained only in terms of new physics (NP) enhancement of electro-weak penguin (EWP) topologies with a large CP-violating NP phase. We will show how, taking into account the isospin breaking effects due to QED corrections, an isospin decomposition of decay amplitudes considering simultaneously SM and possible NP contributions to gluonic penguins can explain the experimental data without further sources of $SU(2)$ breaking beyond the SM.

2 QED Corrections: Scalar Calculation and Application to Experimental Analysis of Rare B Decays

The large amount of data collected so far at B factories has allowed to reach a statistical accuracy on B decays into pairs of pseudoscalars at a level where QED effects cannot be neglected anymore [5]: a correct simulation of the emission of photons in MonteCarlo programs as well as a clear definition of the effective cut on (soft) photon spectra are essential for a consistent definition of the observables. Even though hadronic decays cannot be fully treated in terms of perturbation theory, the universal character of infrared QED singularities allows us to estimate the leading $\mathcal{O}(\alpha)$ contributions to these processes within scalar QED, in the approximation of a point-like weak vertex. The most convenient infrared-safe observable related to the process $H \to P_1 P_2$ is then the photon inclusive width

$$\Gamma_{12}^{\mathrm{incl}}(E^{\mathrm{max}}) = \Gamma(B \to P_1 P_2 + n\gamma)|_{\sum E_\gamma < E_\gamma^{\mathrm{max}}} = \Gamma_{12} + \Gamma_{12+n\gamma}(E_\gamma^{\mathrm{max}}) \,. \quad (2)$$

The infrared cut-off E_γ^{max} is the photon energy below which the state $|P_1 P_2\rangle$ cannot be distinguished from the state $|P_1 P_2 + n\gamma\rangle$. Thanks to general properties of QED, at any order in perturbation theory we can decompose $\Gamma_{12}^{\mathrm{incl}}$ in terms of two theoretical quantities: the so-called non-radiative width, Γ_{12}^0, and the corresponding energy-dependent e.m. correction factor $G_{12}(E_\gamma^{\mathrm{max}})$,

$$\Gamma_{12}^{\mathrm{incl}}(E_\gamma^{\mathrm{max}}) = \Gamma_{12}^0(\mu) G_{12}(E_\gamma^{\mathrm{max}}, \mu) \,. \quad (3)$$

In the limit $E_\gamma^{\mathrm{max}} \ll M_H$ the e.m. correction factor can be reliably estimated within scalar QED. Defining the non-radiative width $\Gamma_{12}^0(\mu)$ as

$$\Gamma_{12}^0(\mu) = \frac{\beta}{16\pi M_B} \left| \mathcal{A}_{B \to h_1 h_2}(\mu) \right|^2 \,, \quad (4)$$

$$\beta^2 = \left[1 - (r_+ + r_-)^2 \right] \left[1 - (r_+ - r_-)^2 \right] \,, \qquad r_i = \frac{M_i}{M_B} \,, \quad (5)$$

namely the tree-level rate expressed in terms of the renormalized (scale-dependent) weak coupling, the function $G_{12}(E_\gamma^{\mathrm{max}}, \mu)$ can be written as [6]

$$G_{12}(E, \mu) = 1 + \frac{\alpha}{\pi} \left[b_{12} \ln\left(\frac{M_B^2}{4E^2} \right) + F_{12} + \frac{1}{2} H_{12} + N_{12}(\mu) \right] \,, \quad (6)$$

where H_{12} and F_{12} represent the finite and energy-independent terms arising respectively from virtual corrections and real emission. The scale dependence in $N_{12}(\mu)$ cancels out in the product $\Gamma_{12}^0 \times G_{12}$ thanks to the corresponding scale dependence of the weak coupling.[1] In principle, in order to quote the results in such a way that radiation effects can be disentangled, it would be necessary to select B candidates with a specified maximum amount of $\mathcal{O}(0.1\,\mathrm{GeV})$ photon energy in the final states, a quantity which is difficult to reconstruct in a tipical analysis of rare B decays at B factories. Instead, one could define the data sample selecting on an observed variable which can be clearly related to the maximum allowed energy for photons E_γ^{max}. The variable $\Delta E = E_B^* - \sqrt{s}/2$, where E_B^* is the reconstructed B candidate energy in the e^+e^- center of mass (CM) frame and $\sqrt{s}$ the total CM energy, clearly suits this scope. The ΔE window chosen for the analysis would then allow for the presence of radiated photons with an energy up to the chosen cut. From a result of such a kind it is then easy to extrapolate the weak couplings employing the theoretical calculation reported above. Calculating R_{c} and R_{n} in terms of the weak coupling estimated through this procedure and correcting for the contribution coming from $R = BR(\Upsilon(4S) \to B^+B^-)/BR(\Upsilon(4S) \to B^0\bar{B}^0) = 1.030 \pm 0.029$, leads to the result $R_{\mathrm{c}} - R_{\mathrm{n}} = 0.05 \pm 0.12$, clearly showing the importance of taking into account radiative corrections when comparing experimental results to theoretical predictions.

3 Isospin Analysis of $B \to K\pi$ Modes

Starting from the effective Hamiltonian $\mathcal{H}_{\mathrm{eff}}^{\Delta F=1}$ we can introduce a minimal set of matrix elements with defined isospin quantum numbers:

$$\langle(\pi K)_{I=1/2}|\mathcal{H}_{\mathrm{eff}}^{\Delta I=0}|B\rangle = \mathcal{P}\,,$$

$$\langle(\pi K)_{I=1/2}|\mathcal{H}_{\mathrm{eff}}^{\Delta I=1}|B\rangle = \mathcal{T}_1\,,$$

$$\langle(\pi K)_{I=3/2}|\mathcal{H}_{\mathrm{eff}}^{\Delta I=1}|B\rangle = \mathcal{T}_3\,. \tag{7}$$

As, due to intial and final states qauntum numbers, operators of any dimension-six $\mathcal{H}_{\mathrm{eff}}$ contributing to $B \to K\pi$ decays carry isospin $I = 0, 1$ even allowing for NP contributions. As $\mathcal{P}$ and $\mathcal{T}_{1,3}$ phases are result of the interplay of the weak phases in $\mathcal{H}_{\mathrm{eff}}$ and the absorptive parts of the amplitudes (strong phases), in general we will need additional complex parameters to describe the $\bar{B}$ decays. This can be seen focusing on the role of EWP operators: the EWP Wilson coefficients hierarchy allows us to neglect $Q_{7,8}$ contributions and to express the remaining matrix elements in terms of those of $Q_{1,2}$, as described in [7]. Defining $\kappa_{EW} = -\frac{3}{2}\frac{C_9+C_{10}}{C_1+C_2}\frac{V_{tb}^*V_{ts}}{V_{ub}^*V_{us}}$, we can thus factorize the

---

[1] For the explicit expressions of F_{12}, H_{12} and N_{12} and a more detailed discussion on the scale dependence μ, we refer the reader to [6].

term $V_{ub}^* V_{us}$ and write out the explicit CKM coefficients dependence of the isospin parameters isolating strong from weak phases:

$$\mathcal{P} = |V_{tb}V_{ts}|P\,\mathrm{e}^{\mathrm{i}\delta_p}\,,$$

$$\mathcal{T}_1 = V_{ub}^* V_{us}[(1 + \kappa_{\mathrm{EW}})t_1^+\,\mathrm{e}^{\mathrm{i}\delta_1^+} + (1 - \kappa_{\mathrm{EW}})t_1^-\,\mathrm{e}^{\mathrm{i}\delta_1^-}]\,,$$

$$\mathcal{T}_3 = V_{ub}^* V_{us}[(1 + \kappa_{\mathrm{EW}})t_3^+ + (1 - \kappa_{\mathrm{EW}})t_3^-]\,, \tag{8}$$

where $t_{1,3}^\pm(\delta_{1,3}^\pm)$ and $P(\delta_P)$ are respectively moduli (strong phases) of the contractions of $H_{\mathrm{eff}}^{\Delta I=1^\pm}$ and $H_{\mathrm{eff}}^{\Delta I=0}$ on the final states, where $\pm$ refers to the different contributions from $Q_\pm$ operators [7]. The CP conjugate $\bar{\mathcal{P}}, \bar{\mathcal{T}}_{1,3}$ needed for $\bar{B}$ decays are obtained from (8) changing sign to the weak phases. Assuming no NP at tree level nor in EWP (i.e. NP does not break $SU(2)$), any NP amplitude can be reabsorbed into the definition of $P, \bar{P}$ and $\delta_{P,\bar{P}}$. Following the approach of [8], we fit to the ten parameters $P, \delta_p, \bar{P}, \delta_{\bar{P}}, t_{1,3}^+, t_{1,3}^-, \delta_1^\pm$ with eight observables (BR, S_{00} and A_{CP}, except for $A_{\mathrm{CP}}(K^0\pi^0)$) and the only (very weak) model-dependent assumption $t_3^-/t_3^+ \leq 0.5$. For the experimental inputs, we take $BR(K^+\pi^-)$ and $BR(K^0\pi^+)$ from BaBar measurements and remove from them the radiation effects; instead, we use HFAG averages for $BR(K^0\pi^0)$ and $BR(K^+\pi^0)$, as the first has no QED corrections and the resolution on π^0 dominates the ΔE radiative tail of the second, not giving any information on E_γ^{max}. The result of the fit in terms of BR's and A_{CP}'s in Table 1 shows that it is possible to fit very precisely the experimental data and in particular the observed asymmetries without invoking any additional source of isospin breaking beyond the SM.

Table 1. Fit results for branching fraction and CP asymmetries. $A_{\mathrm{CP}}(K^0\pi^0)$ is not used in the fit and is predicted $A_{\mathrm{CP}}(K^0\pi^0) = -0.15 \pm 0.07$

	Exp. Input	Fit Output		Exp. Input	Fit Output
$BR(K^+\pi^-)$	19.6 ± 0.9	19.8 ± 0.9	$A_{\mathrm{CP}}(K^+\pi^-)$	-0.12 ± 0.02	-0.11 ± 0.02
$BR(K^+\pi^0)$	12.1 ± 0.8	12.1 ± 0.8	$A_{\mathrm{CP}}(K^+\pi^0)$	0.04 ± 0.04	0.03 ± 0.04
$BR(K^0\pi^+)$	26.4 ± 1.7	26.2 ± 1.6	$A_{\mathrm{CP}}(K^0\pi^+)$	-0.02 ± 0.04	-0.01 ± 0.04
$BR(K^0\pi^0)$	11.5 ± 1.0	11.0 ± 0.9	S_{00}	0.31 ± 0.26	0.36 ± 0.07

References

1. M. Beneke and M. Neubert, Nucl. Phys. B **675** (2003) 333,
2. A. J. Buras and R. Fleischer, Eur. Phys. J. C **11** (1999) 93 A. R. Williamson and J. Zupan, Phys. Rev. D **74** (2006) 014003
3. A. J. Buras *et. al* Eur. Phys. J. C **45** (2006) 701
4. http://www.slac.stanford.edu/xorg/hfag/rare/leppho05/charmless/ index.html

5. B. Aubert [BABAR Collaboration], arXiv:hep-ex/0608003, Y. Chao [BELLE Collaboration], PoS **HEP2005** (2006) 256.

6. E. Baracchini and G. Isidori, Phys. Lett. B **633** (2006) 309

7. M. Ciuchini, M. Pierini and L. Silvestrini, arXiv:hep-ph/0601233.

8. M. Ciuchini *et al.*, JHEP **0107** (2001) 013

Status of the Unitarity Triangle analysis

M. Bona[1], M. Ciuchini[2], E. Franco[3], V. Lubicz[2], G. Martinelli[3],
F. Parodi[4], M. Pierini[5], P. Roudeau[6], C. Schiavi[4], L. Silvestrini[3],
V. Sordini[6], A. Stocchi[6], and V. Vagnoni[7]

[1] Lab. d'Annecy-le-Vieux de Phys. des Part. LAPP, IN2P3/CNRS,
   Univ. de Savoie
[2] Dip. di Fisica, Univ. di Roma Tre and INFN Roma III, Italy
[3] Dip. di Fisica, Univ. di Roma "La Sapienza" and INFN Roma, Italy
[4] Dip. di Fisica, Univ. di Genova and INFN Genova, Italy
[5] Dep. of Physics, Univ. of Wisconsin, Madison, USA
[6] Lab. de l'Accél. Lin., IN2P3-CNRS et Univ. de Paris-Sud, Orsay Cedex, France
[7] Corresponding author, INFN Bologna, Italy

1 Introduction

The analysis of the Unitarity Triangle (UT) and CP violation provides nowadays a unique opportunity to look for New Physics (NP) effects beyond the Standard Model (SM).

In the following, after having shown the results of the UT analysis in the SM, we will introduce a generalized parametrization of $|\Delta F| = 2$ processes and thus present the results of the UT analysis in the presence of arbitrary NP contributions to the mixing amplitudes. With the new measurements from the Tevatron, namely the mass difference Δm_s, the width difference $\Delta \Gamma_s$ and the di-muon asymmetry, it is now possible for the first time to establish significant bounds on NP quantities also in the B_s sector. By restricting the NP scenario to Minimal Flavour Violation (MFV) models, both in the large and small $\tan \beta$ regimes, it is then possible to probe the corresponding NP scales.

All the details not reported in this summary can be found in a series of papers that the **UT**fit Collaboration has already published on the subject [1–4]. Furthermore, the results and the plots presented in this paper can be found at the URL http://www.utfit.org, where they are continuously updated.

2 The Unitarity Triangle in the Standard Model

Several measurements are available to bound the range of the $\bar{\rho}$ and $\bar{\eta}$ parameters nowadays. Here below we report a brief description of the most relevant ones in use in our UT fit:

- The rates of charmed and charmless semileptonic B decays decays which allow to measure the ratio $|V_{ub}| / |V_{cb}|$.
- The mass differences of the $B_q^0 - \bar{B}_q^0$ systems Δm_d and Δm_s.
- The ε_K parameter, which measures CP violation in the neutral kaon system.
- $\sin 2\beta$ from the golden modes $B_d^0 \to c\bar{c}K^0$, that can be determined almost without hadronic uncertainties.
- The angle α, that can be obtained from the $B \to \pi\pi$ and $B \to \rho\rho$ decays, assuming the $SU(2)$ flavour symmetry and neglecting the contributions of electroweak penguins. It can also be obtained using a time-dependent analysis of $B \to (\rho\pi)^0$ decays on the Dalitz plane.
- The angle γ that can be extracted from the tree-level decays $B \to DK$, using the fact that a charged B can decay into a $D^0(\overline{D}^0)K$ final state via a $V_{cb}(V_{ub})$ mediated process. CP violation occurs if the D^0 and the

Table 1. Determination of UT parameters from the SM fit

Parameter	Output	Parameter	Output
$\bar{\rho}$	0.162 ± 0.029	$\bar{\eta}$	0.342 ± 0.017
$\alpha[°]$	92.9 ± 4.4	$\beta[°]$	22.1 ± 0.9
$\gamma[°]$	64.6 ± 4.4	Δm_s [ps^{-1}]	17.4 ± 0.3
$\sin 2\beta$	0.698 ± 0.023	$\mathrm{Im}\lambda_t$ [10^{-5}]	13.7 ± 0.6
$V_{ub}[10^{-3}]$	3.67 ± 0.15	$V_{cb}[10^{-2}]$	4.16 ± 0.06
$V_{td}[10^{-3}]$	8.51 ± 0.28	$|V_{td}/V_{ts}|$	0.208 ± 0.007
R_b	0.379 ± 0.015	R_t	0.904 ± 0.030

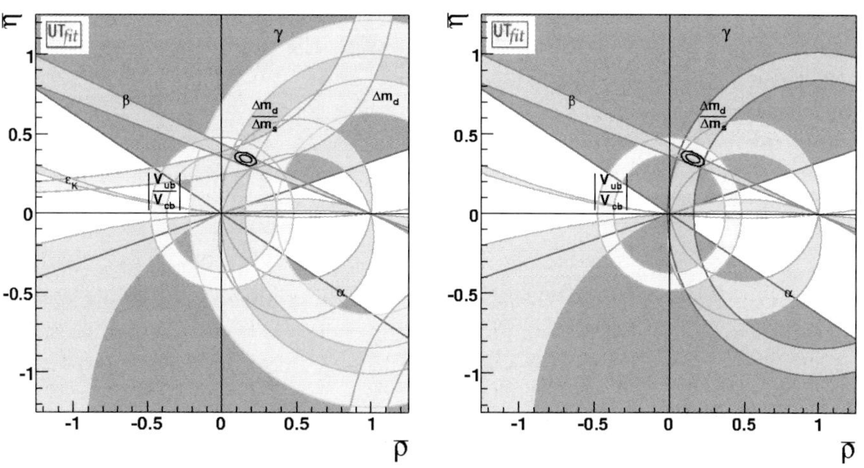

Fig. 1. *Left*: determination of $\bar{\rho}$ and $\bar{\eta}$ from constraints on $|V_{ub}| / |V_{cb}|$, Δm_d, Δm_s, ε_K, β, γ, and α. 68% and 95% total probability contours are shown, together with 95% probability regions from the individual constraints. *Right*: same fit as in *left*, but excluding ε_K and Δm_d (UUT fit)

$\overline{D}^0$ decay to the same final state. The same argument can be applied to $B \to D^*K$ and $B \to D^{(*)}K^*$ decays.

For the values of the relevant input quantities used in the UT fit as well as for details on the statistical treatment see [5, 6]. The output of the SM fit including all the constraints is reported in Table 1 and in Fig. 1. It is worth to mention that in our analysis we find that the experimental value of $|V_{ub}|/|V_{cb}|$ is not in good agreement with the rest of the fit at present, in particular due to a large value of the inclusive determination of $|V_{ub}|$. However, it is a matter of fact that the UT analysis has demonstrated so far the impressive success of the CKM picture in the SM.

3 New Physics Analysis

Starting from a tree-level determination of $\bar{\rho}$ and $\bar{\eta}$, we perform the UT analysis in general extensions of the SM with arbitrary NP contributions to $|\Delta F| = 2$ processes. We assume that NP enters observables in the flavour sector only at the loop level. For this reason the constraints from V_{ub}/V_{cb} and γ from the interference between $b \to c$ and $b \to u$ transitions to DK final states are basically NP-free. Being the mixing processes described by a single amplitude, they can be parameterized without loss of generality in terms of two parameters quantifying the difference of the amplitude with respect to the SM one [7]:

$$C_{B_q}\, e^{2i\phi_{B_q}} = \frac{\langle B_q^0 | H_{\text{eff}}^{\text{full}} | \bar{B}_q^0 \rangle}{\langle B_q^0 | H_{\text{eff}}^{\text{SM}} | \bar{B}_q^0 \rangle}\,, \qquad C_{\epsilon_K} = \frac{\text{Im}[\langle K^0 | H_{\text{eff}}^{\text{full}} | \bar{K}^0 \rangle]}{\text{Im}[\langle K^0 | H_{\text{eff}}^{\text{SM}} | \bar{K}^0 \rangle]}\,. \tag{1}$$

where $H_{\text{eff}}^{\text{SM}}$ includes only the SM box diagrams, while $H_{\text{eff}}^{\text{full}}$ includes also the NP contributions. In the absence of NP, $C_{B_q} = C_{\epsilon_K} = 1$ and $\phi_{B_q} = 0$. With these definitions, NP effects which enter the present analysis are parameterized in terms of 5 real quantities: C_{B_d}, ϕ_{B_d}, C_{B_s}, ϕ_{B_s} and C_{ϵ_K}.

The results of the fit are summarized in Table 2. It is interesting to note that the measurement of Δm_s [8] strongly constrains C_{B_s}, so that C_{B_s} is already known better than C_{B_d}. Finally the introduction of the A_{CH} [9] and $\Delta \Gamma_s$ constraints provides the first relevant bound on ϕ_{B_s}.

In the context of MFV extensions of the SM [10], it is possible to use the so called UUT construction [11] in order to determine the parameters of the CKM matrix independently of NP effects. To this end one has to use all the constraints from tree-level processes and from the angle measurements, as well as the $\Delta m_d/\Delta m_s$ ratio, which in MFV scenarios are NP-free (see Fig. 1). Using the results of the UUT fit and the procedure detailed in [12], we can obtain lower bounds on the MFV scale $\Lambda > 5.9\,\text{TeV}$ at 95% probability for one Higgs doublet or at small $\tan\beta$ and $\Lambda > 5.4\,\text{TeV}$ at 95% probability for large $\tan\beta$.

Table 2. Determination of UT and NP parameters from the NP generalized fit

Parameter	Output	Parameter	Output
C_{B_d}	1.25 ± 0.43	$\phi_{B_d}[^\circ]$	-2.9 ± 2.0
C_{B_s}	1.13 ± 0.35	$\phi_{B_s}[^\circ]$	$(-3 \pm 19) \cup (94 \pm 19)$
C_{ϵ_K}	0.92 ± 0.16		
$\bar{\rho}$	0.20 ± 0.06	$\bar{\eta}$	0.36 ± 0.04
$\alpha[^\circ]$	93 ± 9	$\beta[^\circ]$	24 ± 2
$\gamma[^\circ]$	62 ± 9	$\mathrm{Im}\lambda_t[10^{-5}]$	14.6 ± 1.4
$V_{ub}[10^{-3}]$	4.01 ± 0.25	$V_{cb}[10^{-2}]$	4.15 ± 0.07
$V_{td}[10^{-3}]$	8.3 ± 0.6	$|V_{td}/V_{ts}|$	0.203 ± 0.015
R_b	0.42 ± 0.03	R_t	0.89 ± 0.06
$\sin 2\beta$	0.75 ± 0.04	$\sin 2\beta_s$	0.039 ± 0.004

4 Conclusions

In this paper we have presented an updated analysis of the UT in the SM, using all the relevant measurements available from the B factories and the Tevatron, in particular the recent measurement of Δm_s from CDF.

Then, we have performed a model-independent analysis of the UT in general extensions of the SM with loop-mediated contributions to FCNC processes. We have shown how the great number of measurements nowadays available allow for a simultaneous determination of the CKM parameters, together with the NP contributions to $|\Delta F| = 2$ processes in the K^0, B^0 and B_s^0 sectors.

Finally, we have analyzed in detail the UUT, showing that it is possible to constrain the additional NP parameters very accurately. In this way, we have been able to probe the NP scale in MFV scenarios in the large and small $\tan \beta$ limits up to about 5–6 TeV.

References

1. M. Bona _et al._, JHEP **0507**, 028 (2005) [arXiv:hep-ph/0501199].
2. M. Bona _et al._, JHEP **0603**, 080 (2006) [arXiv:hep-ph/0509219].
3. M. Bona _et al._, arXiv:hep-ph/0606167.
4. M. Bona _et al._, arXiv:hep-ph/0605213.
5. **UT**_fit_ home page, http://www.utfit.org
6. M. Ciuchini _et al._, JHEP **0107** (2001) 013 [arXiv:hep-ph/0012308].
7. Y. Grossman, Y. Nir and M. P. Worah, Phys. Lett. B **407**, 307 (1997)
8. [CDF - Run II Collaboration], Phys. Rev. Lett. **97** (2006) 062003
9. http://www-d0.fnal.gov/Run2Physics/WWW/results/prelim/B/B29/B29.pdf
10. M. Blanke _et al._, arXiv:hep-ph/0604057.
11. A. J. Buras _et al._, Phys. Lett. B **500**, 161 (2001) [arXiv:hep-ph/0007085].
12. G. D'Ambrosio _et al._, Nucl. Phys. B **645**, 155 (2002) [arXiv:hep-ph/0207036].

Non-perturbative inputs for Flavour Physics

Federico Mescia

INFN, Laboratori Nazionali di Frascati, Via E. Fermi 40, I-00044 Frascati, Italy
federico.mescia@lnf.infn.it

1 Introduction

In the pursuit of improving our knowledge of short-distance physics from
weak hadron decay (i. e. the CKM couplings in the Standard Model), reliable
theoretical estimates of QCD non-perturbative effects are needed. For many
decays from K to B physics (see Fig. 1), Lattice QCD remains the only tool
to achieve this task. Once masses heavier than m_W are integrated out, low-
energy QCD effects are encoded into hadronic matrix elements of some local
operator.

At this stage, it is important to emphasize:

- Lattice simulations are not ad hoc QCD models but an estimate of QCD
  path integrals (e. g. Green functions) from first principles of QCD: the
  only parameters entering the computation are those that appear in the
  QCD lagrangian, namely the quark masses and the (bare) strong coupling;
- QCD path integrals are solved numerically on a finite (L) and discrete
  ($L = N \times a$) box (lattice) by using the Monte–Carlo methods which induce
  the statistical errors ($\propto 1/\sqrt{N_{\rm conf}}$).

$$
\begin{pmatrix}
\mathbf{V_{ud}} & \mathbf{V_{us}} & \mathbf{V_{ub}} \\
\pi \to l\nu & K \to \pi l\nu & B \to \pi l\nu \\
 & K \to l\nu & B \to \tau l\nu \\
\mathbf{V_{cd}} & \mathbf{V_{cs}} & \mathbf{V_{cb}} \\
D \to \pi l\nu & D \to K l\nu & B \to D^{(*)} l\nu \\
D \to l\nu & D_s \to l\nu & \\
\mathbf{V_{td}} & \mathbf{V_{ts}} & \mathbf{V_{tb}} \\
\langle B_d | \overline{B}_d \rangle & \langle B_s | \overline{B}_s \rangle &
\end{pmatrix}
\tag{1}
$$

Fig. 1. Gold-plated processes accessible to present lattice studies and relative CKM
matrix elements. The neutral K–$\bar{K}$ mixing is also another gold-plated process,
which gives a constraint on the phase of the CKM matrix

Thus, in principle, computation of the hadronic quantities on the lattice can be made to an arbitrary accuracy: setting a priori the length, L much larger than the Compton length of the lightest particle $m_\pi L \gg 1$ and the lattice spacing, a, much smaller than the Compton length of the heaviest mass $m_b a \ll 1$. In practice, however, limitations in computing resources determine various approximations, each introducing some systematic uncertainty into the final results: statistical errors, finite volume and discretization effects ($L m_\pi$ and $1/(a m_b)$ small), heavy- and light-quark extrapolations (physical m_{ud}, m_b quark masses inaccessible) and sea quark effects (no quark loops). Thanks to theoretical and technical improvements, results are becoming more and more reliable determining an important impact for the phenomenology. For example finite volume and discretization errors are no longer a severe issue, whereas milestone barriers such as "sea quark effects" and "the quark chiral limit" are under investigation from present lattice activities. This final step is indeed demanding in front of new experiments, such LHCb, Super-B and ϕ factory.

In front of this physics demand, the following matrix elements need to be computed on the lattice:

$$\langle \bar{P} | A^\mu_{Qq} | 0 \rangle = i f_P p^\mu_P \,, \quad \{(Q = u, c, b), (q = s, d)\}$$

$$\langle \bar{P}_2 | V^\mu_{Qq} | P_1 \rangle = f_+(q^2)(p^\mu_\pi + p^\mu_K) + f_-(q^2) q^\mu_\pi \,, \quad \{(Q = u, c, b), (q = s, d)\}$$

$$\langle \bar{K}^0 | O_{sd}(\mu) | \bar{K}^0 \rangle = \frac{8}{3} m^2_K f^2_K B_K(\mu) \,, \tag{2}$$

$$\langle \bar{B}^0_q | O_{bq}(\mu) | \bar{B}^0_d \rangle = \frac{8}{3} m^2_{B_d} f^2_{B_d} B_{B_q}(\mu) \,,$$

where $A_{Qq} = (\bar{Q}q)_A$, $V_{Qq} = (\bar{Q}q)_V$, $O_{Qq} = (\bar{Q}q)_{V-A}(\bar{Q}q)_{V-A}$. On the r. h. s. of (2) we recognize the pseudoscalar decay constants f_π (for $\pi \to \ell\nu$), f_K (for $\pi \to \ell\nu$) and $f_{B_{d(s)}}$ (for $B \to \tau\nu$), the form factors $f_+(q^2)$ (for $K \to \pi\ell\nu$) and the familiar "bag" parameters, $B_K(\mu)$ (for ε_K) and $B_{B_{d(s)}}(\mu)$ (for $\Delta_{m_{d,s}}$). All these quantities parameterize our ignorance of QCD effects entering π, K and B processes. Analogue quantities are also defined for D decays. We now collect updated lattice averages for the quantities above [1]:

$$f_K = 153.5(3.2)\,\text{MeV} \,, \quad f_K/f_\pi - 1 = 0.198(3)(^{16}_{5}) \,, \quad f_+(0) = 0.961(8) \,,$$

$$\hat{B}_K = 0.86(6)(14) \,, \quad \hat{B}_{B_d} = 1.34(12) \,,$$

$$f_{B_d} = 216(27)\,\text{MeV} \,, \quad f_{B_s}/f_{B_d} = 1.20(5) \,,$$

$$\xi = (f_{B_s}\sqrt{B_{B_s}})/(f_{B_d}\sqrt{B_{B_d}}) = 1.21(5) \,.$$

These quantities are completely dominated by the systematic uncertainties. At present, these results include rough estimates of chiral log ($m_q/m_s > 0.13$)[1] and unquenched effects ($N_F = 2 + 1$ and $N_F = 2$). Moreover, they still rely on coarse setup ($a^{-1} \sim 2\,\text{GeV}$ and $L^{-1} \sim 80\,\text{MeV}$).

____

[1] Let's remind the physical point m_{ud}/m_s is 0.04.

In what follows, I will mainly discuss on quenched approximation and prospectives to estimate it. For a discussion on the results in (3) and on their phenomenology, I will refer to [1] and the references thereby.

2 Systematic uncertainties: some good news

For long time, lattice simulations have worked in the "so-called" quenched approximations and with heavy quarks. This era seems now finished. Shortly, observables, $\langle O \rangle$, on the lattice are estimated by the following form

$$\langle O \rangle = \int dA\, e^{-S(A)} \prod_{q=1,N_F} \mathrm{Det}[/D + m_q^{\mathrm{sea}}] O(A, m_q^{\mathrm{valence}}) \qquad (3)$$

where dA is the gluonic measure and $S(A)$ the Yang–Mills action, whereas $\prod_{q=1,N_F} \mathrm{Det}[/D + m_q^{\mathrm{sea}}]$ is the residue of the gaussian fermionic integration on N_F flavours and the beasties of lattice calculations for $m_q \ll \mathcal{O}(\Lambda_{\mathrm{QCD}})$. Simulation costs for light masses are very expensive. The Wilson formulation of the lattice QCD action, which used to be the standard one up to 2002, did not allow to get $m_\pi < 500\,\mathrm{MeV}[m_q/m_s < 0.5]$, because of technical issues such as exceptional configurations and algorithms slow down. This barrier, known as "Berlin Wall"[2], is sketched from the Wilson points in Fig. 2. Recent developments however, in both algorithms and new formulations lattice fermions are solving this issue:

- Improved staggered fermions [2]: MILC collaboration from 2002 was available to have, $m_\pi 240\,\mathrm{MeV}$ and $(m_q/m_s \sim 0.08$. These studies with $N_F = 2 + 1$ sea quarks (e.g two degenerate light quarks $m_u = m_d = m_q$

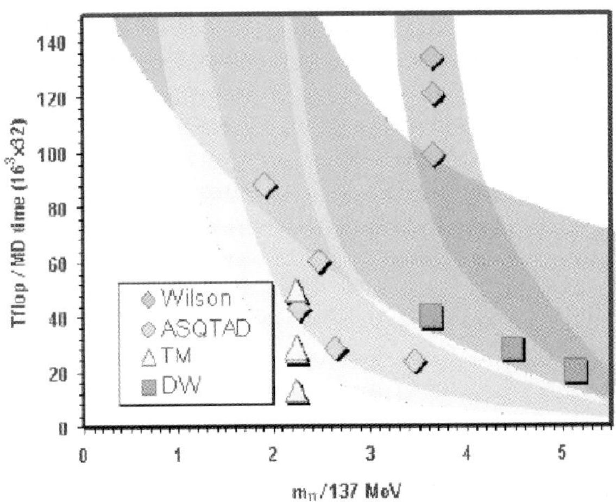

Fig. 2. Computational Cost in Tera Flops units versus the simulated pion mass

---

[2] where it was for the first time shown at 2001 lattice conference in Berlin.

plus a strange quark $m_s > m_q$) are quite advanced and dominate the average in (3). However, even tough fast to simulate, they can leave out some unknown systematics in estimating the dynamical quark effects. More specifically, for any flavored fermion there are four extra copies. Effects from these unphysical tastes require some unclear contrivance:

- a fourth root trick $(\mathrm{Det}[/\!\!\!D + m_q^{\mathrm{sea}}]^{1/4})$ is used to eliminate at finite lattice spacing the degeneracy: no proof that this is correct nor wrong. In perturbation theory, this is correct;
- fit on the observables with tenths of free parameters;

In the end, we have no answers to the following questions: is the continuum limit of staggered fermions QCD? Locality, unitarity?

- Wilson fermions with Schwartz-preconditioning [3]: Luscher in 2005 was available to simulate with a reasonable amount of time $m_\pi \sim 280\,\mathrm{MeV}$ and m_q/m_s 0.16. This formulation is more expensive than staggered fermions but has the QCD physical properties. For the time being, there are just spectrum studies for $N_F = 2$ fermions [4].

- Twisted-mass fermions [5]: This formulation is as good as the Luscher approach but easier to implement and with $\mathcal{O}(a^2)$ [6] scaling. The inconvenience of isospin and parity breaking at $\mathcal{O}(a^2)$ seems fixable. Current simulations are only concentrated on $N_F = 2$ fermions, even tough feasibility studies have already started for $N_F = 2 + 2$ sea quarks.

In conclusion, next year we will see a large amount of lattice results with more physical setup. Yet we cannot expect to work with b and up quarks close to their physical masses in unquenched QCD. Thus, we still have some irreducible uncertainty to take care

- promising strategies for u/d extrapolation: 1) have pions between 200 MeV and 300 MeV with new algorithms; 2) use ChPT functional form in this range with FV effects under control;
- good prospectives for beauty physics: 1) apply recent HQET developments (non-perturbative renormalization understood and physical signal improved) to get results for $m_b \to \infty$: 2) interpolate to B, improved HQET results with QCD above the charm region. Errors are expected to reduce with respect to the present uncertainties: no improvement instead are expected from alternative approaches such as NRQCD (difficulties with higher orders $1/(am_H)$: renormalon shadow, no continuum limit) and FNAL (renormalisation procedure unclear).

I thank the Organizers for their kind invitation and great hospitality in Pavia during the meeting.

References

1. S. Hashimoto, Int. J. Mod. Phys. A **20**, 5133 (2005) [arXiv:hep-ph/0411126]. M. Okamoto, PoS **LAT2005**, 013 (2006) [arXiv:hep-lat/0510113]. C. Daw-

son, PoS **LAT2005**, 007 (2006). P. B. Mackenzie, eConf **C060409**, 022 (2006) [arXiv:hep-ph/0606034]. C. Davies, arXiv:hep-lat/0608018. D. Becirevic, arXiv:hep-ph/0211340. L. Lellouch, Nucl. Phys. Proc. Suppl. **117**, 127 (2003) [arXiv:hep-ph/0211359].

2. C. T. H. Davies *et al.* [HPQCD Collaboration], Phys. Rev. Lett. **92**, 022001 (2004) [arXiv:hep-lat/0304004].

3. M. Luscher, Comput. Phys. Commun. **165**, 199 (2005) [arXiv:hep-lat/0409106].

4. L. Del Debbio, L. Giusti, M. Luscher, R. Petronzio and N. Tantalo, JHEP **0602**, 011 (2006) [arXiv:hep-lat/0512021].

5. T. Chiarappa *et al.*, arXiv:hep-lat/0606011. F. Farchioni *et al.*, Eur. Phys. J. C **47**, 453 (2006) [arXiv:hep-lat/0512017].

6. R. Frezzotti and G. C. Rossi, Nucl. Phys. Proc. Suppl. **129**, 880 (2004) [arXiv:hep-lat/0309157].

Bounds on the supersymmetric flavour space

Valentina Porretti

Dipartimento di Fisica 'E. Amaldi', Università di Roma Tre and INFN, Sezione di Roma Tre, I-00146 Rome, Italy
porretti@fis.uniroma3.it

1 Outline

We introduce the supersymmetric flavour problem (Sect. 2) focusing on mixing processes in the quark sector (Sect. 3) and present updated plots of the bounds on squark mass matrices obtained in the mass insertion approximation (Sect. 4).

2 Flavour issues in supersymmetry

Supersymmetry (SUSY) accomodates and does not explain the very large number of results obtained in flavour experiments up to now. In this respect, it seems to perform worse than the SM, sharing similar problems with around 100 parameters more. However in SUSY the solution to the flavour problem can be postponed to very high energy scales (GUT scales for instance) without destabilizing the electroweak scale. The still unknown physics that breaks SUSY and maybe solves the flavour problem is parametrized by the soft terms in the Lagrangian. Flavour is interesting in SUSY because it implies that this physics must be highly non-generic.

The fact is that SUSY generally fails to reproduce the special circumstances that suppress FC processes in the SM: gauge structure and particle content forbid baryon and lepton number violation and flavour changing neutral currents (FCNCs) at tree level, while fermion masses much smaller than the W mass and an almost diagonal CKM matrix effectively suppress FCNCs in loop processes. The recent measurement of B_s mixing, again in agreement with the SM expectation, confirms that the supersymmetric parameters must be strictly organized. R-parity, universal or aligned scalar mass matrices and partial decouplings, eventually motivated by statistical or anthropic reasons, are the standard tools applied to the soft terms in order to hyde SUSY signatures in flavour experiments [1].

3 Constraints from neutral meson mixings

Supersymmetric flavour changing transitions in the quark sector are potentially much larger than the experimental rates. The main qualitative reason is that they can be mediated by strong interactions. Mixing processes of neutral mesons seem a promising place to look for SUSY effects because in the SM they are four times suppressed: by the loop factor, the weak coupling constant, the GIM-mechanism and the small CKM angles. In generic SUSY models only the loop suppression is present.

Neutral meson mixings change flavour by two units ($\Delta F = 2$) and are mediated by box diagrams where the internal particles can be

- squarks ang gluinos
- charged Higgs (or W) and quarks
- charginos and squarks
- neutralinos and gluinos

The squark-gluino mediation is the most dangerous and should dominate the supersymmetric rate for $\Delta F = 2$ processes. Barring interference effects, the experimental number can be directly interpreted as a bound on the structure of the squark–gluino–quark coupling. It is customary to impose the constraints in the super-CKM basis, where squarks are rotated parallel to quarks and the amount of flavour change is set by the off-diagonal entries of the squark mass matrix in flavour space. The off-diagonal elements divided by the geometric average of the diagonal elements are the mass-insertion parameters which we plan to bound. For instance, K-mixing involves the first two families and the mass-insertions bounded are δ_{AB}^{12} with $A, B = \mathrm{L}, \mathrm{R}$ referring to the helicity of the partners of the squarks running in the loop. The δ's can be kept small increasing the squark masses (decoupling) or suppressing the flavour non-diagonal entries (universality). RG effects generally dilute the bounds going up to the soft scale [2].

4 A next-to-leading order analysis

The computation of the amplitude for neutral meson mixing is done in the effective hamiltonian formalism. It is an operator product expansion that separates the short distance contribution, expressed by the Wilson coefficients, from the long distance physics, contained in the matrix elements of the operators that appear in the calculation after integrating out the heavy particles. The Wilson coefficients must be RG evolved from the supersymmetric scale (close to the squark/gluino masses $\sim 500\,\mathrm{GeV}$) down to the scale of few GeV where the matrix elements are evaluated non-perturbatively in lattice QCD. Large logarithms generated by the gap between the two scales are resummed in the running.

The next-to-leading order (NLO) matrix elements and anomalous dimension matrix for the evolution have been available since [3]. Several partial computation of NLO Wilson coefficients appeared in the meanwhile (see refs. in [4]), while the general case of the MSSM has been recently completed in [4]. We found that the numerical effect of the NLO coefficients with respect to the LO ones is not significant but the dependence on the high energy scale is very reduced (from $\sim 15\%$ to a few %) and leaves the matrix elements as the main source of theoretical uncertainty. Even if the NLO accuracy is obscured by the $\mathcal{O}(10)$ uncertainty in the squark and gluino masses, it could be that at the LHC the supersymmetric spectrum will be measured and in any case the predictions of SUSY models can now be tested more precisely. The improvement with respect to past analysis is mainly due to new experimental bounds and the improved extraction of the CKM parameters allowing for the presence of new physics in the fit of the unitarity triangle.

Equating the experimental rates to the NLO amplitude we derive bounds on the mass insertion parameters. We set to zero all the mass-insertion parameters except the one to be bounded. It is a safe procedure when the mass insertions have different orders of magnitude and no cancelation occurs in the Wilson coefficients, as is usually the case.

We present some significant results in plots (real versus imaginary parts of the δ's), rather than in tables, since they are typically not isotropic in the complex plane and the impact of the single constraints is more transparent. The SM parameters are extracted with gaussian distribution; the ρ, η parameters are obtained from tree-level processes. The gluino and squark masses are set to $350\,\mathrm{GeV}$. The mass insertions are extracted with flat distribution in the complex plane and weighted by $\exp[(O^{\mathrm{th}} - O^{\mathrm{exp}})^2/\sigma_{\mathrm{exp}}^2]$, where O is the observable used as constraint and σ_{exp} the exeprimental error. The size of the squares in the plots is a measure of the p.d.f. of the $\delta's$. The numerical values of the inputs used can be found in [5].

- B_d-mixing, LL case. The experimental constraints are Δm_d (upper,left panel), $\sin 2\beta$ (upper, central panel) and $\cos 2\beta$. All the bounds together are represented in the upper panel on the right, where the dark squares represent the allowed region.
- B_s-mixing, RR case. We added to the $\Delta F = 2$ constraint Δm_s, the $\Delta F = 1$ bound from $B \to s\gamma$, which dominates in the chirality changing cases. In the RR case (lower panel on the left) the impact of the measurement of Δm_s is important.
- K-mixing, LL case. The experimental constraints are ΔM_K (light squares) and ϵ_K (dark squares). We set to zero the SM long distance contribution to ΔM_K. The bound on $\delta_{\mathrm{LR}}^{12}$ is 10 times more stringent than the LL bound presented in the lower central panel.
- D-mixing, LL case. Experiments at the moment only bound the sum of the mixing parameters squared $x^2 + y^2$ ($x = \Delta M_D/\Gamma$, $y = \Delta\Gamma/2\Gamma$). Moreover the SM contribution is by far dominated by the long distance contribution and neither the order of magnitude is assessed. In spite of this, the bounds on D-mixing (lower panel on the right) are already im-

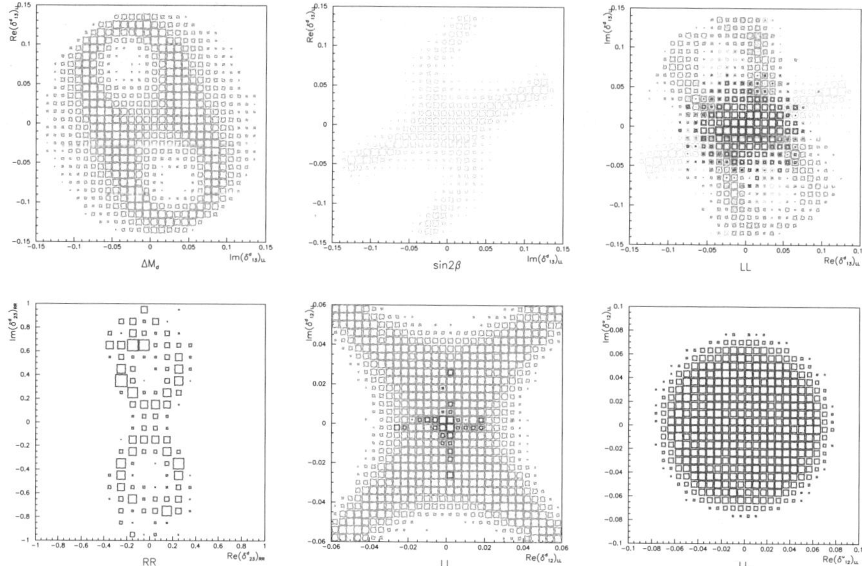

Fig. 1. See text for comment

portant because models with alignment predict large D-mixing rates. This is actually a common feature to many models [6] in which flavour violation in the D sector is somehow complementary to the K sector, where the most stringent bounds apply.

References

1. A recently updated review is S. P. Martin, arXiv:hep-ph/9709356.
2. D. Choudhury, F. Eberlein, A. Konig, J. Louis and S. Pokorski, Phys. Lett. B **342** (1995) 180 [arXiv:hep-ph/9408275].
3. M. Ciuchini, E. Franco, V. Lubicz, G. Martinelli, I. Scimemi and L. Silvestrini, Nucl. Phys. B **523** (1998) 501 [arXiv:hep-ph/9711402]; A. J. Buras, M. Misiak and J. Urban, Nucl. Phys. B **586** (2000) 397 [arXiv:hep-ph/0005183].
4. M. Ciuchini, E. Franco, D. Guadagnoli, V. Lubicz, V. Porretti and L. Silvestrini, arXiv:hep-ph/0606197.
5. http://utfit.roma1.infn.it/
6. A. A. Petrov, eConf **C030603** (2003) MEC05 [arXiv:hep-ph/0311371].

PARALLEL SESSION:
Neutrinos and Cosmic Rays
(E. Lisi and L. Patrizii, conveners)

Neutrino oscillations with artificial sources

Maximiliano Sioli

Dipartimento di Fisica dell'Università di Bologna and INFN Sezione di Bologna,
V.le B. Pichat, I-40127, Bologna, Italy
sioli@bo.infn.it

We review the main experimental results on neutrino oscillations obtained
with man-made neutrino beams. Particular attention will be devoted to past-
generation experiments which were based on conventional neutrino beam
from proton acceleration and from nuclear reactors.

1 Introduction

The experimental discovery of neutrino flavour oscillations provides the first
strong evidence of physics beyond the Standard Model. Under the assumption
that neutrinos have definite masses and that weak and mass eigenstates do not
coincide, the probability that a neutrino $|\nu_\alpha\rangle$ produced in a weak eigenstate
would be detected after a time t in a weak interaction as a $|\nu_\beta\rangle$, oscillates
with a fashion

$$P_{\alpha\beta} = |\langle\nu_\beta||\,\nu_\alpha(t)\rangle|^2 = \left|\sum_{i=1}^{3}\sum_{j=1}^{3} U_{\alpha i}^* U_{\beta j}\langle\nu_j(0)|\,\nu_i(t)\rangle\right|^2 , \tag{1}$$

where $|\nu_i\rangle$ are states with definite mass m_i and U is a matrix which linearly
mix the flavour and the mass bases. This probability can be expressed in
terms of the pathlength L (the "baseline"), the energy E and the squared
mass difference Δm_{ij}^2:

$$P_{\alpha\beta} = \sum_{i=1}^{3}\sum_{j=1}^{3} U_{\alpha j} U_{\beta j}^* U_{\alpha i}^* U_{\beta i}\, e^{-i\frac{\Delta m_{ij}^2 L}{2E}} \tag{2}$$

If we limit for simplicity to a two-neutrino scenario, described by one mixing angle θ and one mass square difference Δm^2, the oscillation probability
assumes the more familiar form (expressed in proper units):

$$P_{\alpha\beta} = \sin^2 2\theta \sin^2 1.27 \frac{L(\text{km})}{E(\text{GeV})} \Delta m^2 (\text{eV}^2). \tag{3}$$

Neutrino oscillations are a typical quantum-mechanical interference phenomenon, hence they are extremely powerful in revealing very tiny effects. $P_{\alpha\beta}$ is called *appearance* probability, and experiments aiming at the detection of ν_β's emerging from an originally pure ν_α beam are called "appearance experiments"; similarly, "disappearance experiments" are those which measure the residual abundance of the original ν_α's after they traversed a baseline L.

Historically, the first indications for neutrino oscillations came from the study of natural sources. The long-standing solar neutrino problem (deficit of ν_e rate with respect to the Solar Standard Model predictions) and the so-called atmospheric anomaly (namely the suppression of ν_μ/ν_e ratio for upwardgoing events) were definitely solved in the last decade with a new generation of experiments. In particular, the Sudbury Neutrino Oscillation experiment [1] provided a model independent proof of the solar $\nu_e \to \nu_{\mu,\tau}$ transition, while new atmospheric neutrino experiments [2–4] showed compelling evidence for muon rate suppression as a function of the energy E and the baseline L, precisely as predicted from (3).

However, the final confirmation of the oscillation hypothesis came from laboratory experiments. The use of artificial neutrinos from reactors (Kam-LAND) and from accelerators (K2K and MINOS) confirmed, respectively, that the solar and atmospheric anomalies were not just an artifact of the poor-knowledge of the source but a fundamental property of the neutrino itself.

In the following we summarize the main steps in the production of reactor and accelerator neutrinos and draw up a (not exhaustive) list of experiments which used them to carry out oscillation studies.

2 Experiments based on artificially-produced neutrinos

Artificial neutrino sources

Reactor neutrinos are produced in β decays of neutron-rich fission fragments. The anti-neutrinos are emitted isotropically from the reactor core with a spectral density of the form $f(E_\nu) = N_f \varrho_f(E_\nu)/4\pi R^2$ where N_f is the number of fission events per second, $\varrho_f(E_\nu)$ is the anti-neutrino spectrum per fission and R is the distance from the reactor core (N_f is related to the thermal power of the reactor). The typical anti-neutrino energy is in the few-MeV range. Therefore, at nuclear reactors only disappearance experiments can be performed, being the interactions under-threshold for creating muon and tau leptons. At the detector site, neutrinos are recorded exploiting the specific signature of the inverse β decay $p\bar{\nu}_e \to ne^+$, where both the positron and the delayed neutron capture are observed.

At accelerators, neutrinos of intermediate energies (10–100 MeV) are produced dumping low energy protons ($\lesssim 1\,\mathrm{GeV}$) onto a target to produce secondary pions. Most of the π^- are absorbed by nuclei, while π^+ de-

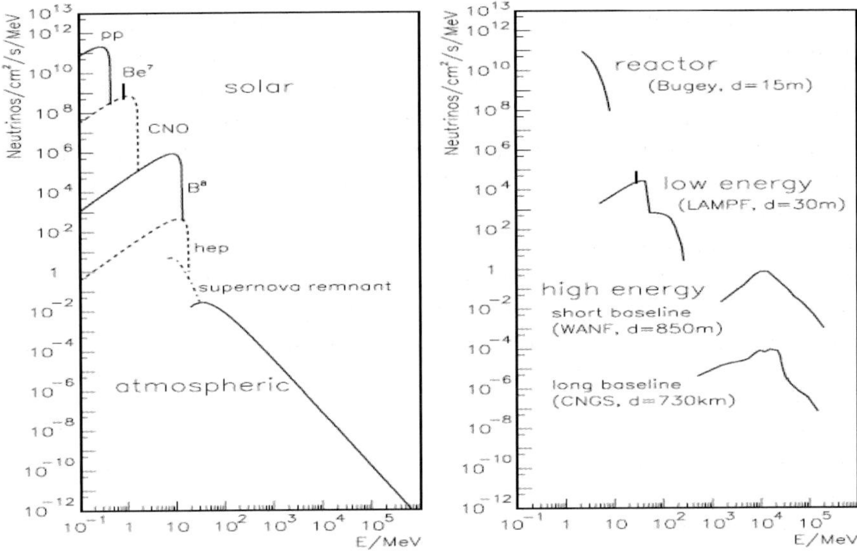

Fig. 1. Comparison of the energy-intensity relation between natural (*left*) and artificial (*right*) neutrino beams [5]. Note that long and short-baseline experiments share the same production mechanism (and hence the same energy range) but they have different intensities (due to the $1/r^2$ reduction factor)

cay at rest (DAR) via $\pi^+ \rightarrow \mu^+ \nu_\mu$; then the muons undergo three-body decays $\mu^+ \rightarrow e^+ \nu_e \bar{\nu}_\mu$.

Finally, high energy neutrino beams ($E_\nu \gtrsim 1\,\mathrm{GeV}$) are produced accelerating protons up to hundreds of GeV and impinging them on a needle-shaped target. The secondary produced (mainly pions and kaons) are momentum-selected by a series lensing devices ("horns") which focus them toward an evacuated decay tunnel of length $\lesssim 1\,\mathrm{km}$. It is possible to switch from ν_μ to $\bar{\nu}_\mu$ simple reversing the current of the magnets.

In Fig. 1 we show the energy-intensity relations for both natural and laboratory neutrino sources. In general we may note the sharp flux decrease as the energy increases; however, this loss is partially compensated by the increse of the neutrino cross-sections at high energy. In the following we briefly report (not in chronological order) some of the laboratory experiments that shed light in the study of neutrino oscillations.

Neutrino detectors

The KamLAND experiment in Japan measures the flux and the spectrum of $\bar{\nu}_e$ produced in several surrounding nuclear reactors. It is located in the Kamioka site at an average distance from the sources of $\sim 180\,\mathrm{km}$. It uses a kton of liquid scintillator surrounded by nearly 2000 PMTs and observes neutrinos via the inverse-beta reaction $\bar{\nu}_e p \rightarrow e^+ n$, where the Cerenkov light produced by both the positron and the delayed neutron capture are detected. KamLAND results [6] showed that terrestrial neutrinos oscillate in the same fashion of solar neutrinos provided that, for the latter, also flavor-conversion phenomena inside the sun (MSW effect) are taken into account.

The K2K experiment in Japan (and more recently MINOS in the USA) observed with an artificial ν_μ beam the same rate suppression of atmospheric neutrino experiments, and also the energy spectrum distortion lead to oscillation parameters laying in the atmospheric allowed region [7].

The CHOOZ reactor experiment ($L \sim 1\,\mathrm{km}, E \sim 3\,\mathrm{MeV}$) looked from $\bar{\nu}_e$ disappearance in the region of atmospheric Δm^2. The negative result of this search excluded $\nu_\mu \to \nu_e$ as the dominant atmospheric oscillation channel and put a stringent bound on the θ_{13} mixing angle [8].

In 1995 the LSND experiment at Los Alamos observed an excess of $\bar{\nu}_e$ events in a predominantly $\bar{\nu}_\mu$ beam [9]. The final analysis reported an excess of $\nu_e = 87.9 \pm 22.4 \pm 6.0$ events, corresponding to an oscillation probability $P(\bar{\nu}_\mu \to \bar{\nu}_e) = (0.264 \pm 0.067 \pm 0.045)\%$. The main concern with the LSND result is that it's allocation in the oscillation scenario requires a fourth sterile neutrino or CPT violation (i.e. different Δm^2 in neutrino and anti-neutrino sectors) or more exotics possibilities (e.g. mass varying neutrinos or violation of the Lorentz Invariance). The (already running) MiniBooNE experiment will unambiguously confirm or refute the 3.8σ effect reported by LSND. First results are expected in the near future.

What is still missing to close the discovery phase is the direct proof the $\nu_\mu \to \nu_\tau$ oscillations at the atmospheric scale, a challenge that is going to be accomplished by the CNGS/OPERA τ appearance program [10]. First neutrinos have already been delivered from CERN and detected in Gran Sasso. The two long baseline experiments OPERA and MINOS would also try a first attempt to measure non-zero θ_{13} below the CHOOZ limit.

References

1. Q. R. Ahmad *et al.*, Phys. Rev. Lett. **87**, 071301 (2001); Phys. Rev. Lett. **89**, 011301 (2002); Phys. Rev. Lett. **89**, 011302 (2002).
2. Y. Fukuda *et al.*, Phys. Rev. Lett. **81**, 1562 (1998).
3. M. Ambrosio *et al.*, Phys. Lett. B **434** 451 (1998); Eur. Phys. J. C **36** 323 (2004);
4. W. W. M. Allison *et al.*, Phys. Lett. B **449**, 137 (1999).
5. A. Geiser, Rep. Prog. Phys. **63**, 1779 (2000).
6. K. Eguchi *et al.*, Phys. Rev. Lett. **90**, 021802 (2003); T. Araki *et al.*, Phys. Rev. Lett. **94**, 081801 (2005).
7. E. Aliu *et al.*, Phys. Rev. Lett. **94**, 081802 (2005).
8. M. Apollonio *et al.*, Phys. Lett. B **466**, 415 (1999).
9. A. Aguilar *et al.*, Phys. Rev. D **64**, 112007 (2001).
10. M. Guler *et al.*, CERN/SPSC 2000-028, SPSC/P318, LNGS P25/2000, July, 2000.

Future measurements of neutrinos from the Sun, Earth and Supernovae

Aldo Ianni

I.N.F.N. Gran Sasso Laboratory, S.S. 17 bis km 18+910, 67010 Assergi (AQ),
Italy aldo.ianni@lngs.infn.it

In this paper a brief review of the research for neutrinos from natural sources,
namely solar neutrinos, neutrinos from the Earth (so-called geoneutrinos) and
Supernova neutrinos is presented. Emphasis is given to established facts and
to future measurements in experiments in operation or planned.

1 Solar neutrinos

The Sun is a natural source of low energy ($< 20\,\mathrm{MeV}$) electron neutrinos, ν_e.
The flux of these neutrinos on Earth is of the order of $60 \times 10^9\,\mathrm{cm}^{-2}\,\mathrm{s}^{-1}$ [1].
Solar neutrinos offer an important opportunity to search for new physics in
the field of both particle physics and astrophysics. As a matter of fact, solar
neutrinos have unique features: low energy (mean energy $\sim 0.4\,\mathrm{MeV}$), long
baseline ($10^{13}\,\mathrm{cm}$) and the possibility to travel through high density matter
($\sim 150\,\mathrm{g/cm}^3$), the Sun's core. This latter fact is of particular importance to
study neutrino–matter interactions. From collected experimental data [2–6]
it has been shown that solar neutrinos undergo matter-enhanced oscillations
(so-called MSW). Using the KamLAND [7] measurement (reactor antineutrinos with a baseline of about $180\,\mathrm{km}$), it is possible to reduce the allowed
range in the space of oscillation parameters, namely $\delta\mathrm{m}_{12}^2 = \mathrm{m}_2 - \mathrm{m}_1$ and
$\sin\theta_{12}$ (see as an example Fig. 5 in [8]). The combination of solar neutrino
observations and KamLAND reactor antineutrino data selects the so-called
Large Mixing Angle (LMA) solution for the phenomenology of neutrinos from
the Sun (assuming CPT). Solar neutrino oscillations are driven mainly by the
parameters $\delta\mathrm{m}_{12}^2$ and $\sin^2\theta_{12}$. The survival probability for electron neutrinos
on Earth depends weakly on θ_{13}. Moreover, solar neutrinos oscillations allow
to establish that $\mathrm{m}_2 > \mathrm{m}_1$, being m_i with $i = 1, 2, 3$ a neutrino mass eigenstate. Yet, at present, it is not known whether m_3 is larger or smaller than
m_1. In the case $\mathrm{m}_1 < \mathrm{m}_2 < \mathrm{m}_3$ we write about "normal hierarchy"; in the
case $\mathrm{m}_3 < \mathrm{m}_1 < \mathrm{m}_2$ we write about "inverted hierarchy". The LMA solution
leads to energy spectrum distortion and regeneration of electron neutrinos on

Earth (so-called day–night effect). Both these effects have not been observed yet by running real-time experiments, SNO and SuperKamiokande, due to low statistics and systematics. The LMA predicts an increase of the survival probability on Earth for electron neutrinos born in the Sun's core. This so-called "up-turn" of the survival probability is expected in the 1–2 MeV range, which is not accessible by neither SNO nor SuperKamiokande.

Solar neutrinos have been measured in real-time only above 5 MeV. Only a tiny fraction (0.01%) of the neutrinos are above 5 MeV. Therefore, future experiments searching for ^{7}Be or pp neutrinos have the opportunity to study in a great detail the Solar Standard Model (as pointed out in [9]) and perform tests for new physics. As a matter of fact, a number of papers [10] suggest that in the 1 MeV region non-standard neutrino–quark interactions may show up. Moreover, in 1 MeV region a measurement of pep and CNO neutrinos could be possible in the case the intrinsic radioactivity of U and Th is at the level of 10^{-17} g/g [11]. For such a measurement it is important that the experimental site is located as deep as possible due to ^{11}C muon-induced background (see [11] for details).

In the coming future (2007) two experiments, Borexino [12] and KamLAND [13], will start measuring low energy (^{7}Be and pep) solar neutrinos; both Borexino and KamLAND make use of a massive organic liquid scintillator. Borexino is getting filled with water at present and before spring 2007 will be filled with scintillator. KamLAND is building a new purification system in order to reduce the internal background due mainly to ^{210}Pb and ^{85}Kr. The expected rate of solar neutrino events in both detectors will allow, as mentioned, to search for new physics, dealing with neutrino–matter interactions, with solar neutrino time variations etc, and to study in great detail the parameters of the Solar Standard Model (see [14]).

Among the planned new experiments to search for solar neutrinos we mention SNO+. SNO+ is the future of SNO: the heavy water will be replaced with 1kton of organic scintillator. This new detector is planned to take data after 2009. Moreover, due to its deep location, SNO+ is well suited to search for pep and CNO neutrinos.

2 Neutrinos from the Earth: geoneutrinos

The measured heat released by the Earth (30–40 TW) is in part due to the radioactivity of long-lived radioactive isotopes such as ^{238}U, ^{232}Th and ^{40}K. At present it is believed that about 40% or more of the heat radiated by the Earth has radiogenic origin [15]. Yet, a detailed knowledge depends on the distribution of U, Th and K in the crust and mantle of the Earth. These quantities are not known and maybe impossible to be measured directly. However, an indirect measurement is feasible by using a massive liquid scintillator detector designed to search for low energy solar neutrinos. As a matter of fact, the radioactivity of U, Th and K produces a natural flux of electron antineu-

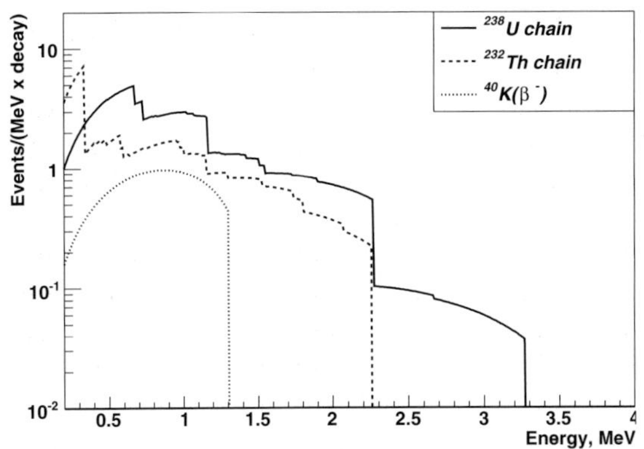

Fig. 1. The geoneutrino energy spectrum from U, Th and K. Spectra are normalized to the number of antineutrinos produced in each decay or chain of decays

trinos. In Fig. 1 we show the expected energy spectrum of these neutrinos, also referred to as geoneutrinos. The total flux of geoneutrinos on the Earth's surface is estimated to be at the level of $10^6 \, \mathrm{cm}^{-2} \, \mathrm{s}^{-1}$, and depends on the location [16]. In fact, the U and Th content in the oceanic crust is much different than that on the continental crust. So, a location by the ocean will see a large fraction of geoneutrinos produced in the oceanic crust; the situation is completely different for a location in the middle of a continent.

As it can be seen from Fig. 1 a fraction of the geoneutrinos has energy above the inverse-beta decay reaction (anti-$\nu_e + p \rightarrow e^+ + n$). In particular, the fraction of geoneutrinos from the U and Th chains above 1.8 MeV are 0.38 and 0.15, respectively. The inverse-beta decay reaction has a well defined signature to search for electron antineutrinos. In particular, in a solar neutrino detector such as Borexino, KamLAND and SNO+ this reaction is almost background-free (see [17] for details of backgrounds and the detection method).

We point out that neutrino oscillations do not play an important role for geoneutrinos. In fact, only the mixing angle enters in the calculation of the expected number of events due to the fact that, given the average neutrino energy and the path between the source and the detector, the oscillation term, that depends on δm_{12}^2, can be average out. Geoneutrino signal allows to distinguish the U and Th contributions. In fact, geoneutrinos from U have larger energy due to the beta decay of ^{214}Bi (end-point energy at 3.72 MeV). Recently, KamLAND has claimed the first observation of geoneutrinos [18]. This measurement, although from a first sample with rather poor statistics, opens a new field of research which aims to establish a link between particle physics and geophysics. In fact, the ultimate goal of the geoneutrino research is to determine the distribution of U and Th within the mantle and test different models of the Earth formation. In the near future (2007–2009) the same experiments reported in the previous Section, namely, Borexino, KamLAND and SNO+, will perform a search for geoneutrinos. In particular, Borexino and SNO+, being far from nuclear reactors, are well-suited for this kind of measurement. However, due to the small number of events expected (in Borexino about 6/yr) and uncertainties on the distribution of U and Th [19], the geoneutrino research will require, in order to say something about the U and Th from the mantle, on the order of 5 years of statistics and data from a few detectors from different locations.

3 Neutrinos from Supernovae

It is believed that massive stars ($M>8M_\odot$) will end in a core-collapse SuperNova (SN) [20]. About 99% of the liberated gravitational energy ($\sim 3 \times 10^{53}$ ergs) goes into neutrinos. At present, only one experimental evidence of supernova neutrinos (SN1987A) exists. This evidence shows a number of puzzling features. It is yet not clear whether these features are due to a poor statistics or to a serious problem with the SN model. Although there is not a well defined SN theory, in the usual calculations of the predicted number of neutrino events in a given detector, $\nu_{\mu,\tau}$ and their antiparticles are expected to be emitted with a temperature $T \sim 8$ MeV, anti-ν_e with $T \sim 5$ MeV and ν_e with $T \sim 3.5$ MeV. Moreover, it is assumed equipartition among the six types of neutrinos and antineutrinos (we refer to this scenario as that for a standard SN). In order to estimate the expected number of neutrino events for a SN burst we usually consider a galactic SN at 10 kpc or at 8.5 kps (the distance of the Sun with respect to the center of the galaxy). We remind that the most likely distance for a galactic SN is predicted to be 11 ± 5 kpc [21].

Due to the fact that the estimated rate for a galactic SN is 40 ± 10 years/SN, the long-term stability of a SN detector is an important factor to be considered. At present SN detectors are: SuperKamiokande (a water Cherenkov); Mini-BooNE, Borexino, KamLAND and LVD (liquid scintillators); AMANDA (long string water Cherenkov). For the next galactic SN these detectors will be able to collect from about 10^2 to 10^4 events by the inverse-beta decay channel. So, a data sample much bigger than the one collected by Kamiokande (11 events) or IMB (6 events) in 1987. Therefore, the next observation of a galactic SN should boost our understanding of a core-collapse explosion.

Liquid scintillator detectors designed to search for solar neutrinos can probe a very promising interaction channel for SN neutrinos, namely the neutrino-proton elastic scattering [22]. This channel is of relevant interest because, in order to measure the temperature and energy of $\nu_{\mu,\tau}$'s and their antiparticles, one needs a spectral signature. The recoils spectrum of protons gives such a signature. However, due to quenching effects the recoil spectrum is pushed toward lower energies. Therefore, a low threshold (at the level of 200–300 keV) detectors is necessary.

Neutrino oscillations affect the SN signal. As a matter of fact, the flux of electron neutrinos, as an example, on Earth is written:

$$F_e^{\text{Earth}} = P_{ee} F_e^{\text{SN}} + (1 - P_{ee}) F_{x=\mu,\tau}^{\text{SN}} \tag{1}$$

where F^{SN} is the flux at the SN and P_{ee} is the survival probability after oscillations within the SN core.

Acknowledgement. I would like to thank the organizers of the neutrinos and cosmic rays parallel session, E. Lisi and L. Patrizii, for the invitation to the conference.

References

1. J.N. Bahcall, *et al.*, Astrophys. J., 555, (2001), 990. J.N. Bahcall, hep-ph/0412068.
2. B.T. Cleveland, *et al.*, Astrophys. J., 496, (2005), 505.
3. Y. Fukuda, *et al.*, Phys. Rev. Lett. 77, (1996), 1683.
4. GALLEX collaboration, W. Hampel, *et al.*, Phys. Lett. B447 , (1999), 127. SAGE collaboration, J.N. Abdurashitov, *et al.*, Zh. Eksp. Teor. Fiz. 122, (2002), 211 [J. Exp. Theor. Phys. 95, (2002), 181].
5. Y. Fukuda, *et al.*, Phys. Rev. Lett., 81, (1998), 1158.
6. Q.R. Ahmad, *et al.*, Phys. Rev. Lett. 87, (2001) 071301. Q.R. Ahmad, *et al.*, Phys. Rev. Lett. 89, (2002) 011301.
7. T. Araki, *et al.*, KamLAND collaboration, Phys.Rev.Lett. 94 (2005) 081801.
8. G.L. Fogli *et al.*, Prog.Part.Nucl.Phys. 57 (2006) 742–795.
9. J.N. Bahcall and C. Pena-Garay, New J.Phys. 6 (2004) 63
10. A. Friedland *et al.*, Phys.Lett. B594 (2004) 347.
11. C. Galbiati *et al.*, Phys.Rev. C71 (2005) 055805.
12. G. Alimonti *et al.*, Borexino collaboration, Astrop. Phys. 16 (2002) 205–234.
13. KamLAND collaboration, Phys.Rev.Lett. 90 (2003) 021802.
14. A. Bandyopadhyay *et al.*, hep-ph/0608323.
15. G. Fiorentini, Phys.Lett. B557 (2003) 139–146.
16. F. Mantovani *et al.*, Phys.Rev. D69 (2004) 013001.
17. M. Balata *et al.*, Borexino collaboration, Eur. Phys. J. C 47 (2006) 21–30.
18. T. Araki *et al.*, KamLAND collaboration, Nature 436 (2005) 499.
19. G.L. Fogli *et al.*, Phys. Scr T127 (2006) 89–94.
20. K. Kotake, K. Sato, K. Takahashi, Rept.Prog.Phys. 69 (2006) 971–1144 (also at astro-ph/0509456).
21. A. Mirizzi, G.G. Raffelt and P.D. Serpico, astro-ph/0604300 v2.
22. J. Beacom *et al.*, Phys.Rev. D66 (2002) 033001.

Cosmology and Neutrinos,
of fixed and variable mass

Marco Cirelli

Department of Physics, Yale University, New Haven, CT 06520, USA
marco.cirelli@yale.edu

Thanks to recent data about the Cosmic Microwave Background (CMB), Large Scale Structures (LSS) and also Type Ia Supernovæ (SNe), cosmology has become the most sensitive probe of some neutrino properties and a very sensitive probe of others, including non standard ones [1–3]. On the issue of ν masses, cosmological observations play the dominant role nowadays: oscillation experiments test squared-mass differences, and other means of probing the absolute ν mass are less sensitive. On the other hand, cosmology may have an even deeper connection with neutrinos, being at the very origin of their mass, according to the proposal of Mass Varying Neutrinos (MaVaNs) [5].

In the first part of this contribution (Sect. 1) I rapidly review the status of the current bounds on ν masses from cosmology, with the use of different techniques, and the prospects of future developments. In the second part (Sect. 2) I introduce the scenario of MaVaNs and mention the constraints on ν masses that it implies if it is realized in Nature [6].

1 Neutrino masses from cosmology

Cosmological probes are sensitive to the mass of the neutrinos in a few different ways. A basic observation is that the mass scale of $\mathcal{O}(\lesssim 1\,\mathrm{eV})$ is such that neutrinos were relativistic at least until just after recombination, when "CMB is formed", so that its effect is expected to be felt on CMB in an indirect way and on LSS formation, which takes place subsequently.

CMB anisotropies spectrum. The well known power spectrum of CMB anisotropies is sensitive to the amount of energy density stored in the relativistic form of neutrinos $\Omega_\nu h^2 = \sum m_\nu/93\,\mathrm{eV}$. The more massive the neutrinos, the larger the contribution to the relativistic energy density and the more delayed the realization of the matter-radiation equality in the Universe. The net effects on the CMB spectrum can be complicated but in first approximation consist of a shift of the peaks "up and to the left", as a consequence of the different evolution of the scale factor and of the reduced integrated Sachs–Wolfe effect. Comparing the modified CMB spectrum with the extremely precise measurements available, it is therefore possible to constrain the ν mass. This was first realized by Ichikawa et al. in 2004 and later applied in a series of studies. After the latest 3-year data release from WMAP [4] we find in [3]

$$\sum m_\nu < 2.2 \text{ eV} \quad \text{at 95\% C.L.} \quad \text{(CMB alone)} \tag{1}$$

Other analysis find similar results (see [3] for a list of references). In the future, the next generation CMB experiments are likely to improve the sensitivity and therefore impose a stricter bound or positively detect the effect of ν masses. E. g. the PLANCK satellite is expected to reach $\sum m_\nu \approx 0.45\,\text{eV}$, while a future experiment with measurements limited only by the intrinsic Cosmic Variance should proceed to $\sum m_\nu \approx 0.15\,\text{eV}$ [2].

Matter power spectrum. The ν mass affects more directly the matter power spectrum of LSS. When relativistic, neutrinos free-stream out of the galactic structures that are being formed by the clustering force of gravity, thus suppressing their growth. When neutrinos become non relativistic, as a consequence of being massive, they can travel smaller distances and therefore their smoothening effect is felt only on the small scales. Thus the ν mass determines the amount and the time (corresponding to the cosmological scale) of the resulting suppression. Confronting with the observed LSS matter power spectrum allows to derive constraints on $\sum m_\nu$. Recent galaxy surveys like 2dFGRS and SDSS measure the matter power spectrum via the distribution of luminous galaxies. This is subject to the assumption that the galactic matter traces the underlying total matter as parameterized by a bias parameter b, that has to be independently determined or modelized. From this set of data we find in [3]

$$\sum m_\nu < 0.73\,\text{eV} \quad \text{at 99.9\% C.L.} \quad \text{(CMB + LSS)}\,. \tag{2}$$

Observing the forest of absorption lines at the (redshifted) Lyα frequency in the light of distant quasars also allows to reconstruct the distribution of the intervening matter (mainly at smaller scales compared to the galaxy surveys). While these measurements are still quite plagued by controversy and uncertainties, taking them into account we find in [3]

$$\sum m_\nu < 0.40\,\text{eV} \quad \text{at 99.9\% C.L.} \quad \text{(CMB + LSS + Ly}\alpha\text{F)}\,. \tag{3}$$

In the future, these techniques are expected to reach sensitivities around $\sum m_\nu \approx 0.1\,\text{eV}$. Other techniques (most notably the galaxy lensing surveys – that will reconstruct the intervening matter distribution from the deformation of galaxy images in the background – and the CMB lensing surveys – that will apply the same concept to the CMB itself) are expected to reach $\sum m_\nu \approx 0.03\,\text{eV}$. If achieved, this "guarantees" a positive detection of ν masses, as the atmospheric ν experiments imply $\sum m_\nu \gtrsim 0.06\,\text{eV} \simeq \sqrt{\Delta m_{\text{atm}}^2}$.

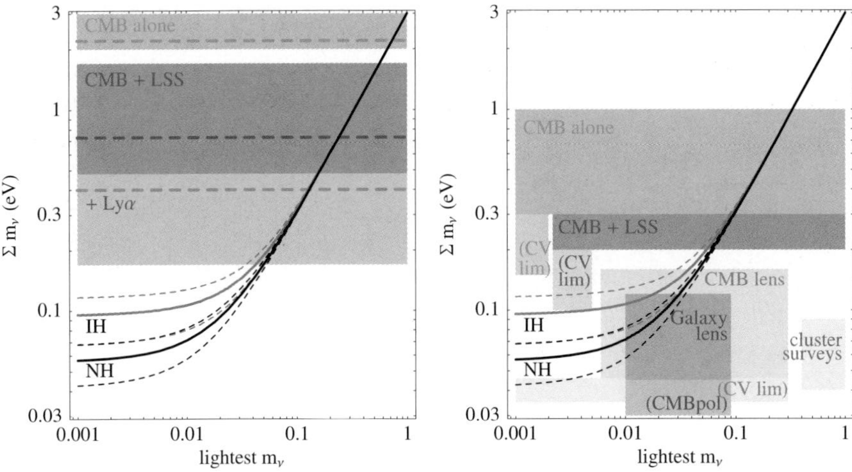

Fig. 1. *Left:* Summary of the current upper bounds on the sum of ν masses $\sum m_\nu$ from cosmology, as a function of the lightest mass m_ν. The *lower* (*upper*) *solid curve* corresponds to normal (inverted) hierarchy, with the errors from the neutrino Δm^2 measurements (*dashed side lines*). Each *shaded rectangle* spans the values found by different analysis and/or authors, while the *horizontal thick dashed lines* are our 99.9% C.L. bounds from [3]. From *top* to *bottom*: bounds from CMB alone, from CMB+LSS and with the addition of Lyα forest data. *Right:* Outlook of the possible future sensitivity. From *top* to *bottom*, the *rectangles* select the ranges predicted by different authors using CMB alone, CMB+LSS, CMB lensing observations, galaxy lensing surveys and cluster surveys. The *paler lower portions* might be explored with hypothetical Cosmic Variance-limited CMB experiments or by CMBpol

2 Mass Varying Neutrinos

In the Mass Varying Neutrino scenario [5], the ν mass arises from the interaction with a scalar field responsible for the acceleration of the Universe, whose effective potential changes as a function of the ν density. This establishes a very intriguing connection between two recent pieces of evidence for New Physics – the indirect observation of Dark Energy and the confirmation of ν masses and oscillations – that are both suggestively characterized by a similar mass scale $\mathcal{O}(10^{-3})\,\mathrm{eV}$. More precisely, in this approach at low energies the effective Lagrangian for m_ν is $\mathcal{L} = m_\nu \bar{\nu}^c \nu + V_{\mathrm{tot}}(m_\nu)$, where $V_{\mathrm{tot}}(m_\nu) = V_\nu(m_\nu) + V_0(m_\nu)$ contains the contribution to the energy density both from the ν backgrounds and from the scalar potential of the scalar field. The neutrino mass m_ν is a dynamical field and the condition of minimization of V_{tot} determines its physical value. In the vacuum, $V_\nu(m_\nu)$ consists of the sole background of cosmic neutrinos ($c\nu b$, with number density $n_{c\nu b}$), so that, choosing an appropriate form for the scalar potential (dictated by the cosmological observation of the scale $\Lambda \approx 10^{-3}\,\mathrm{eV}$

and quasi-cosmological-constant nature of the Dark Energy), this scenario implies

$$m_{\nu,0} \simeq \frac{\Lambda^4}{n_{c\nu b}} . \tag{4}$$

Since the cosmic ν density dilutes with the expansion of the Universe, the neutrino mass in vacuum $m_{\nu,0}$ varies (increases) in time. In a neutrino-rich medium, the potential $V_\nu(m_\nu)$ acquires a new contribution due to the environmental ν density. This is in particular the case for the Sun: the mass m_ν of a solar ν streaming out of the production zone in the center varies as a function of the position r, according to

$$m_\nu = m_{\nu,0} - A(r)\, m_{\nu,0}^2 + \cdots \tag{5}$$

where the factor

$$A(r) \approx \frac{1}{n_{c\nu b}} \frac{n_{\nu,\mathrm{Sun}}}{\langle E_\nu \rangle} \tag{6}$$

depends on the environmental neutrino density $n_{\nu,\mathrm{Sun}}(r)$ and their energy E_ν. It can be computed precisely on the basis of the Solar Standard Model [6]. Finally, since the solar environment is dominated by the ν_e flavor, different mass eigenstates are differently affected and it can be shown that the effective neutrino squared mass difference is modified with respect to the standard Δm_{21}^2, by a function which contains the absolute mass of the lightest eigenstate in vacuum $m_{\nu,0}$[1]. A global fit of solar and KamLAND data with such a modified $\Delta m_{\mathrm{MaVaNs}}^2$ shows that the data adverse the scenario [6] and poses the constraint

$$m_{\nu,0} \lesssim 0.01\,\mathrm{eV} . \tag{7}$$

Summarizing, the results of the solar and KamLAND ν experiments quite generally imply that, if the mechanism of MaVaNs is giving neutrinos their mass, then the absolute ν mass in vacuum is subject to the constraint (7).

References

1. A. Strumia, F. Vissani, arXiv:hep-ph/0606054.
2. J. Lesgourgues, S. Pastor, Phys. Rept. **429**, 307 (2006) [arXiv:astro-ph/0603494].
3. M. Cirelli, A. Strumia, arXiv:astro-ph/0607086.
4. D.N. Spergel *et al.*, arXiv:astro-ph/0603449.
5. R. Fardon, A.E. Nelson, N. Weiner, JCAP **0410**, 005 (2004) [astro-ph/0309800]. See also P. Gu *et al.*, Phys. Rev. D **68**, 087301 (2003) [hep-ph/0307148].
6. M. Cirelli, M.C. Gonzalez-Garcia, C. Pena-Garay, Nucl. Phys. B **719** (2005) 219 [arXiv:hep-ph/0503028].

---

[1] In passing, note that, in this scenario, a quantity relevant for oscillation (Δm^2) depends on the absolute ν mass and can therefore be constrained by oscillation measurements. This is never the case in standard neutrino physics.

Double Beta Decay Experiments

Maura Pavan

Università di Milano Bicocca e sez. INFN di Milano Bicocca
pavan@mib.infn.it

"Double Beta Decay provides a means of finding which form of the beta decay theory is correct, the form postulating ν and $\bar{\nu}$ or the (Majorana) form in which all ν's are the same" (E. Fermi, Nuclear Physics, 1949).

1 Double Beta Decay

After about 70 years since the *Beta Decay Theory* was given to light, after the marvellous successes of the *Standar Model Theory* and about 50 years of neutrino experiments ... we still miss one of the more fundamental information concerning the character of neutrinos, and still *"Double Beta Decay provides"* an almost unique *"means of finding which form of the beta decay theory is correct"*. If Neutrinoless Doube Beta Decay ($\beta\beta(0\nu)$) exists the correct form of the theory is the one in which the ν is a massive Majorana particle identical to its antiparticle. If $\beta\beta(0\nu)$ does not exist – and if this is not the result of accidental "cancellations" – ν is a massive Dirac particle. Double Beta Decay is the only spontaneous decay that lead some unstable nuclei to a lowest energy isomer. In the decay two electrons are emitted transforming two neutrons into two protons. If the decay proceeds like two simultaneous beta decay two neutrinos are emitted toghether with the electrons but if ν is a Majorana particle the decay can proceed also through the exchange of a virtual ν and the emission of just electrons. Unable to detect the eventually emitted neutrinos the only signature that the experimentalist can search, to distinguish between $\beta\beta(2\nu)$ and $\beta\beta(0\nu)$, relies in the 2 electron energy spectrum: the sum of the electron kinetic energies is fixed at the Q-value in the $\beta\beta(0\nu)$ (the nuclear recoil is negligible) while it has a continuum probability distribution, zeroing above the Q-value, in the $\beta\beta(2\nu)$. The half-life predicted for $\beta\beta(0\nu)$ is incredibly high, and a hard fight against an overwhelming background is engaged by experiments that since more than 60 years search for this "rare" process. The comparison among the different $\beta\beta(0\nu)$ experimental results is not trivial, two are the outlet of an experiment: a lower limit for the

$\beta\beta(0\nu)$ half-life ($T^{0\nu}_{1/2}$) and a (corresponding) upper limit for the parameter $|\langle m_\nu \rangle|$. This parameter, named the *Majorana mass of neutrino*, is what – beging the ν a Majorana particle and existing $\beta\beta(0\nu)$ – one would like to measure because of its direct correlation with the ν mass scale [2]. To extract $|\langle m_\nu \rangle|$ from $T^{0\nu}_{1/2}$ one have to know the value of the nuclear matrix element $|M_{0\nu}|$ and of the phase space factor $G^{0\nu}$. Indeed $T^{0\nu}_{1/2} = |\langle m_\nu \rangle|^2 G^{0\nu} |M^{0\nu}|^2$. Unfortunately rather different values of $|M_{0\nu}|$ are provided – for the same nucleus – by the different nuclear models, spanning even an order of magnitude [1]. Therefore generally for each nucleus (and then for each lower limit to $T^{0\nu}_{1/2}$) a range of values for $|\langle m_\nu \rangle|$ is reported, according to a selected sample of $|M_{0\nu}|$ values.

2 Experimental sensitivity and Present Results

The sensitivity of a $\beta\beta(0\nu)$ experiment is usually quoted in terms of the maximum half-life that an experiment would be able to investigate. To project this sensitivity on $|\langle m_\nu \rangle|$ the value of $|M_{0\nu}|$ (and its uncertainties) has to be accounted for. The sensitivity increases with the live time, with the number of candidate nuclei, with the improvement of the energy resolution of the detector (as this measures the capability of bkg discrimination) and with the decrease of the background. In the years different techinques has been used and two dinstinct experimental approaches can be devised. Experiments in which source ($\beta\beta$ candidates) and detector (measuring the two electron energies and often also the electron tracks) are separate ($s \neq d$) and experiments where the detector is "made of" a $\beta\beta$ candidate ($s = d$). In the former, due to the small range of the electrons emitted, only very thin sources can be used. This kind of experiments allow to investigate almost any $\beta\beta$ candidate and exploit the much more defined signature of the $\beta\beta$ event (the tracks of the two neutrinos) but pay the price of a higher difficulty in scaling in mass (because of their complicate structure) and generally use poor energy resolution devices. The latter technique is applyed whenever a detector can be made with a material containing a $\beta\beta$ candidate, generally the only available information in this case is the sum energy of the two electrons and no or poor discrimination technique can be applied to reject background events. The advantage is usually the high energy resolution of the devices employed, a simpler set-up and a higher potentiality in scaling the mass. Both the techniques share the problem of background: any event that in the detector is not distinguishable from a $\beta\beta$ event is a background event that spoil the $\beta\beta(0\nu)$ sensitivity. Radioactive contaminations of materials and cosmic rays are generally troublesome background sources. To reach a significant sensitivity experiments are realized deep underground, they are heavy shielded against γ, n, μ and are constructed with select low-activity materials.

Table 1 summarizes the more recent and most competitive results obtained by present and past $\beta\beta(0\nu)$ experiments. Cuoricino and Heidelberg–

Table 1. Summary table of $\beta\beta(0\nu)$ past, present and future experiments [3]. $|\langle m_\nu \rangle|$ values are computed according to [1]

Experiment	Technique	Mass [kg]	Isotope i.a.	$T_{1/2}[y]$	$|\langle m_\nu \rangle|$ [MeV]
H–M	$s = d$ HPGe	10.9	^{76}Ge 86%	1.9×10^{25}	330–840
IGEX	$s = d$ HPGe		^{76}Ge 86%	1.6×10^{25}	360–920
CUORICINO	$s = d$ TeO$_2$ bol	40.7	^{130}Te 34%	2.4×10^{24}	400–1400
NEMO3	$s \neq d$	6.9	^{100}Mo 100%	4.6×10^{23}	600–2700
NEMO3	$s \neq d$	0.9	^{82}Se 100%	6×10^{22}	1200–3200
GERDA I	$s = d$ HPGe	18	^{76}Ge 86%	2×10^{25}	300–700
GERDA II	$s = d$ HPGe	40	^{76}Ge 86%	2×10^{26}	90–240
Majorana	$s = d$ HPGe	120	^{76}Ge 86%	5.5×10^{26}	61–156
CUORE	$s = d$ TeO$_2$ bol	740	^{130}Te 34%	110×10^{26}	59–208

Moscow (H–M) experiments exploit the $s = d$ technique using the former TeO$_2$ bolometers and the latter Hyperpure Germanium Diodes (HPGe); both are located at Laboratori Nazionali del GranSasso (LNGS – Italy). IGEX used HPGe as well, producing results similar to those of H–M; the experiment was located at Canfranc Underground Laboraotry (Spain). NEMO3, located in the Underground Laboratories of Frejus (France), is on the contrary a $s \neq d$ experiment were different isotopes are studied. NEMO3 and Cuoricino are the only two $\beta\beta(0\nu)$ experiments presently running, they will have a good chance to test the "Klapdor claim" of an evidence for $\beta\beta(0\nu)$ in the H–M data with a half-life of 0.8–1.8×10^{25} y corresponding to a $|\langle m_\nu \rangle|$ value of about 0.11–0.56 eV.

3 Experimental sensitivity and Present Results

The renewed interest in $\beta\beta(0\nu)$ of these years, strongly supported by the recent results on neutrino physics, lead to a proliferation of proposed next generation experiments. These experiments are projected in order to reach a sensitivity on $|\langle m_\nu \rangle|$ in the range predicted for the inverted hierarchy, i. e. ~ 50 MeV. To accomplish these results, huge masses of $\beta\beta$ candidates and extremely low backgrounds are required. Of course the choice of a *favourite* candidate from the point of view of the nuclear transition amplitude ($F_{\rm N}$) could help although within the limits imposed by the nuclear matrix elements uncertainties [4]. GERDA and Majorana are the next generation Ge experiments. Both will use arrays of HPGe diodes, made with ^{76}Geenriched material. The segmentation technique will guarantee a high efficiency to reject multisite events (i. e. a large fraction of bkg). The main differences between the two rely in the set-up design that is much more traditional in the case of Majorana (with groups of Ge diodes placed toghether in an heavy radiopure lead shield and surrounded by thick n-shield) and innovative in the case of GERDA (the naked diodes will be immerged in a LAr/LN filled tank that will act both as low T thermal bath and radioactive shield, the tank being surrounded by water). While the underground site for Majorana is not jet assigned GERDA is an approved, funded experiment that is being constructed in the LNGS. CUORE is a tightly packed array of ~ 1000 TeO$_2$ bolometers. The project is based on the experience of CUORICINO and forsees the realization of the largest array ever projected to work at $10\,\mathrm{mK}$. The array, heavy shielded and mounted in a specially designed dilution refrigerator, forms a highly segmented detector with a good efficiency in rejecting multicrystal events. The experiment is approved, already partially funded and its construction in the LNGS started.

4 Conclusions

$\beta\beta(0\nu)$ experiments are among the most interesting tools to further investigate the unknown properties of the neutrino family. Several new generation experiments have been proposed so far. Their discovery potential will be strictly connected to the background levels that they will be able to guarantee. Most likely in the next ten years one or few experiment will be able to reach the $|\langle m_\nu \rangle|$ interval of few tens of MeV, being therefore able to say something about neutrino mass hierarchy. This will contribute to the reconstruction of the neutrino puzzle, providing complementary information to that produced by the neutrino oscillation factory.

References

1. F. Simkovic, G. Pantis, J.D. Vergados and A. Faessler, Phys. Rev. **60**, (1999) 055502.
   S. Stoica and H.V. Klapdor-Kleingrothaus, Nucl. Phys. A **694**, (2001) 269.
   O. Civitarese and J. Suhonen, Nucl. Phys. A **729**, (2003) 867.
   V.A. Rodin, A. Faessler, F. Simkovic, and P. Vogel, Nucl. Phys. A **766**, 107 (2006).
2. F. Feruglio, A. Strumia and F. Vissani, Nucl. Phys. B **637**, 345 (2002).
   S. Pascoli and S. T. Petcov, Phys. Lett. B **544**, (2002) 239.
   S. Pascoli, S. T. Petcov, Phys.Lett. B **580**, (2004) 280-289
3. NOW2006, "Neutrino Oscillation Workshop", `http://www.ba.infn.it/~now2006/`, to be pubblished on Nucl. Phys. B (Proc. Suppl.).
   Neutrino Physics and Astrophysics 2006, http://neutrinosantafe06.com/, to be pubblished on the Proceedings.
4. V. I. Tretyak and Yu. G. Zdesenko, Atomic Data and Nuclear Data Tables, **80**, (2002) 83.
   S. Elliott and P. Vogel, Ann. Rev. Nucl. Part. Sci.**52**, (2002) 115.
   A. Morales and J. Morales, Nucl. Phys. B (Proc. Suppl.) **114**, (2003) 141.
   O. Cremonesi, Nucl. Phys. B (Proc. Suppl.) **118**, (2003) 287.
   K. Zuber, Acta Polonica B **37**, (2006) 1905.

Some Trends in Theoretical Models for Neutrino Masses

Michele Frigerio

Service de Physique Théorique, CEA-Saclay, 91191 Gif-sur-Yvette Cedex, France
frigerio@spht.saclay.cea.fr

In this talk[1] we discuss (1) which neutrino parameters, to be measured in experiments, have the deepest impact on the theory; (2) what are the possible contributions to neutrino mass from physics beyond the Standard Model (SM); (3) how to introduce a family symmetry which describes lepton flavor; (4) how to embed neutrino data within a grand unified framework.

1 A theoretical perspective on neutrino parameters

The mixing among three light Majorana neutrino mass eigenstates is fully described by the structure of the mass matrix, $m_\nu = U^* diag(m_1, m_2, m_3)U^\dagger$, where m_i are in general complex and the unitary matrix U depends on three mixing angles (θ_{12}, θ_{23} and θ_{13}) as well as one CP violating phase δ. While present oscillation results determine θ_{12}, θ_{23}, $|m_2|^2 - |m_1|^2$ and $||m_3|^2 - |m_2|^2|$, in the future they will have access also to the values of θ_{13}, $sign(|m_3|^2 - |m_2|^2)$ and δ. In particular, $\sin^2\theta_{23} = 0.44(1^{+0.41}_{-0.22})$, $\sin^2\theta_{12} = 0.31(1^{+0.18}_{-0.15})$ and $\sin^2\theta_{13} < 0.04$ [1], which are compatible with $1/2, 1/3$ and 0, respectively. This pattern is known as tri-bi-maximal mixing:

$$\nu_2 \approx \frac{\nu_e - \nu_\mu + \nu_\tau}{\sqrt{3}}, \qquad \nu_3 \approx \frac{\nu_\mu + \nu_\tau}{\sqrt{2}}. \tag{1}$$

A conservative upper bound on the neutrino mass scale, $|m_i| < 0.5\,\text{eV}$, comes from cosmological data, which are sensitive to $\sum_i |m_i|$. The combination of future CMB and lensing experiments may achieve within few years a sensitivity of the order of $\sigma(\sum_i |m_i|) \sim 0.05\,\text{eV}$ [2]. Since $||m_3|^2 - |m_2|^2| \approx (0.05\,\text{eV})^2$, we conclude that cosmological probes will not only improve the upper bound

[1] Preprint SACLAY-T06-106. Further details and references are found in my slides: www.pv.infn.it/~ifae2006/talks/NeutriniRaggiCosmici/Frigerio.pdf. I thank Eligio Lisi for invitation and the Organizers for financial support.

on $|m_i|$, but they will actually determine it. If neutrino masses are hierarchical, they may even distinguish normal ordering ($\sum_i |m_i| \approx 0.05\,\text{eV}$) from inverted ($\sum_i |m_i| \approx 0.1\,\text{eV}$). The neutrinoless 2β-decay experiments are sensitive to one Majorana type CP violating phase, $\arg(m_2/m_1)$, while there are no perspectives at present to access the value of $\arg(m_3/m_2)$.

The priority list to reduce the uncertainties on the structure of m_ν reads: (i) to pin down the neutrino mass spectrum (absolute scale and ordering); (ii) to measure the Majorana phases (at least one); (iii) to determine precisely the mixing angles; (iv) to measure δ.

2 A brief classification of contributions to neutrino mass

Light Majorana neutrino masses are added to the SM via the 5-dimensional operator $C_{ij}L_iL_jHH/M$, where $L_i = (\nu_i\ l_i)^T$ and $H = (H^+\ H^0)^T$ are the lepton and Higgs doublets. When H^0 acquires a vacuum expectation value (VEV) $v = 174\,\text{GeV}$, the mass term $(m_\nu)_{ij}\nu_i\nu_j$ is induced, with $(m_\nu)_{ij} = C_{ij}v^2/M$. A host of new physics candidates may generate such effective operator, so that the neutrino mass matrix observed in experiments should be thought as a sum over several possible contributions, $m_\nu = \sum_i m_\nu^{(i)}$.

A first class of contributions (seesaw mechanism) comes from the tree-level exchange of new particles with mass $M \gg v$. They can be singlet fermions, that is right-handed neutrinos (type I seesaw), isotriplet Higgs bosons (type II) or isotriplet fermions (type III) [3]. Light right-handed (sterile) neutrinos may also contribute to left-handed (active) neutrino mass, in a way proportional to the active-sterile mixing, $m_\nu \sim \sin^2\theta_{\text{as}}m_s$ [4]. A second class of contributions is produced at loop level (radiative mechanism). If we limit ourselves to 1-loop diagrams involving scalars and fermions, we find only four possibilities, shown in Fig. 1. The four external legs represent L and H doublets, while non-standard particles are exchanged in the loop. In such radiative mechanism the induced neutrino mass is suppressed both by the masses of the new heavy particles and by the loop factor. In models where the effective operator is induced only through 2 or more loops, the strong suppression from loop factors makes m_ν sufficiently small even when the new particles in the loops are as light as the electroweak scale and thus may be directly observed.

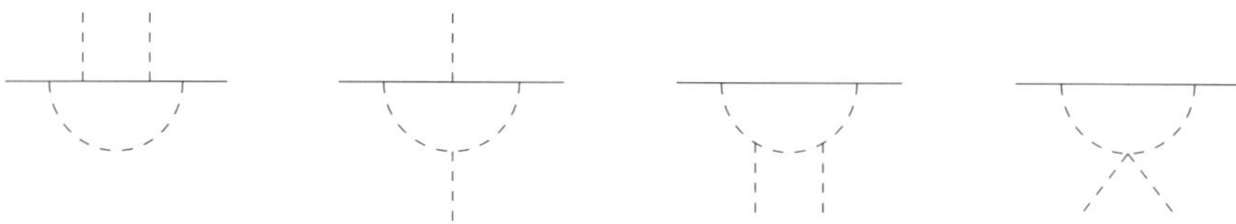

Fig. 1. The four Feynman diagrams which induce the operator $LLHH$ at 1-loop

3 Family symmetries: the case of A_4

The two large mixing angles and the weak mass hierarchy $(\Delta m_{12}^2/\Delta m_{23}^2)^{1/2} \approx 1/5$ indicate that m_ν should be described in terms of order one parameters. To this end, discrete symmetry groups are more flexible than continuos Lie groups, because they have more low dimensional irreducible representations (irreps) to accommodate fermion families. If the family symmetry group is non-abelian and has a 3-dim irrep, all three lepton generations may transform in the same multiplet $\mathbf{3}$, as suggested by the order one lepton mixing angles.

The smallest group with a 3-dim irrep is A_4, the symmetry group of the regular tetrahedron, formed by 12 elements [5]. It has four irreps $\mathbf{1}, \mathbf{1}', \mathbf{1}'', \mathbf{3}$ and the basic tensor product reads $\mathbf{3} \times \mathbf{3} = \mathbf{1} + \mathbf{1}' + \mathbf{1}'' + \mathbf{3_s} + \mathbf{3_a}$. Let us assign lepton and Higgs fields to A_4 representations as follows: left-handed lepton doublets $L_i \sim \mathbf{3}$, right-handed charged leptons $e_j^c \sim \mathbf{1}, \mathbf{1}', \mathbf{1}''$, Higgs doublets $H_k \sim \mathbf{3}$, with VEVs $v_k = \langle H_k^0 \rangle$, and Higgs triplets $\xi \sim \mathbf{1}$, $\xi_k \sim \mathbf{3}$, with VEVs $u = \langle \xi^0 \rangle$, $u_k = \langle \xi_k^0 \rangle$ (naturally small because of type II seesaw). The invariance under A_4 determines the structure of the charged lepton and neutrino mass matrices, generated respectively by the Yukawa couplings $y_{ijk} L_i e_j^c H_k$ and $(c_{ij}\xi + c_{ijk}\xi_k)L_i L_j$. One can verify that tri-bi-maximal mixing (see (1)) is obtained for $v_1 = v_2 = v_3$, u and u_2 non-zero and $u_1 = u_3 = 0$. To realize such vacuum alignment is highly non-trivial, since clearly A_4 is broken by these sets of VEVs in two different directions, respectively in the charged lepton sector and in the neutrino sector.

Several additional ingredients are needed to construct a realistic consistent model. For example [6], consider the Minimal Supersymmetric SM with exact R-parity and $A_4 \times Z_3$ family symmetry. Add 15 gauge singlet scalars: 3 singlets under A_4 and 4 triplets under A_4. Two of these triplets play the role of H_k and ξ_k above. The interplay of all these singlets in the superpotential leads to the correct vacuum alignment and thus to tri-bi-maximal mixing in a *natural* way, in the sense that no fine-tuning is needed among the couplings allowed by the symmetry. This example shows that the A_4 family symmetry helps but it is not sufficient alone to achieve tri-bi-maximal mixing. Also, other simpler models featuring A_4 symmetry may describe correctly lepton mixing even though they do *not* predict exact tri-bi-maximal mixing.

4 Unification: the case of minimal SUSY $SO(10)$

Neutrino mass is automatically generated in grand unification models based on $SO(10)$, where all SM fermions belongs to the same irrep $\mathbf{16}$. The minimal renormalizable SUSY $SO(10)$ model contains only two Higgs multiplets coupling to fermions, $\mathbf{10}_H$ and $\overline{\mathbf{126}}_H$, so that the Yukawa superpotential can be written as

$$W_Y = y_{ij}\mathbf{16}_i\mathbf{16}_j\mathbf{10}_H + f_{ij}\mathbf{16}_i\mathbf{16}_j\overline{\mathbf{126}}_H \ . \tag{2}$$

The four mass matrices of quarks and leptons, M_u, M_d, M_e and m_ν, depend only on the two Yukawa coupling matrices y and f, as well as on the VEVs acquired by two of the $\mathbf{10}_H$ components and four of the $\overline{\mathbf{126}}_H$ components. One realizes, therefore, that it is highly non-trivial to reproduce at the same time the values of 6 quark masses, 4 parameters of the CKM quark mixing matrix, 3 charged lepton masses, 2 neutrino mass squared difference and 2 mixing angles, as well as the upper bounds on θ_{13} and $|m_i|$.

If the VEVs are treated as free parameters, one may obtain a satisfactory fit of all data [7], with the predictions of normal mass hierarchy of the neutrino spectrum as well as $\theta_{13} > 0.1$. A very difficult task in such fit is to control the relative size of type I and type II seesaw contributions to m_ν. When the VEVs are constrained by the consistency between the $SO(10)$ symmetry breaking and the gauge coupling unification, one finds that the type II seesaw contribution is extremely suppressed and even the type I is more than a factor 10 too small to accommodate the largest neutrino oscillation scale $\sim 0.05\,\mathrm{eV}$ [7]. Since the minimal model is on the edge of being ruled out, the search is opened for non-minimal extensions which may rescue the SUSY $SO(10)$ construction *and* still maintain some predictivity. They include (i) naturally small perturbation from a quasi-decoupled $\mathbf{120}_H$ multiplet; (ii) substantial contribution to charged fermion masses from $\mathbf{120}_H$ instead of $\overline{\mathbf{126}}_H$; (iii) effect of non-renormalizable operators; (iv) addition of a second $\mathbf{10}_H$ multiplet; (v) non-SUSY scenario with intermediate scales to achieve gauge coupling unification. Due to lack of space, we refer to [7] for appropriate references.

5 Summary

We illustrated some ingredients needed to identify and test the theory of neutrino mass beyond the SM. We need to know the neutrino mass spectrum (normal or inverted, absolute scale, at least one Majorana phase) and, in the second place, the precise values of mixing angles. Moreover, we need to explore the connections of neutrino parameters with other sectors of the theory, e.g. quark-lepton family symmetries and/or constraints from grand unification.

References

1. G. L. Fogli *et al.*: Prog. Part. Nucl. Phys. **57**, 742 (2006)
2. S. Hannestad, H. Tu, Y. Y. Y. Wong: JCAP **0606**, 025 (2006)
3. E. Ma: Phys. Rev. Lett. **81**, 1171 (1998)
4. A. Y. Smirnov, R. Zukanovich Funchal: Phys. Rev. D **74**, 013001 (2006)
5. E. Ma, G. Rajasekaran: Phys. Rev. D **64**, 113012 (2001)
6. G. Altarelli, F. Feruglio: Nucl. Phys. B **741**, 215 (2006)
7. S. Bertolini, T. Schwetz, M. Malinsky: Phys. Rev. D **73**, 115012 (2006)

Supernova neutrino burst as a probe
of shock waves and matter density fluctuations

Gian Luigi Fogli[1], Eligio Lisi[1],
Alessandro Mirizzi[1](*), and Daniele Montanino[2]

[1] Dipartimento di Fisica and Sezione INFN di Bari
   Via Amendola 173, 70126 Bari, Italy
[2] Dipartimento di Fisica and Sezione INFN di Lecce
   Via Arnesano, 73100 Lecce, Italy

Summary. Future detection of a supernova neutrino burst by large underground detectors would give important information about the supernova matter density profile and unknown neutrino properties, due to the effects of flavor oscillations in the supernova envelope. We mainly discuss the detectability of signatures associated to the shock-wave propagation in observable neutrino signals. We also investigate the effects of possible small-scale stochastic matter density fluctuations in the wake of supernova shock waves. We find that such fluctuations can partly erase the shock-wave imprint on the neutrino event spectra.

1 Introduction

A core collapse supernova (SN) is one of the most spectacular source of ν's in nature. A future detection of supernova neutrino burst by large underground detectors will provide us with important information on the core-collapse explosion mechanism, as well as on neutrino properties. In this context, we aim at studying some time dependent observables related to galactic core-collapse supernovae. The plan of the talk is as follows. In Sect. 2 we briefly describe supernova neutrino oscillations in presence of shock-wave propagation and of (possibly associated) stochastic matter density fluctuations. In Sect. 3 we investigate the signatures of shock waves and of matter fluctuations on observable neutrino time spectra. Conclusions and prospects are given in Sect. 4. This talk is based on the results of [1–3], to which we refer the interested reader for further details.

(*) Speaker. E-mail:`alessandro.mirizzi@ba.infn.it`

2 Neutrino oscillations in "turbulent" supernova shock wave

We assume the standard 3ν oscillation scenario [4]. In this framework, the 3ν squared mass spectrum is parametrized in tems of $\delta m^2 = m_2^2 - m_1^2 \simeq 7.92 \times 10^{-5}\,\mathrm{eV}^2 (> 0)$ and $\Delta m^2 = m_3^2 - m_{1,2}^2 \simeq 2.6 \times 10^{-3}\,\mathrm{eV}^2$. The sign of Δm^2 distinguishes the cases of normal hierarchy (NH: $+\Delta m^2$) and inverted hierarchy (IH: $-\Delta m^2$). As a consequence of the hierarchical mass splitting ($|\Delta m^2| \gg \delta m^2$) and of the smallness of $\sin^2 \theta_{13} \lesssim \mathrm{few}\%$, the relevant electron (anti)neutrino survival probability P_{ee} can be expressed as $P_{ee} \simeq P_{ee}^L \cdot P_{ee}^H$ where P_{ee}^L and P_{ee}^H represent effective electron (anti)neutrino survival probabilities in 2ν subsystems [1]. For the current parameter values, the dependence of P_{ee}^L on δm^2 actually vanishes, and P_{ee}^L equals $\sin^2 \theta_{12} \simeq 0.31$ for ν's and $\cos^2 \theta_{12}$ for $\bar\nu$'s. Instead, P_{ee}^H depends non trivially on Δm^2 and on θ_{13}, and encodes the (shock-wave) matter effects through the neutrino potential $V(x) = \sqrt{2} G_F N_e(x)$, where N_e is the electron density at the supernova radius x.

In a seminal paper [5], Schirato and Fuller noted that the propagation of the forward shock wave produces a time-dependent potential $V(x,t)$ that could induce peculiar modifications of the probability P_{ee}^H a few seconds after the core bounce. In addition to forward shock effects, it has been recently pointed out [6] that a second ("reverse") shock front, is also expected to propagate behind the forward one.

In addition, recent core-collapse SN simulations show turbulent convective motions creating a fluctuating density field in the region behind the shock front. In absence of a clear description of such phenomena, we assume that stochastic matter density fluctuations arise only *after* the passage of the shock-wave. We also assume that fluctuations arise at relatively small length scales (say, $L_0 \simeq 10\,\mathrm{km}$), as compared with the neutrino oscillation length in matter. This implies that the fluctuations can be considered as a δ-correlated noise. The amplitude of the fluctuations (ξ) is taken as few % of the local matter potential ($V \to V(1+\xi)$).

We find that the main effect of the fluctuations is to suppress neutrino flavor conversion [3]. For strong damping one gets the limit $P_{ee}^H \to 1/2$, corresponding to a sort of complete "flavor depolarization".

3 Shock-waves and turbulence signatures

In this Sect. we study possible signatures of shock-wave propagation and stochastic density fluctuations that might be seen in next-generations large volume water Cherenkov detectors. We assume as detector fiducial mass $0.4\,\mathrm{Mton}$. We assume a galactic supernova ($d = 10\,\mathrm{kpc}$) and consider the shock signatures in the absolute spectra of the high statistic sample of inverse beta decay events, $\bar\nu_e + p \to n + e^+$. Inverse beta decay events can be

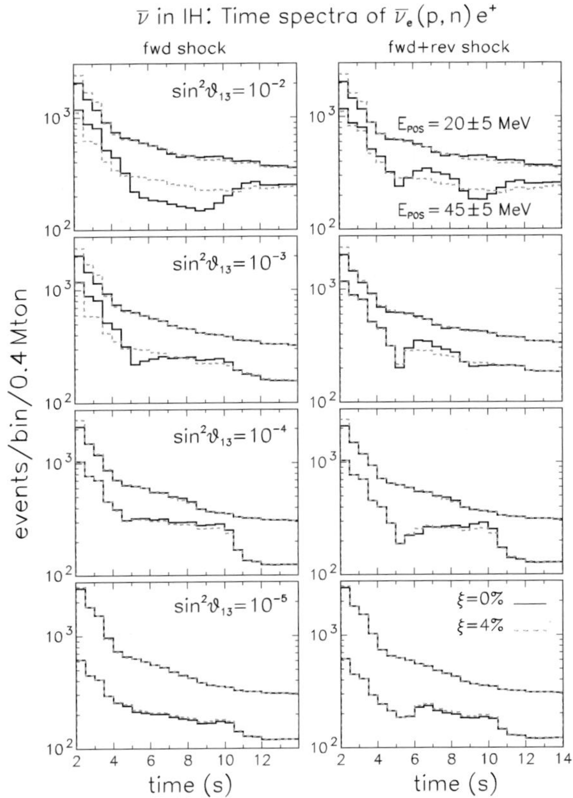

Fig. 1. Absolute time spectra of positron events induced by $\bar{\nu}_e$ in IH in a 0.4 Mton water-Cherenkov detector, in the presence of forward shock (*left panels*) and forward plus reverse shock (*right panels*). The *solid histograms* refer to the case of no fluctuation ($\xi = 0$), while the *dashed ones* refer to the case of fluctuations with amplitude $\xi = 4\%$

sensitive to shock-wave effects on P_{ee}^H only in the case of inverted hierarchy, on which we focus in the following discussion. Figure 1 shows absolute time spectra of events for the cases of forward shock only (left panels) and forward plus reverse shock (right panels), for the different representative values of $\sin^2\theta_{13}$. We consider both $\xi = 0$ (no fluctuations, solid histograms) and $\xi = 4\%$ (dashed histograms). In the absence of stochastic density fluctuations ($\xi = 0$ in Fig. 1), the shock-induced time variations in the representative positron energy bin $E_{\mathrm{pos}} = 20 \pm 5\,\mathrm{MeV}$ are small, and could be hardly characterized a priori as non-monotonic: We are around the "critical energy" E_c ($E_c \simeq 20\,\mathrm{MeV}$ for the our adopted emission model) where the initial $\bar{\nu}_e$ and ν_x spectra are approximatively equal, and so flavor transitions effects are largely cancelled. Conversely, in the bin $E_{\mathrm{pos}} = 45 \pm 5\,\mathrm{MeV}$ no cancellation is expected and the time spectra show strong signatures of shock-wave effects for inverted hierarchy, in the form of non-monotonic time variations of the event rate, especially at relatively high values of $\sin^2\theta_{13}$.

In the presence of stochastic density fluctuations ($\xi = 4\%$ in Fig. 1), the spectra for the low-energy positron bin are basically unaffected, since all flavor transition effects (modified or not by fluctuations) are small. The time spectra for the high-energy positron bin ($E_{\mathrm{pos}} = 45 \pm 5\,\mathrm{MeV}$) are instead significantly smoothed out by fluctuation effects for $\sin^2 \theta_{13} \gtrsim \mathcal{O}(10^{-3})$. Therefore, even at small amplitude ($\xi = 4\%$), fluctuations can suppress the shock imprint on the time spectra, and make them qualitatively similar to those in normal hierarchy (where they are expected a priori to be smooth and monotonic).

4 Conclusions

Future observations of a SN neutrino burst might be an useful probe of neutrino flavor transitions and of explosion mechanism. In this context, we have studied the time spectra of the inverse beta decay events in next-generation Mton-class water Cherenkov detectors, in order to extract signatures of the shock-wave propagation in the stellar envelope after the supernova explosion.

We have also studied the impact of stochastic density flucutations possibly arising after the shock front passage. We find that stochastic fluctuations may significantly suppress the imprint of SN shock waves on observable neutrino signal, the more the larger $\sin^2 \theta_{13}$. Therefore, in the presence of stochastic fluctuations, it might be difficult to "monitor" the shock-wave in real time in a future galactic SN explosion, or to find unmistakable signatures of inverted mass hierarchy effects. Therefore, a better theoretical understanding of turbulences behind the shock front [7] would be of great benefit for the interpretation of future SN neutrino events.

Acknowledgement. A.M. thanks the organizers of IFAE 2006 for the kind hospitality in Pavia.

References

1. G. L. Fogli, E. Lisi, A. Mirizzi and D. Montanino: Phys. Rev. D **68**, 033005 (2003).
2. G. L. Fogli, E. Lisi, A. Mirizzi and D. Montanino: JCAP **0504**, 002 (2005).
3. G. L. Fogli, E. Lisi, A. Mirizzi and D. Montanino: JCAP **0606**, 012 (2006)
4. G. L. Fogli *et al.*: hep-ph/0608060.
5. R. C. Schirato, G. M. Fuller: astro-ph/0205390.
6. R. Tomas, M. Kachelriess, G. Raffelt, A. Dighe, H. T. Janka and L. Scheck: JCAP **0409**, 015 (2004).
7. A. Friedland and A. Gruzinov: astro-ph/0607244.

Leptonic CP violation

Davide Meloni

INFN, Sezione di Roma1, P.le Aldo Moro, 2 – 00185 Roma – Italy
`meloni@roma1.infn.it`

We briefly review the concept of CP violation in the leptonic sector, focusing on several theoretical aspects concerning the possibility of measuring the CP phase δ at future neutrino facilities.

1 Introduction

The atmospheric and solar sector of the PMNS leptonic mixing matrix have been measured with quite good resolution by SK, SNO and KamLand. These experiments measure two angles, θ_{12} and θ_{23}, and two mass differences, Δm_{12}^2 and Δm_{23}^2. The present bound on θ_{13}, $\sin^2\theta_{13} \leq 0.04$, is extracted from the negative results of CHOOZ and from three-family analysis of atmospheric and solar data. The PMNS phase δ is totally unbounded as no experiment is sensitive, up to now, to the leptonic CP violation. The main goal of next neutrino experiments will be to measure these two, still unknown, parameters. The best channel for measuring (θ_{13}, δ) is the $\nu_e \to \nu_\mu$ appearance channel [1] (and/or its T and CP conjugate ones). Unfortunately, this measure is severely affected by the presence of an eightfold degeneracy [2,3] and from our present imprecise knowledge of atmospheric parameters, whose uncertainties are far too large to be neglected when looking for such tiny signals as those expected in appearance experiments driven by the $\nu_\mu \to \nu_e$ and $\nu_e \to \nu_\mu, \nu_\tau$ oscillation probabilities [4].

2 CP violating effects

The CP-violating term of the neutrino transition probability can be shortly written in the following way:

$$P_{\mathrm{CP-odd}} = 4\, c_{12}\, c_{13}^2\, c_{23}\, s_{12}\, s_{13}\, s_{23}\, \sin\delta \times$$

$$\times \left[\sin\left(\frac{\Delta m_{12}^2\, L}{2\, E}\right) + \sin\left(\frac{\Delta m_{23}^2\, L}{2\, E}\right) - \sin\left(\frac{\Delta m_{13}^2\, L}{2\, E}\right) \right] \quad (1)$$

in which $c_{ij}\, (s_{ij})$ stands for $\cos\theta_{ij}\, (\sin\theta_{ij})$.

From (1), the conditions to have observable CP-violation can be easily determined: we need the CP-phase δ different from $0, 180°$, the baseline not too small (the oscillation terms would give vanish contributions) nor too large (otherwise we should average over fast oscillating terms, giving zero) and each mass difference and mixing angle different from zero. The last requirement is satisfied for the atmospheric and solar mixing angles but for θ_{13} we do not know: it could be either very small or null, thus preventing the measurement of the CP phase, since θ_{13} and δ are strongly correlated in appearance probabilities. In fact, it was originally pointed out in [2] that the appearance probability $P_{\alpha\beta}$ for neutrinos with a fixed Baseline/Energy (L/E) ratio and input parameters $(\bar{\theta}_{13}, \bar{\delta})$ has no unique solution. Indeed, the equation

$$P_{\alpha\beta}(\bar{\theta}_{13}, \bar{\delta}) = P_{\alpha\beta}(\theta_{13}, \delta) \tag{2}$$

has a continuous number of solutions. The locus of the (θ_{13}, δ) plane satisfying this equation is called *equiprobability curve* (see left plot in Fig. 1).

On the other hand, if we consider at the same time the equiprobability curves for neutrinos $(+)$ and antineutrinos $(-)$, the previous system of equations has to be modified to

$$P^{\pm}_{\alpha\beta}(\bar{\theta}_{13}, \bar{\delta}) = P^{\pm}_{\alpha\beta}(\theta_{13}, \delta) \tag{3}$$

Two intersections appear (see right plot in Fig. 1): the input pair $(\bar{\theta}_{13}, \bar{\delta})$ and a second, L/E dependent, point, which is called the *intrinsic clone* solution [2]. Then one needs to add more informations to solve the intrinsic degeneracy.

It can be done in two ways: i) using independent experiments (i. e. different L/E) and/or ii) using independent oscillation channels. In case i) one

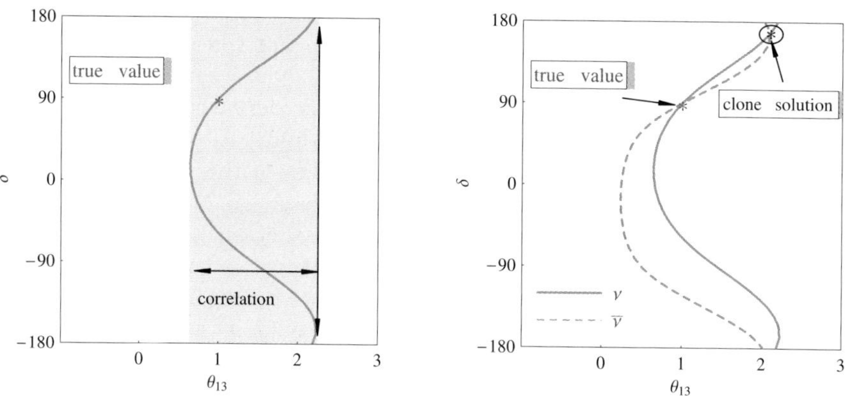

Fig. 1. Correlation of θ_{13} and δ. *Left:* if only neutrinos (or antineutrinos) are measured (infinite degeneracy). *Right:* if neutrinos (*full line*) and antineutrinos (*dashed line*) are measured (twofold degeneracy)

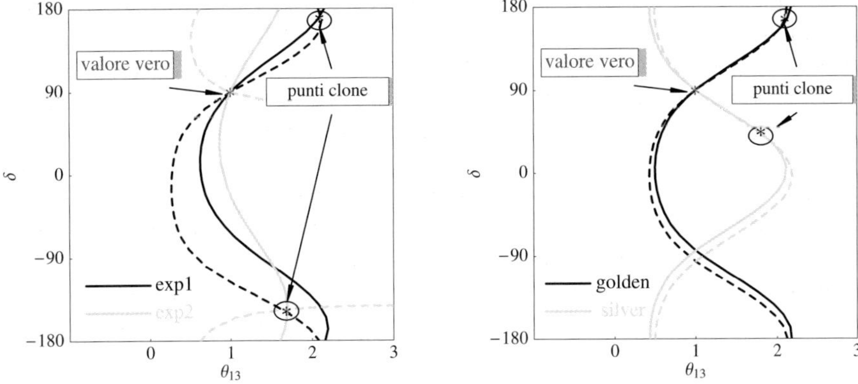

Fig. 2. *Left*: solving the intrinsic degeneracy using the same oscillation channel but two different values of L/E. *Right*: solving the intrinsic degeneracy using the same L/E but two different oscillation channels (i. e. golden and silver)

can think to observe the same neutrino oscillation channel using neutrino (antineutrino) beams with different L/E. In the left plot of Fig. 2 one can see that experiments with different L/E present clone solutions in different regions of the (θ_{13}, δ) parameter space. If the clones are well separated one can solve the degeneracy. Another possibility, case ii), is to fix L/E and to use contemporaneously two different oscillation channels (like for example $\nu_e \to \nu_\mu$ and $\nu_e \to \nu_\tau$). In the right plot of Fig. 2 one can see how the intrinsic clones for the two channels appear in different locations and so the intrinsic degeneracy can be solved. This procedure has been shown to work satisfactory [5].

Then the best way for solving the degeneracies is to add all the possible available informations: different baselines, different energy bins (i. e. different L/E) and different channels in order to localize clones in different points of the (θ_{13}, δ) parameter space [6].

Unfortunately, the appearance of the intrinsic degeneracy is only a part of the "clone problem". As it was pointed out in [3], two other sources of ambiguities arise due to our ignorance of the sign of the atmospheric mass difference, $s_{\mathrm{atm}} = \mathrm{sign}[\Delta m_{23}^2]$, and of the θ_{23} octant, $s_{\mathrm{oct}} = \mathrm{sign}[\tan(2\theta_{23})]$. These two discrete variables (to be measured together with θ_{13} and δ) assume the values ± 1, depending on the sign of Δm_{23}^2 ($s_{\mathrm{atm}} = 1$ for $m_3^2 > m_2^2$ and $s_{\mathrm{atm}} = -1$ for $m_3^2 < m_2^2$) and of the θ_{23}-octant ($s_{\mathrm{oct}} = 1$ for $\theta_{23} < \pi/4$ and $s_{\mathrm{oct}} = -1$ for $\theta_{23} > \pi/4$). From these considerations it follows that (3) has to be replaced by the systems of equations (each for any possible choice of the s_{atm} and s_{oct} signs):

$$P_{\alpha\beta}^{\pm}(\bar{\theta}_{13}, \bar{\delta}; \bar{s}_{\mathrm{atm}}, \bar{s}_{\mathrm{oct}}) =$$
$$P_{\alpha\beta}^{\pm}(\theta_{13}, \delta; s_{\mathrm{atm}} = \pm\bar{s}_{\mathrm{atm}}; s_{\mathrm{oct}} = \pm\bar{s}_{\mathrm{oct}}). \qquad (4)$$

Solving the four systems of (4) will result in obtaining the true solution plus additional *clones* to form an eightfold degeneracy. These eight solutions are respectively: the true solution and its *intrinsic clone* (when the right s_{atm} and s_{oct} signs are used in (4)), the Δm_{23}^2-*sign clones* (when $s_{\mathrm{atm}} = -\bar{s}_{\mathrm{atm}}$ is used), the θ_{23}-*octant clones* (when $s_{\mathrm{oct}} = -\bar{s}_{\mathrm{oct}}$ is used) and finally the *mixed clones* (when simultaneously $s_{\mathrm{atm}} = -\bar{s}_{\mathrm{atm}}$ and $s_{\mathrm{oct}} = -\bar{s}_{\mathrm{oct}}$ are used).

Several ideas have been proposed to face the eightfold degeneracy problem (and then measure the CP violating phase), many of them based on the mentioned necessity of combining as much informations as possible to decrease the statistical importance of the clone points. Among them, β-beams [7], Superbeams [8] and Neutrino Factories [9] seem to be the most promising new neutrino facilities for solving the problem of multiple solutions. Up to now, a unique and well defined strategy to attack the clones has not been identified and the question is still open: several combination of accelerators and detectors could produce similar results depending on the capability of producing intense neutrino beams and building large detectors, required to access leptonic CP violation. What is also important, one should take into account *cost estimates* which are one of the major difficulty in making such large facilities.

References

1. A. Cervera *et al.*, Nucl. Phys. B **579** (2000) 17 [Erratum-ibid. B **593** (2001) 731].
2. J. Burguet-Castell *et al.*, Nucl. Phys. B **608** (2001) 301; M. C. Gonzalez-Garcia and C. Pena-Garay, Phys. Rev. D **68**, 093003 (2003); H. Minakata and H. Nunokawa, JHEP **0110** (2001) 001; G. L. Fogli and E. Lisi, Phys. Rev. D **54** (1996) 3667; V. Barger *et al.*, Phys. Rev. D **65** (2002) 073023.
3. G. L. Fogli and E. Lisi, Phys. Rev. D **54** (1996) 3667; H. Minakata and H. Nunokawa, JHEP **0110** (2001) 001; V. Barger *et al.* Phys. Rev. D **65** (2002) 073023.
4. A. Donini, D. Meloni and S. Rigolin, Eur. Phys. J. C **45**, 73 (2006) [arXiv:hep-ph/0506100].
5. A. Donini *et al.* Nucl.Phys.B **646**, 321 (2002); D. Autiero *et al.*, Eur. Phys. J. C **33**, 243 (2004).
6. A. Donini *et al.*, JHEP **0406**, 011 (2004).
7. P. Zucchelli, Phys. Lett. B **532** (2002) 166; A. Donini *et al.*, Phys.Lett. B **621** (2005) 276; M. Mezzetto *et al.*, AIP Conf. Proc. **721**, 37 (2004); J. Burguet-Castell *et al.*, Nucl. Phys. B **695**, 217 (2004).
8. See for example P. Huber *et al.*, Nucl. Phys. B **645** (2002) 3, and references therein.
9. S. Geer, Phys. Rev. D **57**, 6989 (1998) [Erratum-ibid. D **59**, 039903 (1999)].

Neutrino astronomy with km^3 underwater and under ice

Giorgio Riccobene

Laboratori Nazionali del Sud – INFN, Via S. Sofia 62, I-95123, Catania, Italy
`riccobene@lne.infn.it`

1 High energy neutrino sources and expected fluxes

High energy neutrinos are considered optimal probes to identify the sources
of high energy cosmic rays. Many indications suggest, indeed, that cosmic
objects, where acceleration of charged particles take place are the sources
of the detected UHECRs, the same sources should also produce high energy
neutrino fluxes. Indeed, $p\gamma$ or pp interactions responsible for $>$ TeV neutrinos
and γ rays fluxes, are expected to occur in several astrophysical environments.
In Supernova Remnants (SNRs) protons, accelerated through Fermi mechanism, can interact with gas in dense SN shells or molecular clouds, producing
both neutral and charged pions. Decay of neutral pion originates gamma
rays. The observation of $\simeq 10$ TeV gamma rays from Supernova Remnant
RXJ1713.7-3946 claimed by CANGAROO [1], seems to validate this hypothesis, but the result is still under discussion. The neutrino flux, produced by
charged pion decay, is expected to be similar to the hadronic high energy
gamma rays one. Results from the HESS air Čerenkov Telescope, concerning the observation of TeV gamma ray from the Sgr A* region (the Galactic centre), show an $\propto E^{-2.2}$ γ ray spectrum that could be originated by
hadronic interactions, an interpretation which also implies the production of
intense high energy neutrino fluxes [2]. Photo-meson ($p\gamma$) interactions can occur in astrophysical environments that show dense low energy photons fields:
microquasar jets, AGNs and in GRBs are examples. AGNs and GRBs are
particularly relevant since they are the candidate sources of UHECR. Starting from this hypothesis Waxman and Bahcall shown that there is an upper
bound for the high energy diffuse neutrino flux that reaches the Earth of
$E_\nu^2 \Phi_\nu \simeq 10^{-7.5}$ GeV $/$ (cm^2 ssr) [3]. This limit, object to discussion and crit-
icism, sets however a reference value for the dimensions of future neutrino
telescopes to about 1 km^2 of effective detection area.

2 Detection

High energy neutrinos are detected by observation of the Čerenkov radiation from secondary particles produced by neutrinos interacting inside large volumes of water/ice instrumented with a lattice of photomultiplier tubes (PMTs). In the case of muon neutrino interactions, muons range out over kilometeres at TeV energy, to tens of kilometres at EeV energy, generating also showers along their track. The Čerenkov light wave-front (radiated by the showers and by the muon) is reconstructed using the information of photon hits and PMT positions. Apart from muon tracks, cascades (mainly initiated by ν_e or oscillated ν_τ neutrinos) can also be detected. With a typical size of tens of meters in ice or water, cascades can be considered as an intense quasi-point-like light sources compared to the typical spacing of PMTs (50–100 m) and can be detected as contained events. When an upward going Čerenkov wavefront is reconstructed, this is a signature of neutrino event, being the atmospheric upgoing muon background completely filtered by the Earth. This is the reason why it is often said that neutrino telescopes *look downward*.

3 Under-ice neutrino detectors

The first generation of underwater/ice neutrino telescopes: BAIKAL [4] (a neutrino detector with an effective area of $\simeq 10^5\,\mathrm{m}^2$ for TeV muons, built in Lake Baikal, Siberia) and AMANDA, despite their limited size, have already set first constraints on TeV neutrino production astrophysical models. AMANDA consists of 677 optical modules (OM) pressure resistant glass vessel hosting downward oriented PMTs and readout electronics. OMs are arranged in 19 vertical strings, deployed in holes drilled in ice between 1.3 and 2.4 km depth. Due to ice optical properties for blue-UV light (absorption length $\simeq 100\,\mathrm{m}$, effective light scattering length $\simeq 20\,\mathrm{m}$) AMANDA is a good calorimeter for astrophysical events, with an energy resolution 0.4 in $\log(E)$ for muons and 0.15 in $\log(E)$ for electron cascades, on the contrary the angular resolution is rather poor $\simeq 3°$ for muons and $\simeq 30°$ for cascades. AMANDA data have permitted to measure for the first time the upgoing atmospheric neutrino spectrum in the energy range from few TeV to 300 TeV. The atmospheric neutrino spectrum recorded by AMANDA-II was then used to set a 90% C.L. upper limit on a diffuse muon neutrino flux of $E_{\nu_\mu}^2 \Phi_{\nu_\mu} < 2.6 \times 10^{-7}\,\mathrm{GeV\,cm^{-2}\,s^{-1}\,sr^{-1}}$ ($100 < E_\nu < 300\,\mathrm{TeV}$). For point sources the detector reached a sensitivity $E_{\nu_\mu}^2 \Phi_{\nu_\mu} \simeq 7 \times 10^{-8}\,\mathrm{GeV\,cm^{-2}\,s^{-1}}$ calculated over 807 days live time (years 2000–2003) [5]. The success of AMANDA opened the way to the construction of IceCube. When completed (approximately on 2010), IceCube will consist of 4800 downward looking PMTs arranged in 80 strings. Simulations run by the IceCube Collaboration show that in three years of live time, the detector will reach a sensitivity

$E^2_{\nu_\mu} \Phi_{\nu_\mu} = 4.2 \times 10^{-9} \, \mathrm{GeV \, cm^{-2} \, s^{-1} \, sr^{-1}}$ for a diffuse E^{-2}_ν neutrino spectrum and a sensitivity $E^2_{\nu_\mu} \Phi_{\nu_\mu} = 2.4 \times 10^{-9} \, \mathrm{GeV \, cm^{-2} \, s^{-1}}$ for a point-like source [6].

3.1 The Mediterranean km^3

The scientific community is supporting the construction of another neutrino telescope, in the Northern Earth Hemisphere, to allow contemporary observation of the full sky. Moreover neutrino events detection from the Northern Hemisphere is required to observe the Galactic Centre, not seen by IceCube. In the Northern Hemisphere a favourable region is offered by the Mediterranean Sea, where several abyssal sites at depth $> 3000\,\mathrm{m}$ (at these depths the atmospheric muon background is reduced by a factor ≥ 5 with respect to $2000\,\mathrm{m}$ depth, where IceCube is located) are present. Few of these are relatively close to the coast and to scientific and industrial infrastructures. Since water properties may change from site to site, the selection of an abyssal region with optimal oceanographic and optical parameters is, therefore, a major task for the three Mediterranean Collaborations: NESTOR, ANTARES and NEMO.

NESTOR proposes to deploy a modular detector at $3800\,\mathrm{m}$ depth in the Ionian Sea, near the Peloponnese coast (Greece). An array of 12 PMTs was connected to the shore, and transmitted data for about one month, allowing a rough reconstruction of the downgoing muon spectrum [7]. ANTARES will be a *demonstrator* neutrino telescope with an effective area of $0.1\,\mathrm{km}^2$ for astrophysical neutrinos. It will be located in a marine site near Toulon (France), at $2400\,\mathrm{m}$ depth. The whole detector installation is scheduled to be completed in 2007 [8]. The first *line* of the detector is taking data since spring 2006. Data recorded by optical modules show an unexpectedly high optical background ranging from 60 to several hundreds kHz, well above the one produced by ^{40}K decay, then probably due to bioluminescence.

The NEMO Collaboration is carrying out an R&D towards the km^3 neutrino telescope. The activity has been mainly focused on the search and characterization of an optimal site for the installation and on the development of a feasibility study of the detector. NEMO has intensively studied the oceanographic and optical properties in several deep sea sites close the Italian coast. Results indicate that a large region located $80\,\mathrm{km}$ SE of Capo Passero (Sicily) is excellent for the installation of the km^3 detector. The bathymetric profile of the region is extremely flat over hundreds km^2, with an average depth of $\simeq 3500\,\mathrm{m}$. Average deep sea currents are $3\,\mathrm{cm\,s^{-1}}$ and never stronger than $12\,\mathrm{cms^{-1}}$. The measured value of blue light absorption length is $66 \pm 5\,\mathrm{m}$, close to the value of optically pure sea-water. The optical background noise on $10''$ PMTs (0.5 s. p. e.) is 20–$30\,\mathrm{kHz}$, compatible with the one expected from ^{40}K decay, with negligible contribution of bioluminescence. NEMO proposes innovative structures to host OMs: the NEMO *towers*. They are designed to deploy, during a single operation, a large number of PMTs (≥ 60) arranged in a 3-dimensional shape in order to locally permit event trigger and track reconstruction. A proposed detector geometry consisting of a squared array of 9×9 NEMO towers equipped with 5832 optical modules ($10''$ diameter PMTs). The distance between the towers is $140\,\mathrm{m}$. Using this geometry the detector can reach an effective area $> 1\,\mathrm{km}^2$ at muon energy of about $10\,\mathrm{TeV}$, with an angular resolution lower than 0.1 degrees. Results of simulations also show that the expected sensitivity to a point like source $E_{\nu_\mu}^{-2}$ is about $1.2 \times 10^{-9}\,\mathrm{GeV\,cm^{-2}\,s^{-1}}$, obtained for a search bin of 0.3 degree [9]. These results indicate that such a detector may reach a better sensitivity and smaller search bin than IceCube, allowing a better identification of point-like sources. As an intermediate step towards the km^3, the Collaboration is installing a technological demonstrator (the so called NEMO Phase 1): a mini-tower equipped with 16 PMTs connected to the shore through a Junction Box and a $25\,\mathrm{km}$ long electro-optical cable. The detector is under final assembly at LNS-INFN and will be deployed $25\,\mathrm{km}$ offshore Catania within 2006. Phase 1 will permit to test all the electronics and mechanical elements of the proposed NEMO detector.

4 Conclusions

The forthcoming km^3 neutrino telescopes are *discovery* detectors that could widen the knowledge of the Universe. These detectors have high potential to solve questions as the detection of UHECR sources, the investigation of hadronic processes in astrophysical environments or massive dark matter. The quest for the construction of km^3 size detectors has already started: in the South Pole the IceCube neutrino telescope is under construction; in the Mediterranean the ANTARES and NESTOR Collaborations are installing small scale neutrino detectors, NEMO has proposed Capo Passero as an optimal site for the detector installation and is conducting an R&D for the km^3 underwater telescope. Since 2006 the three Collaborations joined KM3NeT [10], an EU-funded Design Study which will define, within 2009, a technical design report for the Mediterranean km^3 telescope.

References

1. R. Enamoto et al. *Nature*, **416:823**, (2002).
2. F. Aharonian, the H.E.S.S. Collaboration *Astron. Astroph*, **225**, L13-L17 (2004)
3. E. Waxman & J. Bahcall, *Phys. Rev.*, **D64**, 023001, (1999).
4. R. Wischnewski *et al.*, Proc. of *29th ICRC*, Pune, India, 2005, astro-ph/0507715
5. M. Ackermann *et al.*, *Phys. ReV.*, **D077102**, (2005).
6. J. Ahrens *et al.*, *Astrop. Phys.*, **20**, 507, (2004).
7. G. Aggouras *et al.*, *Astrop. Phys.*, **23**, 377, (2005).
8. J. Aguilar *et al.*, astro-ph/0606229
9. C. Distefano *et al.*, Proc. of *BCN06*, Barcelona, Spain, 2006, astro-ph/0608053
10. http://www.km3net.org

Ultra High Energy Cosmic Rays:
Observations and Theoretical Aspects

Daniel De Marco

Bartol Research Institute, University of Delaware, Newark, DE 19716, U.S.A.
ddm@bartol.udel.edu

Summary. We present a brief introduction to the physics of Ultra High Energy
Cosmic Rays (UHECRs), concentrating on the experimental results obtained so far
and on what, from these results, can be inferred about the sources of UHECRs.

1 Introduction

Since the discovery of cosmic rays (CRs) by Victor Hess in 1912 there has
been a constant search for the end of the cosmic-ray spectrum. This end
has long been thought to be determined by the highest energy the cosmic
accelerators might be able to achieve, but despite several decades of research
no end of the spectrum was in sight until 1966. In 1966, right after the
discovery of the cosmic microwave background (CMB), it was understood [1]
that protons with sufficiently high energy would interact inelastically with
the photons of the CMB and produce pions. This process rapidly degrades
the proton energy and, if the sources of CRs are homogeneously distributed,
it produces a drastic suppression in the CR flux around 10^{20} eV, where the so-
called photo-pion production starts to be kinematically allowed. The physical
reason for this suppression is that at energies around 10^{19} eV the loss length
for protons propagating in the CMB is of the order of a Gpc and we are
receiving particles from almost all the visible universe, whereas at 10^{20} eV
the loss length is about $100\,\mathrm{Mpc}$ and we are receiving particles only from
a tiny fraction of the universe. This suppression is usually called the *GZK
cutoff* [1]. After several more decades of experimental activity we have now
experiments exploring the energy region around 10^{20} eV and beyond, but the
end of the spectrum is still eluding us and whether the GZK cutoff is present
or not in the observed spectra is still an open question.

In Sect. 2 we briefly review the present UHECR data sets and the issues
they raise and in Sect. 3 we briefly discuss how we can improve our understanding of the sources of UHECRs using the new data that is now being
collected by the Pierre Auger Observatory (PAO).

2 UHECRs: present

Until a year ago the two largest experiments measuring UHECRs were AGASA and HiRes. In 2005 the PAO [2] reported the results of the first year of data taking [3], but, since those results are still preliminary and the error-bars are still quite large, in the following discussion we will concentrate on AGASA and HiRes. We are considering particles of ultra high energy, around and above 10^{19} eV, and CRs of such high energies, entering the earth atmosphere, interact with it producing extensive showers of secondaries that propagate in the atmosphere close to the speed of light. The two above mentioned experiments use two complementary techniques to detect these extensive air showers (EAS): AGASA used an array of detectors on the ground that sampled the lateral distribution of the EAS when it hit the ground while HiRes uses a telescope to observe the fluorescence light produced by the shower while it propagates in the atmosphere. For a review of the detection techniques see [4].

Despite the fact that two completely different methods were used, the two experiments report somehow similar results for what concerns the energy spectrum at low energy, with some conflicts at high energy. At low energy, where the number of detected events per energy bin is big the two experiments report fluxes that differ by about a factor 2, but this discrepancy can be accounted for by correcting for the systematic errors on the energy determination reported by the two collaborations, about $\pm 15\%$. Doing this shift the two experiments agree perfectly in this energy range ($\leq 10^{20}$ eV) [5, 6].

At $E > 10^{20}$ eV, where the statistics of events is very sparse due to the steepness of the spectrum of CRs, the two experiments report opposite results: AGASA claims [7] to have observed a continuation of the spectrum beyond the expected cutoff whereas HiRes claims [8] to have observed the expected suppression. The statistics of events above 10^{20} eV is however really small and the discrepancy between the two experiments is just about 3σ. Taking into account the systematic errors as in the low energy region this discrepancy is reduced to about 2σ [5]. Recently the AGASA collaboration revised down their energy assignments [9] by about 10% further reducing the alleged discrepancy.

Even if nowadays the presence of the GZK suppression in the spectrum seems more plausible than its absence the fact still remains that events with energies above 10^{20} eV have been measured several times by different experiments. Where did those particles come from? For astrophysical accelerators it is extremely challenging to accelerate particles to such high energies [10] and in the few plausible models the sources are usually too far away from us for the particles to be able to propagate to the earth without suffering sensible energy losses. Indeed no suitable sources have been found within reasonable distances around the arrival directions of the highest energy events. For a review on the origin of UHECRs see [11] and references therein.

From the measurement of an EAS we can basically obtain three informations about the primary particle: its energy, its direction and its nature or chemical composition. Each one of these informations is important and can provide useful clues about the sources. Some experimental techniques are better suited to measure one of them or another one [4], but in general all the experiments are in the end reconstructing those three quantities. We already discussed the energy spectrum, we will skip the discussion of the chemical composition (for a review of the experimental results see [12] while for discussions of the interesting problem of the galactic–extra-galactic transition see [13, 14]) and we will now concentrate on the arrival directions of those UHE events.

AGASA reported [15] the presence of clustering in its set of events with energies above 4×10^{19} eV. While on large scales the arrival directions of these events appear to be isotropic, on small scales they appear to arrive in clusters. AGASA observed 6 doublets and 1 triplet with angular separation less than $2.5°$ on a set of about 70 events. These data point in the direction of astrophysical point sources with a density of about 10^{-5} Mpc^{-3}, with large error bars of about one order of magnitude. Combining this result with the energy spectrum it is possible to obtain information about the luminosity of the sources themselves [16, 17]. The significance of this result is however still debated. First of all because the statistical significance of the clustering signal, that in the beginning was quite high, turned out to be lower in subsequent analyses [18] and also because HiRes did not see any anisotropy in its data set [19] though in this case too the statistical significance of the absence of clustering is not very high [20]. Moreover it seems that the AGASA data set itself presents some internal inconsistency and the probability of reproducing the AGASA result on the spectrum is reduced by a large factor when taking into account a source density of 10^{-5} Mpc^{-3} [5].

3 UHECRs: (near) future

The PAO, being built in Argentina, is a new kind of experiment that combines the two above-mentioned measuring techniques [2]. It consists of four fluorescence telescopes overlooking a ground array of 1600 surface detectors covering an area of 3000 km^2. The ground array exposure, above 10^{19} eV, after 10 years of data-taking will be $70\,000$ km^2 sr yr, to be compared for example with 1645 km^2 sr yr that was the AGASA exposure after 10 years of operation. About 10% of the detected events will be *hybrid events*, detected at the same time by the ground array and by the fluorescence telescopes. The Auger data set will help tremendously in our understanding of UHECRs and of their sources. First of all because measuring hybrid events it will be possible to solve the discrepancy in the energy assignments between fluorescence and surface detectors. Moreover, with the huge statistics of events it will collect at high energy, the spectrum in the GZK region will no longer

be dominated by statistical fluctuations as in the AGASA and HiRes case
and we will be able to observe the presence or absence of the GZK feature in
the spectrum and maybe the end of the CR spectrum [5]. The huge statistics
will be even more important to study the anisotropies in the event arrival
directions. The PAO will be able, already after a few years of operations, to
detect the presence of anisotropies both on large [21] and small [17] scales.
For example it will be able to distinguish between a uniform distribution of
sources and a discrete distribution of sources with a given density already
after 5 years if the density is smaller than $10^{-3}\,\mathrm{Mpc}^{-3}$, whereas in order to
distinguish between different densities 15 years of operations are required or
even a bigger experiment (for example Auger North).

Acknowledgement. This research is funded in part by NASA APT grant ATP03-
0000-0080 at University of Delaware.

References

1. K. Greisen: Phys. Rev. Lett. **16** 748 (1966)
   G.T. Zatsepin and V.A. Kuzmin: Sov. Phys. JETP Lett. **4** 78 (1966)
2. http://www.auger.org/
3. P. Sommers *et al.* (Auger Collaboration): Proceedings of ICRC 2005, Pune,
   India. astro-ph/0507150
4. M. Nagano and A. A. Watson: Rev. Mod. Phys. **72** 689 (2000)
5. D. De Marco, P. Blasi and A.V. Olinto: Astropart. Phys. **20** 53 (2003)
   D. De Marco, P. Blasi and A.V. Olinto: JCAP01(2006)002
6. D. De Marco and T. Stanev: Phys. Rev. **D72** 081301 (2005)
7. M. Takeda *et al.*: Phys. Rev. Lett. **81** 1163 (1998)
   N. Hayashida *et al.*: Astron.J. **120** 2190 (2000)
8. R.U. Abbasi *et al.* (HiRes Collaboration): Phys. Lett. **B619** 271 (2005)
9. M. Teshima: Proceedings of CRIS 2006, Catania, Italy
10. A. M. Hillas: Ann. Rev. Astron. Astrophys. **22** 425 (1984)
11. P. Blasi: Mod. Phys. Lett. A **20** 3055 (2005)
12. A. A. Watson: Nucl. Phys. Proc. Suppl. **151** 83 (2006)
13. R. Aloisio, V. Berezinsky, P. Blasi, A. Gazizov, S. Grigorieva and B. Hnatyk:
    astro-ph/0608219
14. D. Allard, E. Parizot and A. V. Olinto: astro-ph/0512345
    D. Allard, E. Parizot, E. Khan, S. Goriely and A. V. Olinto: astro-ph/0505566
15. N. Hayashida *et al.*: Astrophys. J. **522** 225 (1999)
16. P. Blasi and D. De Marco: Astropart. Phys. **20** 559 (2004)
17. D. De Marco, P. Blasi and A.V. Olinto: JCAP07(2006)015
18. C.B. Finley and S. Westerhoff: Astropart. Phys. **21** 359 (2004)
19. C.B. Finley and S. Westerhoff (HiRes Collaboration): Proceedings of ICRC
    2003, Tsukuba, Japan
20. H. Yoshiguchi, S. Nagataki and K. Sato: Astrophys. J. **614** 43 (2004)
21. A. Cuoco, R. D. Abrusco, G. Longo, G. Miele and P. D. Serpico:
    JCAP01(2006)009

Recent Results in Gamma Ray Astronomy with IACTs

Vincenzo Vitale

Dipartimento di Fisica Università di Udine
vitale@fisica.uniud.it

1 Gamma-ray Astronomy with Cherenkov Telescopes

Very high energy particles interact with Earth's atmosphere (28 radiation length of thickness) producing the so-called extensive air showers. Energetic particles in the showers generate Cherenkov radiation, which can be collected by an optical reflector and focused on a camera of photo-multipliers. In such a way the atmosphere and the telescope (reflector + camera) work as an imaging calorimeter.

In order to select γ ray induced events a dominant background of hadronic (nuclei in the cosmic radiation) events should be rejected. The imaging technique consists in exploiting the differences in the images produced by electromagnetic showers and hadronic cascades. With such a technique a rejection of 99.9% of the background events and a selection of 50% of γs can be achieved.

The Cherenkov radiation light-pool at the ground has a radius of $\approx 10^2$ m, then the geometric collection area of a telescope is in the order of $10^4\,\mathrm{m}^2$. The large geometric area and the selection technique allow the detection of γ ray fluxes in the order of few % of the Crab emission (the standard candle of the TeV γ ray field) in tens of hours.

The first detection of TeV γ rays was in 1989 from the Crab Nebula [14]. After that few blazars (a sub-class of active galactic nuclei) were detected, among which Mkn 421 [12] and Mkn 501 [13]. With the new generation of telescopes the detection of sources and new classes of TeV γ ray emitters has accelerated drastically. At the present time also radio-galaxies, pulsar-wind nebulae, micro-quasars, Bc binary sistems, shell type supernova remnants and unidentified sources are among the detected objects.

It is interesting to note that the majority of the sources and also of the unidentified ones cluster around the galactic plane and were discovered with a scan survey [1].

In the following selected recent results of galactic and extra-galactic observations with Imaging Atmospheric Cherenkov Telescopes (IACT) are reported.

2 Selected Results

The Galactic Center. The Galactic Center (GC) has been established as a strong VHE γ ray emitter. The MAGIC telescope in 2005 [5] measured a flux compatible with that reported by the HESS collaboration [3]. The differential flux from GC can be described by a single power law $(2.9 \pm 0.6)10^{-12} (E/\text{TeV})^{-2.2 \pm 0.2} \text{cm}^{-2} \text{s}^{-1} \text{TeV}^{-1}$. The TeV γ emission is steady and consistent with the SgrA position as well as the supernova remnant SgrA East. The nature of the emission is not yet understood. Nevertheless astrophysical processes associated with the GC rich environment or the central super-massive black hole are favoured. The scenario of γ radiation from dark matter annihilation is not ruled out but less likely. Furthermore diffuse γ ray emission correlated spatially with a complex of giant molecular clouds in the central 200 pc of the Milky Way has been discovered from the Galactic Center Ridge [4]. This emission might be the signature of cosmic rays interacting with interstellar gas.

Microquasars. Microquasars are binary systems composed by ordinary stars and a compact object (black hole, neutron star). They have an accretion disk surrounding the compact companion and a pair of radio jets. Because of their structure they are considered scaled down version of quasars. Microquasars are very important for the study of relativistic jets. The jets are formed close to the black hole, and timescales near the black hole are proportional to the mass of the black hole. Therefore, ordinary quasars take centuries to go through variations a microquasar experiences in one day.

VHE γ radiation has been detected from LS 5039 [2] and LS I +61 303 [6], therefore microquasars have been established as a new class of VHE γ emitters. For LS I +61 303 Six orbital cycles were recorded with the MAGIC telescope. Several detections occur at a similar orbital phase, which suggests that the emission is periodic. The strongest gamma-ray emission is not ob-

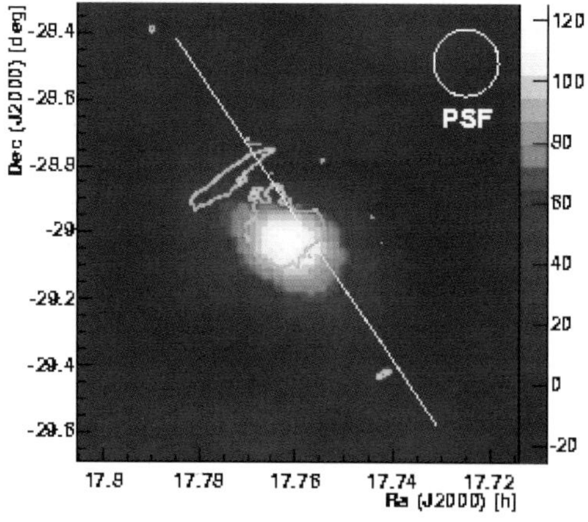

Fig. 1. Smoothed sky map of γ-ray excess, background subtracted, in the direction of the Galactic Center for an energy threshold of about 1 TeV, obtained with the MAGIC telescope. *Overlayed* are contours of 90 cm VLA radio data [11], the *line* shows the galactic plane

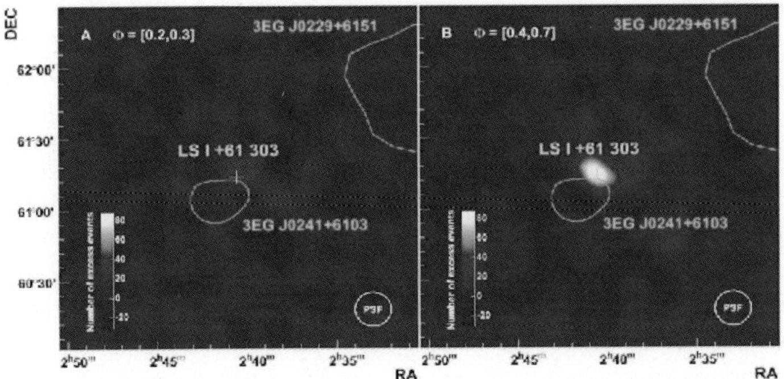

Fig. 2. Smoothed maps of γ ray excess events above 400 GeV around LSI +61 303, obtained with the MAGIC telescope. **A** 15.5 hour exposure at periastron, orbital phase 0.2–0.3. No significant excess is detected. **B** 10.7 hours exposure at orbital phase between 0.4 and 0.7. An excess with a significance of 9.4σ is clearly detected in the location of LSI +61 303

Table 1. List of extra-galactic VHE γ ray sources

Source	Redshift	Type
M87	0.004	FRI
Mkn 421	0.031	BL Lac
Mkn 501	0.034	BL Lac
1ES 2344 +514	0.044	BL Lac
Mkn 180	0.045	BL Lac
1ES 1959	0.047	BL Lac
PKS 2005 -489	0.071	BL Lac
PKS 2155 -304	0.116	BL Lac
H 1426 +428	0.129	BL Lac
H 2356 -309	0.165	BL Lac
1ES 1218+304	0.182	BL Lac
1ES 1101-232	0.186	BL Lac
PG 1553+113	<0.78	BL Lac

served when the two stars are closest to one another, implying a strong orbital modulation of the emission or absorption processes.

Blazars The number of AGNs detected as VHE γ ray emitters is fastly increasing with the time, as can be seen from the list in Table 1. It shoud be noted that M87, which is not of BL LAC type, has been also detected in the VHE domain. Among the new VHE blazars MAGIC discovered Mkn180 [10], 1ES1218+304 [8] and co-discovered PG 1553+113 [9].

Gamma ray bursts The long-duration GRB050713a was observed by the MAGIC telescope, just 40 seconds after the burst onset [7] . The observa-

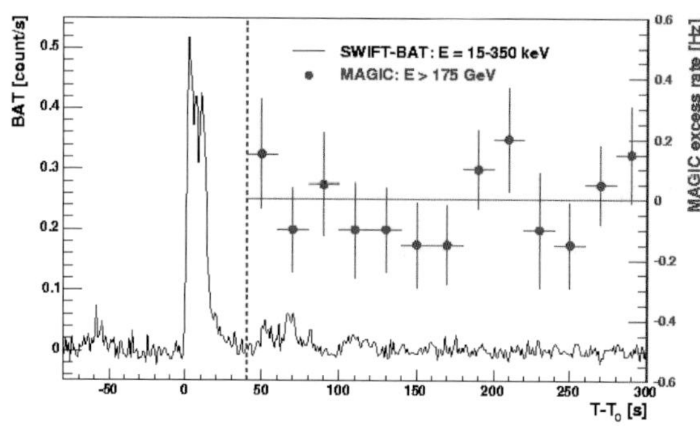

Fig. 3. MAGIC events rate compared with SWIFT-BAT observations. The *vertical line* indicates the start of the MAGIC observations

tion energy threshold was $\approx 175\,\mathrm{GeV}$. No evidence of γ ray signal was found. As the redshift of the GRB was not measured directly, the flux upper limit was given as function of the assumed redshift. Such upper limit is in the order of the γ flux from Crab Nebula.

References

1. Aharonian, F., *et al.* 2005, Science 307, 1938
2. Aharonian, F., *et al.* 2005, Science 309, 746
3. Aharonian, F., *et al.* 2004, A&A 432, 25
4. Aharonian, F., *et al.* 2006 Nature 439, 695
5. Albert, J. *et al.* 2006, ApJ 638, 101
6. Albert, J. *et al.* 2006, Science 312, 1771
7. Albert, J. *et al.* 2006 ApJ 641, 9
8. Albert, J. *et al.* 2006 ApJ 642, 119
9. Albert, J. *et al.* 2006 submitted to ApJ
10. Albert, J. *et al.* 2006 accepted by ApJ
11. LaRosa, T.N. *et al.* AJ 119, 207
12. Punch, M. 1992 Nature 358, 477
13. Quinn, J. 1996 ApJ 456, 83
14. Weekes, T.C., *et al.* 1989, ApJ 342, 379

Gamma-ray Astronomy
with full coverage experiments

Paola Salvini

Istituto Nazionale di Fisica Nucleare, Sez.di Pavia, Italy
Paola.Salvini@pv.infn.it

1 Introduction

Full coverage experiments in gamma astronomy mean detectors observing
most of the shower particles by means of a nearly continuous active area: in
this way one may be able to get a lower energy threshold respect to that typical of the EAS sampling arrays (which are detecting only a small percentage
of the particles reaching the ground). The Full Coverage detectors typically
cover an energy range from about a few hundreds of GeV up to tenths of
TeV, partially overlapping the usual range of the Cherenkov technique. The
main topics of interest in this Very High Energy (VHE) range are the study
of supernova remnants, the search for Active Galactic Nuclei (AGNs) and the
high energy emission from gamma ray bursts (GRBs).

As the Air Cherenkov Telescopes (ACTs) and the full coverage Extensive Air Shower (EAS) detectors cover a similar energy range in gamma-ray
astronomy, it is useful to compare the performance of the two techniques
in order to understand their complementarity. The ACTs have a very good
energy resolution and a good gamma/hadron discrimination, that's where
their very good sensitivity comes from: however they have a low duty cycle
($\approx 10\%$) and a small field of view ($\approx 20\,\mathrm{msr}$), partially compensated by the
possibility of changing the detector orientation. The full coverage appara-
tuses, on the contrary, have a field of view of about $2\,\mathrm{sr}$ and a duty cycle that
in principle can reach 100%, i. e. an uncomparably large exposure time to any
source in their wide field of view: nevertheless their gamma/hadron discrimination is usually bad compared to that of the ACTs. For these reasons one
can say that the full coverage apparatuses are specially good in discovering
new sources, particularly extended ones (where the small field of view may
hinder the detection, [1]), and in looking for transient phenomena, such as
GRBs.

The GRBs spectra, currently well studied by experiments on satellite
in the keV–MeV range, are still quite unknown at higher energies, where
constraints on different production mechanisms are hoped to be found. GRBs

are detected from isotropically distributed directions in the Universe and they are among the most energetic phenomena known in it (estimates say that their emitted energy is comparable with that of Supernovae). The observation of their optical afterglow emission allowed to measure, in some cases, the redshift and tags them as cosmological objects [2]. The time duration of the detected keV–MeV emission shows a bimodal distribution peaked around $\approx 0.3\,$s (short bursts) and $\approx 30\,$s (long bursts) [3]. While for these latter a largely accepted model for the production mechanism has already been developed, much less is known on the short GRBs. It appears that the search for high energy emission from GRBs is one of the most important goals for an apparatus with a wide field of view and high duty cycle. Up to now, very few experimental data exist concerning the GRBs high energy emission: one observation by EGRET of a $18\,$GeV photon detected 90 minutes after the main event of GRB940217 [4] and one candidate of a high energy emission ($> 650\,$GeV) from GRB970417a obtained with the Milagro test apparatus (Milagrito) [5].

At present, two full coverage experiments are taking data: Milagro, a water Cherenkov detector placed in an artificial pond at $2600\,$m a. s. l. in New Mexico (USA), and ARGO-YBJ, a "carpet" of resistive plate chambers located at $4300\,$m a. s. l. on a plateau in Tibet (P.R. China). It has to be mentioned a proposal for a $(200 \times 200)\,\mathrm{m}^2$ water Cherenkov detector to be placed at high altitude (for a description see [6]). It has to be stressed that the high altitude location of most ground-based experiments is obviously motivated by the purpose of lowering the energy threshold of the detected gamma showers by reducing the atmosphere thickness over the apparatus.

2 Milagro

Milagro is a double layer of photomultipliers dipped into a water reservoir $\approx 5000\,\mathrm{m}^2$ of area, completely shielded from the solar light, where the PMTs observe the Cherenkov light produced by the shower particles crossing the pond (see [7] for a short description). Moreover, from 2002, the experiment is surrounded by an array of 175 PMTs in water tanks, covering a total sampling area of about $\approx 40\,000\,\mathrm{m}^2$.

The experiment has an angular resolution of about 0.75^0 around TeV γ-rays [9] and a good background rejection (based on the water depth reached by the shower particles, in particular muons) which allows to discard 91% of the protons keeping about 50% of gamma-rays.

The Milagro sky map, after five years of data taking, can be found in [8]: it shows the Crab at $\approx 10\sigma$, the source Mrk421 at $\approx 6\sigma$ and, most remarkably, an excess in the Cygnus region discussed in [9]. This latter represents the first observation of emission in the TeV region from an extended source near the galactic plane. The emission spectrum from the galactic plane depends on the different interaction models between cosmic rays and interstellar medium.

In [9] the flux measured by Milagro is compared with an extrapolation of the EGRET spectrum between 1 and 30 GeV in the same region of the Galaxy.

Furthermore, in [10] the Milagro collaboration reports a negative search for high energy emission from 45 GRBs detected by satellite experiments between 2002 and 2005.

3 ARGO-YBJ

ARGO-YBJ (Astrophysical Radiation with Ground-based Observatory at YangBaJing), has about $6500\,\mathrm{m}^2$ of total detector area, $5500\,\mathrm{m}^2$ of them with full coverage (see [11] and references therein for a complete description of the apparatus). The experiment consists of a single layer of Resistive Plate Chambers (RPCs) with about 93% of active area. Argo is currently taking data with the complete central carpet, and the foreseen detector ring installation, together with a lead converter layer 0.5 cm thick, will improve its sensitivity.

Argo has two possible detection modes: the "shower mode", where a reconstruction of the primary particle direction is performed by a fit to the space-time shower front, and the "scaler mode",where the rate for four different multiplicities ($\geq 1, \geq 2, \geq 3$ and ≥ 4) on each detector unit (i. e. a cluster of 12 RPCs large $7.64 \times 5.72\,\mathrm{m}^2$) and for a time window of 0.5 s is recorded. Thanks to its high altitude location, the RPC good time resolution and efficiency and its high spatial granularity (a space-time "pixel" of $56 \times 62\,\mathrm{cm}^2$ and at low energy the possibility to detect the position of the few particles reaching the ground by strips of $6 \times 62\,\mathrm{cm}^2$), ARGO-YBJ has a good angular resolution ($\approx 0.5^0$ at about 1 TeV) and the possibility to perform a continuous sky survey in the celestial latitude range $(-10^0, +70^0)$ with a threshold of a few hundreds of GeV in shower mode. In scaler mode, where the reconstruction of the primary particle direction is not performed, the energy threshold is around few GeV.

Concerning the "shower mode", some preliminary results have been reported [12], obtained with the first 6 months of data taken with only a third of the complete carpet (about $1900\,\mathrm{m}^2$). These first data were mainly used to check hardware and software instruments on several cosmic rays distributions, as described in [13]. The first results of a gamma-ray sky survey has been reported as well in [12] where, as expected from the MonteCarlo sensitivity estimate, no significant excess has been claimed.

The "scaler mode" allows the detection at low energy of transient phenomena (seen as a non-statistical fluctuation of the cosmic ray background) such as GRBs or solar flares. The use of four different scalers allows a rough energy estimate of the phenomenon. The counting rate fluctuations are corrected for the influence of environmental parameters, such as atmospheric pressure and detector temperature, in order to maintain the statistical behavior of the detector. A search for VHE emission from GRBs has been presented in [14].

No significant emission has been detected up to now and fluency upper limits $\approx 10^{-4}-10^{-5}\,\mathrm{ergcm}^{-2}$ were set in the 1–100 GeV energy range. The 3σ upper limits were estimated for GRBs detected by satellite experiments in the ARGO-YBJ field of view, using the extrapolation of the spectral index determined by satellites, and a model for the Extragalactic Background Light (EBL) $\gamma\gamma$ absorption [15] when the redshift had been measured.

4 Conclusions

The full coverage air shower arrays are showing-up as powerful instruments for the detection of extended sources [9] or transient phenomena at energies $> 10\,\mathrm{GeV}$ [10, 16]. Even if the knowledge of the VHE region gamma sources has greatly improved in the last few years, due to the small field of view of the ACTs and to the low sensitivity of the EAS sampling arrays, there is still much to be done concerning the study of extended sources, the observation of temporal variations of known sources gamma-ray emission and the discovery of new ones. At last, it is worth to mention that the full coverage detectors have an unique opportunity to look for high energy emission of GRBs. Therefore the contribution of a full coverage detector in complementing observations made by ACTs in this energy range, could be really important.

References

1. XXIX ICRC Conf.Proc. (Pune,India), 10(2006)227
2. C.Firmani, V.Avila-Reese, G.Ghisellini and G.Ghirlanda, submitted to Mon.Not.R.Astron.Soc. (astro-ph/0605267)
3. L.G.Balazs *et al.*, Astron.Astrophys.401(2003)129
4. K.Hurley *et al.*, Nature 371(1994)652
5. ApJ Lett.,533(2000)L119
6. "High energy gamma-ray astronomy" (Heidelberg,2004)Conf.Proc. 745 (2005) 234
7. Nucl.Instr.Meth.A449(2000)478
8. XXIX ICRC Conf.Proc.(Pune,India), 4(2006)81
9. Phys.Rev.Lett.95(2005)251103
10. XXIX ICRC Conf.Proc.(Pune,India), 4(2006)463, 467
11. XXIX ICRC Conf.Proc.(Pune,India), (2006)
12. XXIX ICRC Conf.Proc.(Pune,India), 4(2006)375
13. XXIX ICRC Conf.Proc.(Pune,India), Bernardini
14. G. Di Sciascio and T. Di Girolamo, Proc. of Workshop "The Multi-Messenger Approach To High Energy Gamma-Ray Sources", Barcelona (Spain), July 4–7, 2006 (astro-ph/0609317)
15. T.M.Kneiske *et al.*, Astronomy and Astrophysics 413(2004)807
16. XXIX ICRC Conf.Proc.(Pune,India), 4(2006)431

PARALLEL SESSION:
Detectors and New Technologies
(A. Cardini, M. Michelotto and V. Rosso,
conveners)

The Liquid Xenon calorimeter of the MEG experiment

Fabrizio Cei

INFN and University of Pisa, Largo B. Pontecorvo, 3 – 56127 Pisa (Italy)
`fabrizio.cei@pi.infn.it`

1 Introduction

In the Standard Model (SM) of electroweak interactions the Lepton Flavour
Violating (LFV) processes are forbidden at all; however almost all SM extensions predict processes which do not conserve the lepton number. In particular the supersymmetric models predict branching ratios for LFV reactions at
level of $10^{-(13 \div 16)}$, which should be experimentally observable. The $\mu \to e\gamma$
process is one of the golden channels for the LFV observation [1, 6]. The
present limit for the branching ratio $(\mu \to e\gamma)/(\mu \to ALL)$ is 1.2×10^{-11} [6]
and the aim of the MEG experiment at Paul Scherrer Institute (PSI) is to
improve this limit by a factor ~ 100. This challenging goal requires refined
and innovative detection techniques in order to reach very high resolutions
in the measurement of γ and e^+ energy, momentum and timing. One of the
key elements is the Liquid Xenon (LXe) e.m. calorimeter, the main subject
of this paper.

2 Overview of the MEG experiment

A layout of the MEG experiment [7] is shown in Fig. 1. The muon beam
($3 \times 10^7 \, \mu^+$s, the most intense muon continuos beam line in the world) will
stop in a polyethilene target, slanted by $22°$ with respect to the beam direction to minimize the positron multiple scattering within the target. The
positrons will be detected by a magnetic spectrometer composed by a superconducting magnet (the central value of the field is $1.26 \, \mathrm{T}$) and by 16 drift
chambers (DC). The magnetic spectrometer has a gradient field, which allows a fast expulsion of low longitudinal momentum positrons and a momentum selection roughly independent on the positron emission angle. The DCs,
aligned at $10°$ intervals, are organized in 2 staggered arrays and equipped
with wires and cathodic kapton foils. The positron timing will be measured

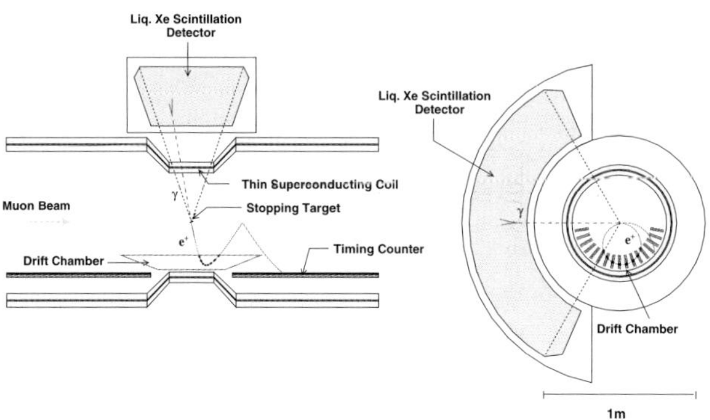

Fig. 1. Layout of the MEG experiment

by a double array (curved fibers and horizontal rods) of scintillation counters (Timing Counter, TC): the inner layer will be mainly devoted to trigger purposes, while the outer layer will measure the positron timing and impact point. The LXe calorimeter for measuring the γ energy, momentum and timing will be described in the next sections. Table 1 shows the required resolutions (FWHM) of MEG sub-detectors in the measurement of $e^+\!-\!\gamma$ relative timing, relative angle and energy, compared with that of previous $\mu \to e\gamma$ experiments. We stress that some of them (e. g. the TC timing resolution) were obtained in accelerator beam tests under various conditions (i. e. with or without magnetic field).

3 The Liquid Xenon calorimeter

The LXe calorimeter has an active volume of $\approx 800\,\mathrm{l}$ and the LXe scintillation light is observed by ≈ 850 two inches PMTs. The main properties of LXe as scintillation medium are: high density ($2.95\,\mathrm{g\,cm^{-3}}$), short radiation length ($2.77\,\mathrm{cm}$), fast response ($\tau \approx 22\,\mathrm{ns}$ and $\approx 45\,\mathrm{ns}$) and high light yield ($\approx 80\%$ anthracene). Since a so large volume of LXe was never used in experiments, a refined R&D work was needed to ensure that the required performances can be reached and maintained for a couple of years. Here we discuss the improving of PMT response, the development of purification system and the results obtained with a calorimeter prototype. Note that the building of the LXe detector is almost completed: the cryostat should be delivered at PSI in November and the PMT mounting and LXe filling should start immediately after the delivery. **PMT R&D.** The PMT behaviour in LXe was studied by means of a cryogenic facility in Pisa [8]. In the original project, the Hamamatsu R6041 PMTs were selected; however, when operated in intense photon background conditions (simulated by a LED flashing at tens-of-KHz rate) their response to a fixed reference signal exhibited strong variations, in correspondence with the switching on and off of the crowding source. This behaviour (due to the increase of photocathode resistivity at

Table 1. Required performances of MEG sub-detectors (all FWHM) compared with that obtained by previous $\mu \to e\gamma$ experiments

Experiment	Year	ΔE_e	ΔE_γ	$\Delta t_{e\gamma}$	$\Delta\theta_{e\gamma}$	Stop rate	Duty Cycle	Upper limit	Ref.
SIN	1977	8.7%	9.3%	1.4 ns	–	$5 \times 10^5\ \mu^+/s$	100%	$< 1.0 \times 10^{-9}$	[2]
TRIUMF	1977	10%	8.7%	6.7 ns	–	$2 \times 10^5\ \mu^+/s$	100%	$< 3.6 \times 10^{-9}$	[3]
LANL	1979	8.8%	8%	1.9 ns	37 mrad	$2.4 \times 10^5\ \mu^+/s$	6.4%	$< 1.7 \times 10^{-10}$	[5]
LANL	1986	8%	8%	1.8 ns	87 mrad	$4 \times 10^5\ \mu^+/s$	6–9%	$< 4.9 \times 10^{-11}$	[4]
MEGA	2002	1.2%	4.5%	1.6 ns	17 mrad	$2.5 \times 10^8\ \mu^+/s$	6–7%	$< 1.2 \times 10^{-11}$	[6]
MEG	2008	0.8%	4%	0.15 ns	19 mrad	$3 \times 10^7\ \mu^+/s$	100%	$< 1.0 \times 10^{-13}$	

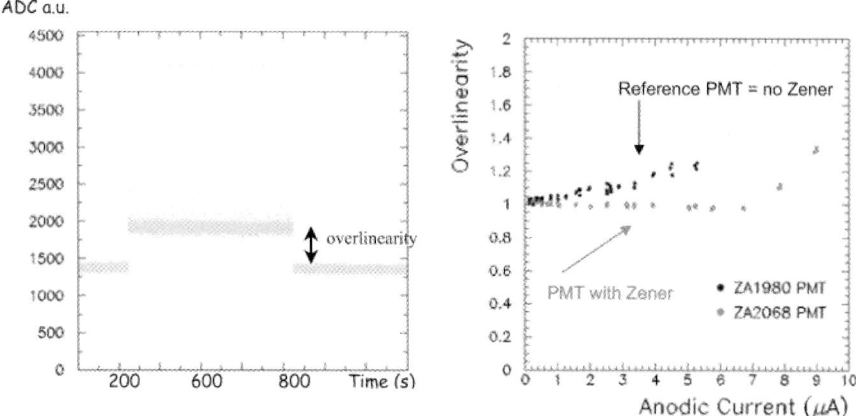

Fig. 2. *Left*: the overlinearity effect. *Right*: Dependence of overlinearity on anodic current for two representative PMTs, one with and one without Zener diodes

cryogenic temperatures) was eliminated by changing the photocatode composition (from Rb–Cs–Sb to K–Cs–Sb mixture) and by adding Aluminum strips in front of it (PMT types R9288 and R9869, the latter with a double density of Al strips). Moreover, an overlinearity effect was observed in presence of the intense γ background because of the changes in PMT gain induced by an average anodic current comparable with the base current. This second effect was reduced by inserting two Zener diodes in the last two stages of the base amplification chain of R9869 PMTs. Figure 2 shows the overlinearity effect (left) and the dependence of the overlinearity on the anodic current (right) for two R9869 PMTs, one with and one without Zener diodes. Since the working point for MEG PMTs should be about 2–$3\,\mu$A, the behaviour of Zener corrected R9869 PMTs looks satisfactory.

The purification system. The LXe calorimeter performances are strongly affected by processes which absorb the scintillation light. Because of the scintillation mechanism [9], the Xenon is almost transparent to its own scintillation light, but possible contaminants can be very "opaque"; therefore, a complex purification system was developed, which removes water vapor and Oxygen contaminants by multiple passages through purifier cartridges filled with molecular sieves. The system has been repeatedly improved in speed and efficiency; in the present configuration the Xenon is circulated in liquid phase by a Barber-Nicols cryogenic pump, at a rate of $\approx 100\,$l/hour. The scintillation light absorption length after the purification is $\lambda_{\mathrm{abs}} > 1\,$m.

Results on calorimeter prototype. A $\approx 70\,$l LXe calorimeter prototype (Large Prototype, Fig. 3, left) was built and operated in a beam test at PSI to detect γ's from charge exchange reaction $\pi^- \mathrm{p} \to \pi^0 \mathrm{n}, \pi^0 \to \gamma\gamma$. By using a back-to-back coincidence with a NaI detector, $54.9\,$MeV and $82.9\,$MeV photons were selected; the reconstructed $54.9\,$MeV γ-line is shown in Fig. 3, right. The

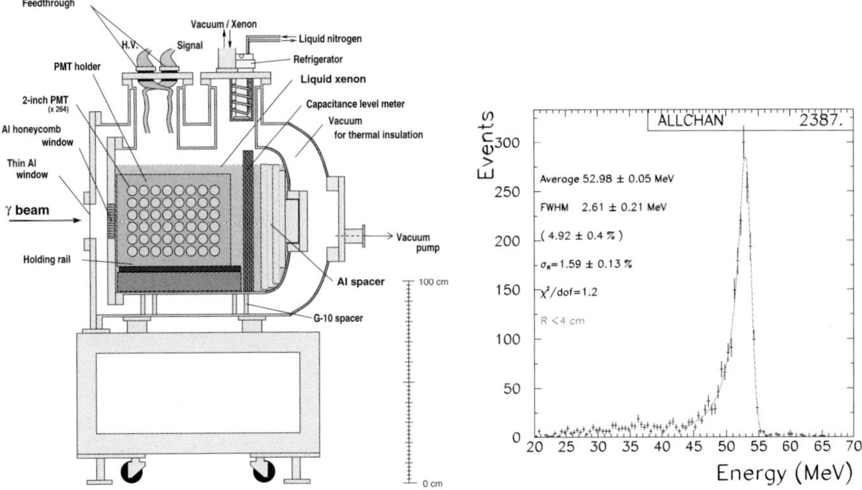

Fig. 3. *Left*: The Large Prototype (LP). *Right*: 54.9 MeV γ-line seen in the LP

measured energy, timing and position resolutions at 54.9 MeV were $\approx$ 4.9%, 140 ps and 8.9 mm, close to the required values (Table 1).

4 Conclusions

The ambitious goal of the MEG experiment (improve the present limit in the $(\mu \to e\gamma)/(\mu \to ALL)$ branching ratio by a factor ~ 100) demands detectors with challenging resolutions, most of which were experimentally obtained in dedicated beam tests. For the LXe calorimeter, a long R&D work was needed to develop PMTs with good Q.E. and stable behaviour at cryogenic temperatures and a fast and reliable Xenon purification system. The beginning of data taking for the MEG experiment is foreseen for the end of 2006.

References

1. E.P. Hincks and B. Pontecorvo: Can. J. Res. A **28**, 29, (1950)
2. A. Van der Schaaf *et al.*: Nucl. Phys. A **340**, 249, (1980)
3. P. Depommier *et al.*: Phys. Rev. Lett. **39**, 1113, (1977)
4. R.D. Bolton *et al.*: Phys. Rev. D **38**, 2077, (1988)
5. W.W. Kinnison *et al.*: Phys. Rev. D **25**, 2846, (1982)
6. M. Ahmed *et al.*: Phys. Rev. D **65**, 112002, (2002)
7. MEG proposal to INFN, available at `http://meg/ps.ch/docs/prop_infn/nproposal.pdf`
8. A. Baldini *et al.*: approved by NIM, (2006)
9. A. Baldini *et al.*: Nucl. Inst. and Meth. A **545**, 753, (2005)

The Silicon Vertex Trigger Upgrade at CDF

Alberto Annovi for the CDF Collaboration

Istituto Nazionale di Fisica Nucleare, Via E. Fermi 40, I-00044 Frascati, Italy
annovi@lnf.infn.it

Motivations, design, performance and upgrade of the CDF Silicon Vertex Trigger are presented. The system provides CDF with a powerful tool for online tracking with offline quality in order to enhance the reach on B-physics and large P_t-physics coupled to b quarks.

1 The SVT: online tracking with offline quality

SVT is the Silicon Vertex Trigger [1] of the CDF experiment [2] at FNAL. SVT uses data from COT, the Central Tracking Chamber, and SVX, the Silicon Vertex Detector. SVX consists of six adjacent barrels of high spatial resolution detectors surrounding the beam pipe. Each barrel consists of five detector layers. SVX is naturally subdivided into 12 azimuthal sectors, 30 degrees each, called phi sectors or wedges.

SVT gives important contributions to several triggers. However, it was specifically designed for selecting B hadronic decays for both B-physics and high P_t-physics strongly coupled to b quarks. At an hadron collider B-physics can benefit of a large B production cross section. Moreover, the high energy can produce new particles decaying into b quarks. On the other hand, the QCD background is overwhelming. For these reasons, one definitely needs to improve the signal to background ratio at trigger level. For this purpose, one can rely on the long lifetime of the B mesons, by detecting the decay vertex. However, this is technically challenging, since one needs to track with offline resolution at high event rates. SVT detects large impact parameters of decay tracks, with respect to the proton-antiproton collision point.

At which trigger level can we use our online tracker in CDF? At level three it is too late, because most physics would be already lost. On the other side, the silicon signals are digitized only after the level 1 accept. Thus, the right place is the level 2, with an average latency of about 20 microseconds. Within this time, SVT must achieve offline resolution for a full event with thousands of hits, where offline CPUs usually take hundreds of milliseconds.

SVT takes in input the COT tracks reconstructed with low resolution at level 1, track phi, track P_t and the digitized pulse heights from the silicon vertex. With these inputs, SVT makes high resolution tracks in few stages. First, it implements a hit cluster finding on the SVX data in order to get high resolution hit coordinates. Then SVT divides the tracking problem in two steps. First, it solves the pattern recognition problem at low resolution. Adjacent detector channels are logically ORed together into super-bins. The hit super-bins are associated in order to form candidate tracks called roads. At this level, the COT track parameters can be logically treated as the other hits. The resolution is coarse (a few hundred µm) in order to limit the combinatorial problem. Few residual ambiguities, such as those of multiple hits in the same super-bin, are resolved in the next step, but most of the pattern recognition is done at this point. The final step consists in fitting the full resolution hits to determine the track parameters. The SVT output consists of the reconstructed track parameters, phi, P_t and most important the impact parameter. The reconstruction is bi-dimensional on the transverse plane.

The pattern recognition is based on a pattern matching algorithm [3]. The set of all super-bin combinations compatible with the passage of a particle is calculated in advance and stored in a large memory called pattern bank. During the run, for every event each stored pattern is compared with the event looking for matching patterns. Of course, the pattern bank is large and we don't have the time to compare the patterns sequentially. We developed a dedicated device in order to achieve the maximum parallelism: the associative memory (AM). The working principle is similar to that of the bingo game. Each player has its own scorecard with a unique combination of detector super-bins. While the detector hits are being read out, each player puts a mark on matching super-bins on his scorecard. Then, each bingo corresponds to a track candidate and it is sent out of the device. The pattern recognition is performed during the detector readout: without additional computing time.

Once a track is confined to a road, fitting becomes easy. We use a linear expansion in the hit positions of both chi square and track parameters. The fit is so reduced to a number of scalar products which allow a very fast evaluation in FPGAs. The expansion constants can be evaluated in advance from detector geometry. Corrections and feedbacks for mechanical alignments proceed through linear algorithms which are intrinsically fast and stable.

From the abstraction to the real world: SVT uses ten 9-units VME crates in the CDF counting room. Each crate hosts the hardware for two phi sectors, plus two crates for track fitting and two crates for fan-in and fan-out. The data flow through the main boards pipeline: the hit finder, the associative memory and the track fitter.

The SVT performance at a luminosity of 5×10^{31} cm^2 s1. The latency distribution, which includes SVX digitization and readout, nicely peaks around 24 µs. The impact parameter distribution has a sigma smaller than 50 µm, which is a convolution of the spatial resolution (about 35 µm) and the beam spot dimension (about $33\,\mu\mathrm{m}$). The efficiency is about 90% for fiducial offline tracks with 4 silicon hits. It is worth to mention that SVT can solve the pattern recognition without inputs from the tracking chamber. With SVX data only, the impact parameter resolution is degraded to about $87\,\mu\mathrm{m}$, because of the shorter lever arm in measuring the track P_t. But, this result is interesting in view of possible future applications to all-silicon trackers.

SVT continuously monitors the beam position in an actual run, in order to subtract it when determining impact parameters with respect to the beam axis. If the beam moves from the nominal position, all tracks will have an impact parameter with respect to the nominal position. The impact parameter as a function of phi for a displaced beam follows a sinusoidal law and the sinusoid parameters are in a straightforward relation with the beam position.

The trigger requirements for B selection. At level 1 we basically require two tracks with some P_t threshold. At level 2 we typically require the impact parameter on both tracks and a positive B decay length. It is worth to underline the big advantage that comes form SVT. For instance, let ús consider the case of the B^0 decaying into two charged mesons. This signal is extremely hard to be extracted from the QCD background, but it is possible thanks to SVT that gives a background rejection of about three orders of magnitude.

Thanks to SVT, CDF has an hadron-hadron mass distribution with a nice peak of the B^0 mass. Before SVT was first used for data taking, many physicists were skeptical about the possibility of just seeing this decay at an hadron collider because of the huge QCD background. Nowadays we even use a nice D0 peak as an on-line monitor for data quality.

The Tevatron luminosity is ever increasing. This is the main reason for upgrading SVT. Especially the latency time suffers at higher luminosities. In fact, the average latency is fine, but the problem comes from the distributions tail due to crowded events. Because of historical reasons, during level 2 the DAQ uses only 4 event buffers, which can be saturated by fluctuations in the processing time, thus generating dead time.

We proposed the SVT upgrade in 2003. To cope with the Tevatron schedule it was necessary to complete it within 2 years. The idea for a fast implementation was to replace boards without changing the SVT data-flow and infrastructure. We replaced 4 different kinds of boards for a total of 60 boards, in less than 30 months. The upgrade consists in making faster boards and, in order to reduce the number of fake combinations, thinner roads.

The latter required to implement a larger pattern bank. The core of the new pattern bank is based on a new standard cell AM chip [4] with a $0.18\,\mu\mathrm{m}$ technology. The new chip contains 5000 patterns. The chips have been simulated with a clock rate up to $50\,\mathrm{MHz}$ and actually tested at $40\,\mathrm{MHz}$. 3000 chips were produced in April 2005 with a yield of 70%. The first AM boards, loaded with 64 AM chips each, have been installed in August 2005.

The new pattern bank (512k patterns/wedge) works with a resolution of $240\,\mu\mathrm{m}$, while the original bank (32k patterns/wedge) had a typical resolution of $600\,\mu\mathrm{m}$. With the improved resolution, the number of ambiguities left after the SVT pattern recognition, measured as the total number of fits to be performed is decreased by a factor ≈ 2.5. A strong reduction in the tail of the number of fits distribution is also achieved.

Three other kind of boards have been re-built for this upgrade. They are the AMSRW (AM sequencer and road warrior), the hit buffer, and the track fitter. We implemented these function with Pulsar boards [5]. They are flexible programmable boards developed for other CDF upgrades. They have 3 powerful FPGAs on board and all the CDF DAQ connectors. The RAM available on board was not sufficient for the demanding SVT application. It has been provided with custom RAM mezzanines. Pulsars have been a key element to minimize the effort and time required for hardware development. This allowed us to develop and install these boards in less than 18 months.

Another key element of the SVT upgrade is the road warrior [6] that solves the common problem of ghosts removal. This problem arise in SVT when majority logic (4 out of 5) is used for pattern matching. A large number of ghost roads are found and have to be fitted. The road warrior identifies and removes them before fitting. This function is implemented as a small associative memory within an FPGA.

With the SVT upgrade the available L1 trigger bandwidth increased from about 18 kHz to 25–30 kHz. The speed of the upgraded system is only slightly sensitive to increasing instantaneous luminosity. The same L1 bandwidth is now available up to peak luminosity.

The next challenges in this field are at LHC and other future machines, especially in moving the reconstruction of all-silicon detectors down to the level 1 trigger. We are working on this R&D, for instance Fast Track is an ongoing study for applications to the LHC triggers [7,8]. The SLIM5 project is R&D for a new pixel detector designed along with a dedicated L1 trigger.

In summary, the design and construction of SVT was a significant step forward in the technology of fast track finding. Performance of SVT is as expected. CDF is triggering on impact parameter and collecting data leading to significant physics results. We described the SVT upgrade for operation at high instantaneous Tevatron luminosity. We put emphasis on the upgrade features and the key ideas that allowed fast and effective implementation.

References

1. B. Ashmanskas *et al:* Nucl. Instrum. Meth. A **518**, 532-536, (2004)
2. F. Abe *et al.*, Nucl. Instr. and Meth. A **271**, 387 (1988)
3. M. DellOrso and L. Ristori, Nucl. Instr. and Meth. A **278**, 436 (1989).
4. A. Annovi *et al:* NSS2005 Conference Record **1**, 259 (2005)
5. http://hep.uchicago.edu/thliu/projects/Pulsar/
6. J. Adelman *et al:* IEEE Trans.on Nucl. Sci. **53**, 648 (2006)
7. A. Annovi *et al.*, IEEE Trans. on Nucl. Sci. **48**, 575 (2001).
8. http://www.pi.infn.it/~orso/ftk/

Monolithic Active Pixel Sensors
in a 130 nm Triple Well CMOS Technology

V. Re[1,2], C. Andreoli[2,3], M. Manghisoni[1,2], E. Pozzati[2,3], L. Ratti[2,3],
V. Speziali[2,3], G. Traversi[1,2], S. Bettarini[4], G. Calderini[4], R. Cenci[4],
F. Forti[4], M. Giorgi[4], F. Morsani[4], N. Neri[4], G. Rizzo[4].

[1] University of Bergamo, I-24044 Dalmine (BG)
[2] INFN Pavia, I-27100 Pavia
[3] University of Pavia, I-27100 Pavia
[4] INFN Pisa and University of Pisa, I-56100 Pisa

1 Introduction

The attention of several groups in the particle physics community has been
drawn to monolithic active pixel sensors (MAPS) in CMOS technology as
promising candidates for charge particle tracking at the future high luminosity colliders. Their working principle, based on the diffusion of minority
carriers in a lightly doped, thin epitaxial layer, might be exploited in the fabrication of highly granular, light detectors possibly satisfying the resolution
constraints set by next generation experiments at the International Linear
Collider and at the Super B-Factory [1].

In this work, the large scale of integration and the triple-well feature of
deep submicron CMOS processes have been exploited to develop an integrated sensor with high functional density in the elementary cell and with
the potential for thin detector fabrication, therefore incorporating the main
characteristics of hybrid pixels and standard MAPS in a monolithic crystal. Since an N-well with a deep junction is used as the collecting electrode,
the proposed device will be called deep N-well monolithic active pixel sensor (DNW-MAPS). The main features of such a detector will be highlighted
in the following section. Experimental results from the characterization of
the front-end electronics for DNW-MAPS belonging to two prototype chips,
Apsel0 and Apsel1, will then be presented and discussed.

2 Deep N-well pixel sensor

In standard CMOS-MAPS, use of PMOS devices in the design of the front-
end electronics is avoided as the N-well they are integrated in might subtract
charge to the collecting electrode leading to potentially serious efficiency loss.

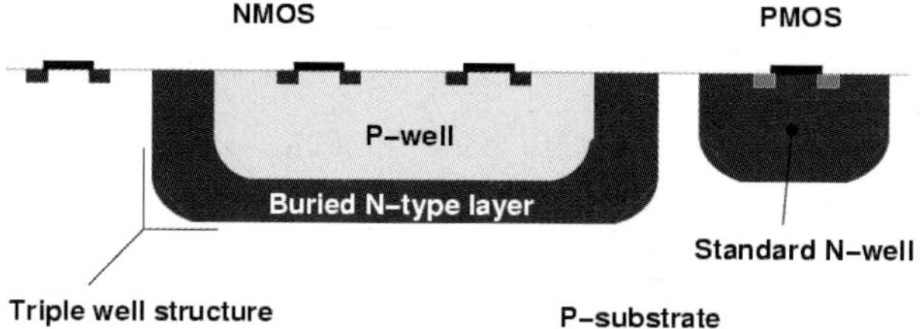

Fig. 1. Cross-sectional view of a CMOS process including a triple-well structure

In the deep N-well MAPS discussed in this paper, the problem has been circumvented by using a buried N-type layer to design a comparatively larger electrode than it is found in common CMOS sensors (Fig. 1). If the dimensions of the collecting electrode are chosen to be sufficiently large, N-type wells can be allowed in the vicinity with no significant degradation of the collection efficiency. The front-end electronics is based upon a standard processing chain for capacitive sources and includes a charge sensitive amplifier, a first order, semigaussian unipolar (RC-CR) shaper with programmable peaking time (0.5, 1 and 2 μs), a comparator and a NOR latch. The analog and the digital sections of the readout processor are separated from each other, in order to minimize cross talk issues.

3 First prototype chip

The first prototype test chip (Apsel0) was fabricated in a 130 nm triple-well, epitaxial, CMOS process provided by STMicroelectronics. A detailed description of its features has been given in a previous paper [2]. The purpose of this first prototype was to demonstrate the feasibility of the DNW-MAPS sensor and its capability to detect ionizing radiation. The availability of single cells with different sensor area was meant to help study charge collection and signal–to–noise properties of the device. In the Apsel0 chip, the forward block of the preamplifier is a cascode stage with a cascode load. In the input NMOS element, the channel width $W = 3\,\mu m$, and length $L = 0.35\,\mu m$, were optimized for a detector capacitance of 100 fF and a drain current of 1 μA. The power dissipated by the analog section and the threshold discriminator is about 10 μW. In Table 1, the charge sensitivity is displayed for all of the three channels with the injection capacitance at the preamplifier input at the three available peaking times. In the ch5 pixel structure, the charge preamplifier was connected to the relevant deep N-well sensor (featuring a 270 fF capacitance) to form a DNW-MAPS. No electrical connection to the collecting electrode was laid out instead in the case of ch2, nor in the case of ch1, which consists of the bare readout channel. Nevertheless, for both of them, the NMOS devices belonging to the analog section were integrated in a deep N-well covering the same area ($830\,\mu m^2$) as in the ch5 single pixel. The significant change in the charge sensitivity featured by ch5 is to be blamed on

Table 1. Charge sensitivity [mV/fC] and ENC [e^- rms] at the three available peaking times for the structures with integrated calibration capacitance (Apsel0)

Peaking time [µs]	Charge sens. [mV/fC]/**ENC** [e^- rms]		
	ch1 ($C_D = 0$)	ch2 ($C_D = 100\,\text{fF}$)	ch5 ($C_D = 270\,\text{fF}$)
0.5	610/**38**	580/**80**	460/**150**
1	590/**40**	550/**77**	450/**153**
2	530/**39**	520/**73**	430/**149**

the small loop gain in the preamplifier, which was designed to process signals from a 100 fF detector. Equivalent noise charge, also shown in Table 1 was found to be independent of the peaking time, a hint of the fact that $1/f$ noise in the preamplifier input device is the dominant contribution to the overall channel noise performances.

4 Front-end electronics characterization

The second prototype (Apsel1) includes five single pixel cells, all provided with 60 fF injection capacitance. One of them consists of the standalone readout circuit not connected to the DNW sensor, while the other four are DNW-MAPS with different sensor area. An 8 by 8 pixel matrix was also integrated, capable of generating a trigger signal as the wired OR of all of the latch outputs and featuring a row by row, 8-line parallel readout. In the design of the front-end electronics for the new prototype, the noise and gain issues raised by the Apsel0 chip were addressed by implementing a folded cascode circuit with active load stage in the charge preamplifier. Moreover, the input NMOS transistor dimensions, $W/L = 16/0.25$, were optimized for a sensor parasitic capacitance C_D of 460 fF. The drain current in the input element was set to 30 µA so as to improve significantly the front-end noise performances. Simulations predicted a power dissipation of about 60 µW per channel. The equivalent noise charge and the charge sensitivity, measured for the same device (channel 1), is shown in Table 2. ENC was found to be

Table 2. ENC, charge sensitivity and ENC slope for the single DNW-MAPS with the same layout as the matrix pixels (channel 1, $C_D = 460\,\text{fF}$, Apsel1)

Peaking time [µs]	ENC [e^- rms]	Charge sens. [mV/fC]	dENC/dC_D [e^-/pF]
0.5	41	466	70
1	39	432	68
2	39	406	68

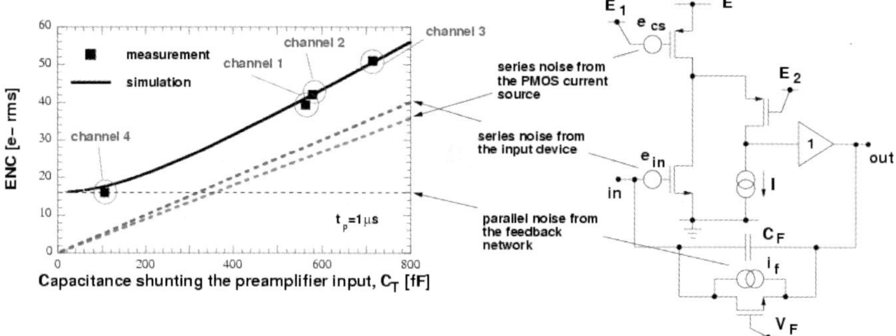

Fig. 2. Equivalent noise charge as a function of the capacitance shunting the preamplifier input terminal, C_T. Measurement data are compared to simulation results. Noise contributions to the total ENC are specified in a simplified schematic representation of the preamplifier

about 40 electrons, irrespective of the peaking time, in good agreement with post layout simulations. In Fig. 2, measurement results are compared to simulation data. The main noise contributions are represented by the voltage sources e_{in} and e_{cs} and by the current source i_f in a simplified schematic of the charge preamplifier.

5 Conclusions

In this paper, a novel concept of CMOS monolithic active pixel sensor was presented. The proposed MAPS takes advantage of the deep N-well structure available in deep submicron CMOS processes to collect the charge released in the epitaxial layer and to increase the complexity and the functional density of the readout electronics integrated at the pixel level. Measurements performed on the first prototype, Apsel0 demonstrated that the device is capable to detect ionizing radiation. The characterization of the second prototype, Apsel1, indicates that the gain and noise issues raised by the Apsel0 chip have been correctly addressed. Design and submission of a full size MAPS, with hybrid-pixel-like functionalities and implementing data sparsification, is planned for the end of 2006.

References

1. H.S. Matis *et al.*, *IEEE Trans. Nucl. Sci.*, **50** 1020 (2004).
2. L. Ratti *et al.*, *2005 IEEE Nuclear Science Symposium Conference Record*, Vol. 2, October 23-29, 2005, pp. 969–973.

The external scanning proton microprobe in Florence: set-up and examples of applications

Lorenzo Giuntini

Dipartimento di Fisica dellUniversitàe Sezione INFN Firenze, Italy

1 Accelerator

During 2003, at the LABEC laboratory (labec@fi.infn.it), Firenze, Italy, a new 3 MV tandem accelerator was installed, mainly for Accelerator Mass Spectrometry (AMS) and Ion Beam Analysis (IBA) measurements. Our system is equipped with a cesium sputtering ion source for AMS and two sources for IBA applications, i.e. a duoplasmatron and a second cesium sputtering; duoplasmatron is usually preferred for micro beam applications, because, working with this source, it is possible to deliver more intense beam on target. The ion beam is focussed by two electrostatic lenses EL1 and EL2 and mass/energy analyzed by a $\sim 90°$ dipole magnet; beam waist is some 0.4 m after the exit port of this magnet. Here a remotely controlled aperture is installed, to allow beam intensity regulation, and a beam profile monitor BPM1, to control shape, dimensions and intensity of the beam as it is transmitted through the slits; it is also possible to measure the beam current after the slits with a Faraday cup (FC1). The beam is then focussed by the electrostatic lenses EL3 and EL4 just at the entrance of the accelerator, where BPM2 allows monitoring proper beam shaping and FC2 is used to measure the current. The beam is then focalized in the high voltage terminal, inside the stripping canal; as the accelerating column has a strong focussing action, at the entrance of the column, a pre-acceleration electrode is installed, so that, increasing the speed of the ions, the particles are less sensitive to the column focussing. Coming out of the accelerator, the beam passes through an electrostatic quadrupole doublet, the action of which is to focus the ions some 2 m downstream of the accelerator, where BPM3 allows verifying the action of the quadrupole doublet; the beam then enters a slits system which regulates beam dimensions and intensity, as it can be monitored by BPM4. The combined action of a switcher magnet SW and a triplet of electrostatic quadrupoles allows beam deflection and focalization along any of the five working beamlines; selecting the microbeam line $(-30°)$, the beam is focussed some 2 m downstream of the SW, where OS slits defining the object of the microbeam lens are installed. Working with a 3 MeV proton beam, as it is usual for IBA measurement, the natural size of the beam at the object slits OS is some mm^2 with a current of the order of some tens of μA.

2 Microbeam [2–4]

Downstream of the objects slits OS, a second slits system CS is installed, to limit beam divergence and, as a consequence, aberrations of the strong focussing lens, constituted by a magnetic quadrupole doublet MQD. As in an optical system, to have an image much smaller than the object, it is necessary to have the object-to-lens distance p much greater than the lens-to-image distance q; in our set up we have $p \sim 6$ m and $q \sim 0.2$ m. With an OS aperture of $\sim 60 \times 300\,\mu m^2$, we have, in vacuum, $\sim 7 \times 5\,\mu m^2$ beam spot on sample with a ~ 1 nA current. X Y steering magnets ST1, roughly halfway between SW and OS, allow us to adjust particle trajectory to match the OS aperture; a second X Y steering magnets halfway between OS and MQD allow us to set the beam along the axis of the MQD lens. Immediately downstream of OS, a diagnostic station allows us to measure the current transmitted by the OS by means of a Faraday cup and to evaluate position, intensity and dimensions of the beam both by BPM5 and a quartz system, composed by a quartz, which can be inserted along the beam path in vacuum, and a TV chamber + monitor, to view beam fluorescence on the quartz. Finally, BPM6, between CS and MQD, allows us to verify beam alignment along the axis of the MQD lens. The beam is then extracted outside the vacuum line; an external beam set-up has many advantages compared to the conventional one with the target in vacuum: ease of handling and moving the target, whichever its size, much smaller risk of damage, reduced charging and warming effects or no effect at all. As a drawback, working in an external environment, implies a widening of the beam dimensions and of the particle energy distribution, due to ion scattering along the external path; in our set up the beam scattering due to the external configuration is minimized thanks to the use of: 1) an extra thin (100 nm) Si3N4 beam exit window [1, 2], 2) a beam path in atmosphere as short as possible (2 mm, tipically), 3) flowing a light gas (He) along the particle path. In our typical working conditions (3 MeV proton beam, 100 nm Si3N4 exit window and 2 mm path in helium) beam spot on sample comes out to be $\sim 9 \times 8\,\mu m^2$ FWHM, with currents of the order of some nA [4]. With our external microbeam it is possible to characterize samples of interest by means of some of the ion beam analysis techniques, i. e. PIXE, PIGE and BS; the first two rely respectively on the detection of characteristic X rays and γ rays produced in the sample by the ion bombardment, while the BS technique is based on the detection of beam particles elastically back-scattered by the target ions. Thanks to the extremely high X rays production cross section values, with the PIXE, elements (Z > 10) are detected down to trace levels. To measure the number of projectiles hitting the target, we implemented a system which exploits the yield of Si X-rays emitted from the Si3N4 window during beam bombardment. The number of counts of Si X rays collected, even with very weak currents (down to the pA range) and short measuring times (minutes), is large enough to produce small errors. Our facility is equipped with a scanning system, which allows us to sweep the beam over a static target and move the target relative to a fixed beam. The system allows us to obtain concentration maps from any user-defined area, with spatial resolution of the order of the beam dimensions, i. e. better than 10 µm.

3 Examples of applications

1 – CMS Pitch adapter analysis [5]

A pitch adapter (PA) is a device situated between the silicon sensor and the front-end hybrid, to make the micro bonding connections possible between the input of the readout chips and the silicon sensor (they have different pitches, see Fig. 1). It consists in a fan of metallic strips, engraved on a glass support; the strips nominally consist in a $400\,\mu g\,cm^{-2}$ aluminium coating over a $70\,\mu g\,cm^{-2}$ chromium layer. On both sides, each strip ends with pads $(50 \times 250\,\mu m^2$ fan out side and $80 \times 340\,\mu m^2$ fan in side) to make the micro bonding contacts possible. Micro-bonding tests on the first batch of adapters showed both mechanical and electrical problems and the hypothesis was made that this unsatisfactory behaviour might be due to contaminations in the aluminium deposition on the PA. Our micro beam analysis pointed out a relevant Cu contamination of the strips; the mean copper thickness came out to be $190\,\mu g\,cm^{-2}$, almost 50% of the expected aluminium areal density. This contamination is evident in Fig. 1, where the Cu maps of the pads are reported, together with an optical image of the same area.

2 – Iron-gall inks in ancient manuscript [6]

The "imaging" information deduced from elemental maps allows selecting the correct areas where to perform quantitative analysis, avoiding inhomogeneities mainly present in the paper substrate. As an example, we show some

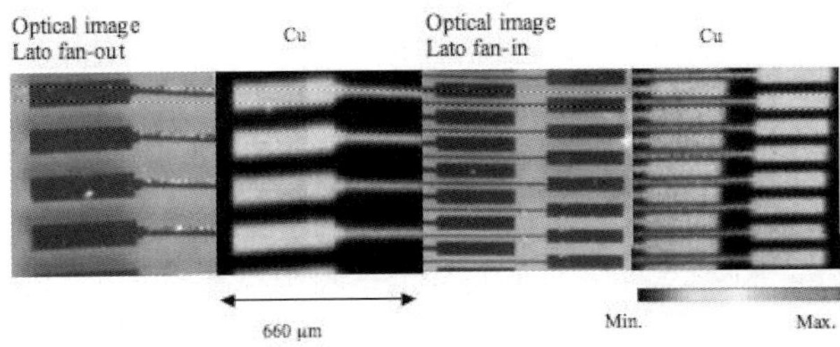

Fig. 1. Cu maps for both fan-out side and fan-in side. The optical images of analyzed areas are shown aside

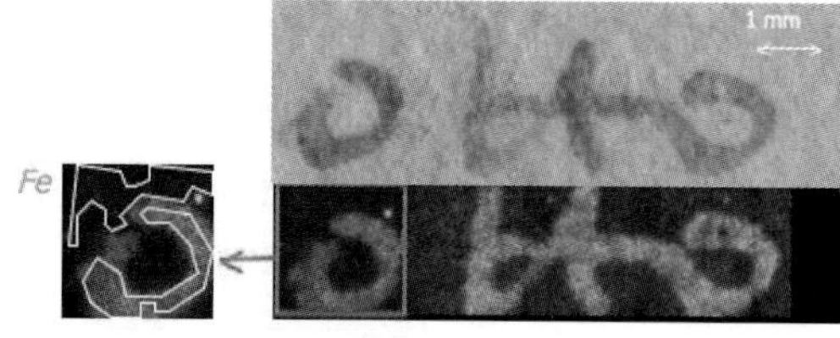

Fig. 2. Optical image, Fe map of the same area and the zone over which the selected scan was done

measured areas from a page of a document of the Florence State Archive; Fe elemental map is shown, together with a photograph of the scanned area.

Spectra of the selected regions can be constructed off line and quantitative analysis is thus obtained with a more significant knowledge than from single point measurements; ink composition is deduced after subtraction of the paper contribution from an equivalent area. Significantly lower currents can be used, still keeping sufficiently high statistics, just because all the "good" areas can be exploited.

3 – Metal point drawings analysis [6]

The knowledge of the material used both in the preparation and in the track is crucial to solve conservation problems of these precious and frail artworks. The non-uniformity of the track left by the stylus makes point analysis problematic, especially when the same elements are present also in the paper preparation. An imaging, non destructive technique is necessary, with a spatial resolution of 100 µm at least. Preliminary measurements on test samples of lead point drawings on a Hg-Pb prepared paper were successful and confirmed the power of the scanning approach to this kind of problems.

Acknowledgement. I wish to thank all the people I work with and, in particular, N. Grassi and P.A.-Mandò. Special thanks to Mirko Massi, who shared with me all the joys of our microbeam

References

1. T. Calligaro *et al.*, Development of an external beam nuclear microprobe on the Aglae facility of the Louvre museum, NIM B 161 163 (2000) 328 333
2. M. Massi, Progetto e realizzazione di una facility di microfascio all'acceleratore KN3000 di Firenze, degree thesis, 2000, `http://labec.fi.infn.it`
3. M. Massi *et al.*, The external beam microprobe facility in Florence: Set-up and performance, NIM B190 (2002) 276-282
4. M. Massi, Development of the external proton microbeam line at the new 3MV Tandetron accelerator in Florence, Ph. D thesis, 2005, `http://labec.fi.infn.it`
5. M. Massi *et al.*, Use of micro-PIXE analysis for the identiƥcation of contaminants in the metal deposition on a CMS, NIM B 219-220 (2004) 722-726
6. N. Grassi *et al.*, Scanning-mode Ion Beam Analysis in Cultural Heritage problems, submitted to NIM B

Infrastructure of the ATLAS Event Filter

Andrea Negri

University of California Irvine
andrea.negri@cern.ch

1 The Atlas Trigger and Data Acquisition system

The ATLAS detector [1] is a high energy physics experiment designed to exploit the full physics potential provided by the Large Hadron Collider (LHC), which will provide pp collisions at a centre-of-mass energy of 14 TeV and a luminosity of $10^{34}\,\mathrm{cm}^{-2}\,\mathrm{s}^{-1}$. The corresponding 40 MHz bunch crossing rate, the huge amount of detector channels ($\sim 10^8$) and a storage data flux of about 300 MB/s outline the challenge of the Trigger and Data Acquisition (TDAQ) system: select every second the most interesting hundred events out of millions. The TDAQ system is organized in three different trigger levels (Fig. 1). The first one (LVL1), realized in hardware by custom electronics, reduces the data rate from the 40 MHz collision rate to about 75 kHz. The High Level Triggers (LVL2 and Event Filter) [2], implemented on two different commodity component farms, provide a further reduction factor of $\sim 10^3$. The LVL2 operates on the full granularity data inside Regions of Interest

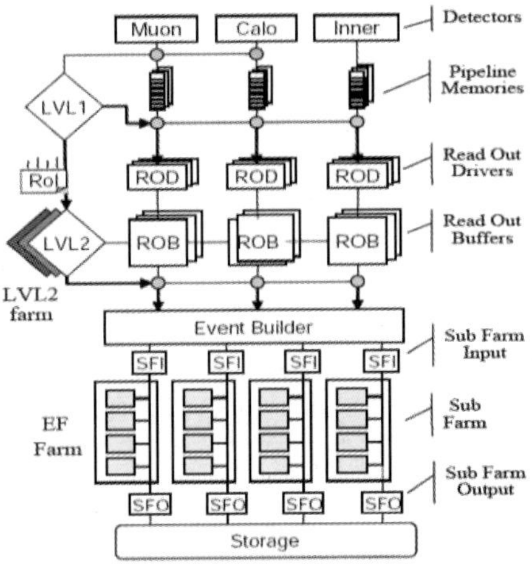

Fig. 1. Block diagram of the Atlas trigger and data acquisition system

(RoIs) identified by LVL1 and it reduces the event rate down to $\sim 2\,\text{kHz}$. For events accepted by LVL2, the full event data is assembled by the Event Builder system (EB) and shipped to the EF which performs the last selection (a rejection of a factor ~ 10) with a latency of the order of a second.

2 The Event Filter system

The EF primary function is the reduction of data flow and rate to a value acceptable by the mass storage operations and by the subsequent off-line data reconstruction and analysis steps. It also provides initial event sorting into streams for off-line production and global physics and detector monitoring, essential to ensure the quality of recorded data.

The computing instrument of the EF system is organized as a set of independent processor farms (sub-farms), each connected to one output port of the EB switch (Sub Farm Inputs). The SFIs perform the building operation, while the SFOs (Sub Farm Output) are the interfaces to the storage system. The final EF farm will be composed of about 2000 multi-core machines.

The running environment for the trigger algorithms is the HLT event selection software framework [3], which is based on the ATLAS off-line reconstruction and analysis environment ATHENA [4]. The reconstruction, scheduled by the HLT Steering, is carried out in a stepwise fashion and the outcome of one algorithm is used as a seed for the subsequent one. At every step one or more algorithms are executed and they validate the triggering embodied by a Trigger Element (TE). The execution of one algorithm sequence is driven by a static configuration that informs the Steering what particular algorithm must be executed in case a trigger condition (TE) is active. At the end of every step the Steering takes the decision whether to reject an event or not. This is done by matching the expected combinations of satisfied trigger conditions (signatures) against the actual outcome of event processing.

3 EF Infrastructure Design

The design of the ATLAS EF system [5] is object oriented (C++) and it uses multi-thread programming techniques, which permit a better exploitation of multi processors architectures.

Each processing host manages its own connection with the SFI and SFO elements and implements the client part of the communication protocol. Therefore processing nodes can be transparently added to a running sub-farm and software or hardware failures in a node do not affect the operation of other sub-farm elements. Similarly complete sub-farms, even geographically distributed, can be easily hot-plugged in a running system.

Fault tolerance and data security, which are fundamental requirements for any on-line architecture, become even more critical in the ATLAS selection

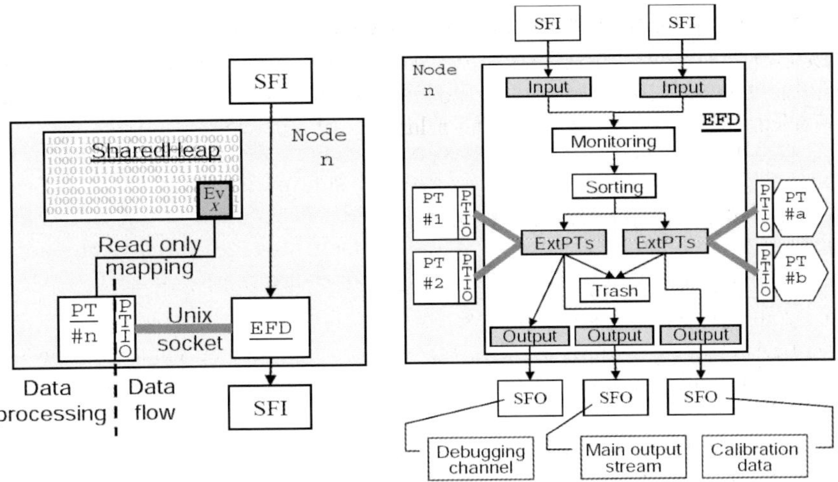

Fig. 2. *Left*: EFD design, decoupling between data processing operations and data flow functionalities. *Right*: implementation example of internal EFD data flow

environment, because EF algorithms are not specially developed for on-line environment, but they are inherited from the off-line ones. Therefore, in order to provide additional reliability in case of algorithm crash and ensure data security and fault tolerance, in each node the data processing operations are completely decoupled from the data flow functionalities. The latter ones are provided by the EF Data-flow process (EFD) that manages the communication with SFI and SFO elements and makes the events available to the Processing Tasks (PTs), which are in charge of data processing and event selection (Fig. 2). The PTs are implemented by separated processes running the ESS in the standard ATLAS off-line framework [3]. The events are made available to PTs via a shared memory (called SharedHeap), which stores the events during their transit in the processing node.

The EFD function is divided into different specific tasks that could be dynamically interconnected to form a fully configurable EF data-flow network (Fig. 2). A worker thread, starting execution from the first task, drives the event references through the internal data-flow network.

A plug-in interface (PTIO library) allows PTs to access the data-flow part opening a connection with the UNIX domain socket server implemented by the EFD. When a PT requests an event, the PTIO library transmits the request to the EFD, obtains the offset and size of the SharedHeap portion containing the event to be processed and maps this region in memory. The returned memory pointer is used by the PT to access the event and process it. Because the map is read only, the PT cannot corrupt the event or the SharedHeap structure. PT problems are reliably handled by the EFD that can identify PT crashes via socket hang-ups and PT dead locks by means of configurable processing timeouts. In both cases, the EFD, which owns the event, can assign it to another PT or send it directly to SFO. The processing operation produces a filtering decision, used to steer the internal EFD data-flow, and additional reconstructed information, which are appended by the EFD to the raw event.

The SharedHeap is implemented as a memory mapped file hosted in the local file system. This solution provides a safe and elegant event recovery mechanism. Indeed in case of EFD crash, the events can be recovered from the file system at EFD restart. The OS itself manages directly the actual write operations avoiding useless disk I/O over-heading.

4 Functional tests and deployments

The EF architecture has been validated on various testbeds of different sizes, it has been deployed in test-beam environments and it is being used for detector commissioning. The design showed the required robustness and fault tolerance: no major failures has been observed and expressly provoked crashes have been recovered without any data loss. System scalability has been proved both inside a node (correct exploitation of multiprocessors architectures) and at the farm level. Large scale scalability tests [6], carried out on 700 dual processor nodes, allowed to identify potential scalability problems related to concurrent access to the databases; they have been fixed and the changes will be validated in the next large scale scalability test scheduled for the end of the year. During test-beam runs, the possibility to use geographically distributed sub-farm implementations has been tested: remote farms located in Canada and Poland has been integrated in the running DAQ environment. Even if the primary EF function is the event selection, the good flexibility and modularity of the EF design allowed the use of the EF infrastructure also for calibration and monitoring purposes.

References

1. ATLAS Collaboration, *Technical Proposal for a General- Purpose pp Experiment at the LHC*, CERN/LHCC/94-43, 1994.
2. ATLAS Collaboration, *ATLAS High-Level Triggers, DAQ and DCS Technical Proposal*, CERN/LHCC/2000-17, 2000.
3. M. Elsing *et al.*, *Analysis and Conceptual Design of the HLT Selection*, ATLAS Internal Note, ATL-DAQ-2002-013, 2002.
4. ATLAS Collaboration, *Athena: User Guide and Tutorial*, Available at: http://ATLAS.web.cern.ch/ATLAS/GROUPS/SOFTWARE/OO/architecture
5. A. Negri *et al.*, *Design, deployment and functional tests of the on-line Event Filter for the ATLAS experiment at LHC*, proceedings of "2004 IEEE Nuclear Science Symposium", Rome, Italy, 16-22 Oct 2004
6. D. Burckhart-Chromeck *et al.*, *Testing on a large scale: running the Atlas data acquisition and high level trigger software on 700 pc nodes*, proceedings of "Computing In High Energy Physics CHEP 2006", Mumbai, India, 13–17 Feb 2006

The CMS High-Level Trigger

Pietro Govoni

Università e INFN di Milano-Bicocca
pietro.govoni@mib.infn.it

Introduction

At the LHC, the Large Hadron Collider that will start operating the next year at CERN, proton beams will collide with energies of 7 TeV each. The beams will be composed by several bunches, that will collide in the interacting points with a frequency of about 40 MHz. The CMS experiment (Compact Muon Solenoid), which is one of the two multi-purpose experiments that will take data, will have to cope with such a high interaction rate and reduce it to the order of 100 Hz, that is the constraint given by the persistency on tape. This reducing rate of a factor of $O(10^6)$ of single events of 2 MB size will be achieved in two separate stages: the level one trigger (L1) [1], implemented by means of hardware devices, and the high-level trigger (HLT) [2], implemented in a software way on a farm of commercial PCs.

1 The CMS HLT general structure

The HLT process benefits from the flexibility of its software implementation, that can be easily updated and changed, while profiting from the up-to-date computing technologies available, with the low cost and maintenance typical of market products. The CMS events will be built by *bulider units*, that merge the L1 output and dispatch in an asynchronous mode bunches complete events to the single machines that compose the DAQ farm, the *filter units*, which in turn select the good events to be saved by *storage managers* on local buffers disks. Their content is then flushed into the tapes of the persitency.

The HLT software implementation performs a quick and raw reconstruction of physical objects by using the data selected by the L1 and selects the events to be saved by applying physical cuts on the reconstructed objects parameters. It is designed to be efficient on the channels of the CMS physics programme; to be as general as possible to save unexpected events; to be the less sensitive is possible to the precise knowledge of the detector conditions, like the alignment and the calibration; to make use of the off-line reconstruction algorithms when possible; to be monitored by saving a small fraction of the discarded events as well. Its main limitations are the CPU time necessary to the algorithms to run, the saving rate on tape, the precision of the calibration and alignment of the detector.

To optimize their performances, the reconstruction algorithms are designed by following some general rules:

- *regional reconstruction*: the algorithms run on the portion of the detector flagged as interesting by the L1 trigger;
- *partial reconstruction*: the reconstruction of the physics events is performed only to the precision necessary to select each event;
- algorithms are subdivided into sub-levels, to stop the reconstruction as soon as an event has to be discarded;
- the good events undergo all the recosntruction algorithms, to be divided in different physics streams after the selection.

2 The HLT Physics events recostruction

The reconstruction patterns follow the different possible particles triggered by CMS. Some of them will be dedicated to single particle identifications, while some other selections will aim at tagging coule of physiscs objects.

The electrons reconstruction is as example of a single particle reconstruction stream. It starts from the energy deposit in the electromagnetic calorimeter, where the energy deposited in the crystals is calculated by means of a dynamic shape clustering algorithm that recovers the energy lost by bremsstrahlung in the tracker material. A first threshold is applied to the collected energy. In the subsequent passage, hits in the pixel detector (the innermost layers of the tracker) that match the track direction with the ECAL energy barycenter are looked for. In case no tracks are found, the energy deposit is associated to a photon and a higher threshold applied to select the events. When a matching is found, the object is identified as an electron and the track is reconstructed by using all the tracker layers available.

3 The HLT performances

The parameters used to evaluate the performances of the HLT are the available trigger rate the efficiency in the channels identification on one side, the CPU time needed for the processing on the other. Montecarlo studies, performed on some benchmark channels, show the results listed in Table 1 for the trigger rates, Table 2 for the efficiencies and Table 3 for the CPU times respectively.

Table 1. the CMS HLT trigger rates for some benchmark channels, as resulting from Montecarlo studies

Trigger	Threshold (GeV) ($\epsilon = 90-95\%$)	Indiv. rate (Hz)	Cumul. rate (Hz)
1e, 2e	29, 17	34	34
1γ, 2γ	80 (40×25)	9	43
1μ, 2μ	19, 7	29	72
1τ, 2τ	86, 59	4	76
Jet $\times$ Miss-E_T	180×123	5	81
1-jet, 3-jet, 4-jet	657, 247, 113	9	89
e+ jet	19×52	1	90
Inclusive b-jets	237	5	96
Calibation/other		10	**105**

Table 2. the CMS HLT trigger efficiencies for some benchmark channels, as resulting from Montecarlo studies. The values relative to the whole detector are reported, together with the ones relative to the fiducial regions, where no edge effects affect the data taking

Channel	Efficiency (for fiducial objects)
H(115 GeV)$\to \gamma\gamma$	77%
H(160 GeV)$\to$WW$^*\to 2\mu$	92%
H$\to$ZZ$\to 4\mu$	92%
A/H(200 GeV)$\to 2\tau$	45%
SUSY (~ 0.5 TeV sparticles)	$\sim 60\%$
With R$_P$-violation	$\sim 20\%$
W$\to$eν	67% (fid: 60%)
W$\to$eμ	69% (fid: 50%)
Top$\to \mu$X	72%

Table 3. the CMS HLT CPU times for some benchmark channels, as resulting from Montecarlo studies

Trigger	CPU (ms)	Rate (kHz)	**Total (s)**
1e/γ, 2e/γ	160	4.3	**688**
1μ, 2μ	710	3.6	**2556**
1τ, 2τ	130	3.0	**390**
Jets, Jet + Miss-E_T	50	3.4	**170**
e+ Jet	165	0.8	**132**
B-jets	300	0.5	**150**

The average CPU time required is about 300 ms/event on 1 GHz PentiumIII CPU. At the low luminosity that will be achieved during the first months of data taking, to satisfy the expected output rate of 50 Hz a farm of 2000 CPU will be assembled.

Conclusions

The CMS HLT will handle an unprecedented rate of events that has to be reduced by a factor of $O(10^3)$. This result will be accomplished by suited software tools running on a farm of commercial PCs, with algorithms that will exploit an efficient, inclusive and stable selection of the events. Its intrinsically flexible architecture will nicely fit the requirements of the CMS startup, when detailed studies will be performed to set-up and fine-tune all the selections.

References

1. The Trigger and Data Acquisition project, Volume I: The Level 1 Trigger. Technical Design Report. **CERN/LHCC 2000-038**
2. The Trigger and Data Acquisition project, Volume II: Data Acquisition and Hihg-level Trigger. Technical Design Report. **CERN/LHCC 2002-26**

WLCG Service Challenges
and Tiered architecture in the LHC era

Daniele Bonacorsi and Tiziana Ferrari (on behalf of INFN SC group)

INFN-CNAF, Viale B. Pichat 6/2, 40127, Bologna, Italy
Daniele.Bonacorsi@cnaf.infn.it, Tiziana.Ferrari@cnaf.infn.it

The implementation of the computing infrastructure for LHC experiments requires graduate and accurate verification of performance of hardware resources (CPU, storage, network), of the middleware services provided by Grid projects and of the LHC experiment-specific software applications. The realization of the Computing Models of the LHC experiments needs high reliability of resources and services, and the demonstration of their functionality, robustness and effective usability in addressing real experiment use-cases. The data management and workload management sector of the Computing Models of LHC experiments are tested both in experiment-specific "Data Challenges" and in more general "Service Challenges" of increasing complexity, the latter being driven by Worldwide LHC Computing Grid (WLCG). The objective is to improve the quality and reliability of the distributed Grid services to address the computing needs in the LHC era.

1 Worldwide LHC Computing Grid

The purpose of the Worldwide LHC Computing Grid (WLCG) project is to provide the computing resources needed to process and analyze the data gathered by the LHC experiments. The LCG project [1] is assembling at multiple computing centres the main offline data-storage and computing resources needed by LHC experiments, and operating these resources in a shared grid-like manner. The main goal is to provide common tools and implement uniform means of accessing resources.

The path to follow is stated in the WLCG Memorandum of Understanding (MoU) [2] for collaboration in the Deployment and Exploitation of the WLCG Grid between the computing Tiers. After LCG Phase 1 (technology development and tests leading to a production prototype), the WLCG MoU governs the execution of the LCG Phase-2 (deployment and exploitation of LCG as a Service), by defining the program of work, distribution of duties and responsibilities to the involved parties as well as the Computing Resources

Levels they will offer to the LHC exps (also organizational, managerial and financial guidelines).

1.1 Computing Tiers in WLCG

The "Parties" in the MoU are CERN with the Tier-0 (T0) and all Institutions participating in the provision of the WLCG with a Tier-1 (T1) and/or Tier-2 (T2) computing centre (federations included).

The CERN T0 receives raw data from experiments' online computing farms and records them on permanent MSS, performs first pass reconstruction and distributes them to Tier-1's. In addition, the CERN hosts an Analysis Facility with the functionality of a combined T1-T2 centre, except that it does not offer permanent storage of back-up copies of raw data. The Tier-1's (T1) provide a distributed permanent back-up of the raw data, storage and management of data needed during the analysis process, and offer a grid-enabled data service data intensive analysis and reprocessing. The Tier-2's (T2) provide well-managed, grid-enabled disk storage and concentrate on tasks such as simulation, end-user analysis and high-performance parallel analysis. The technical participation of the ÒPartiesÓ is defined separately in terms of: *i)* Computing Resources levels that they pledge to provide to one/more LHC exps that they serve; *ii)* Computing Services levels that they pledge to the WLCG collaboration.

2 WLCG Service Challenges

The WLCG Service Challenges (SC) are a mechanism by which the readiness of the overall LHC computing infrastructure to meet the experiments' requirements is measured and – if/where necessary – corrected. It is a joined effort with Tiers community to trigger resources deployment, to test them on realistic use patterns and to drive activity planning towards the LHC days. It is a long path, done in several steps, as outlined below.

The first two Service Challenge phases (SC1 and SC2) were focussed on building up the necessary data management infrastructure to perform reliable T0-T1 transfers and permanent production services with the appropriate throughput. They basically helped to build up the T0-T1 infrastructure and services to handle production transfers and data flows. No T2 were involved, no experiment-specific software and no offline use-cases. SC2 met its throughput goal (100 MB/s/site, 500 MB/s sustained out of CERN), but not its service goals.

The subsequents steps (SC3 and SC4) are instead focussed on bringing T2s into the loop and fully address experiments use-cases towards full production services: a specific "service" section was e. g. added to the set-up and throughput goals, and focus was on data management, batch production and real data access. The focus of the following sections of this paper is on the latter two phases.

3 Service Challenge (phase 3)

The SC3 phase is the first one with experiment-oriented objectives. Its time slot was July–December 2005: T0, all T1s and a small nb of T2s participated. In July 2005 a set-up ("throughput") phase was addressed, with throughput targets of 150 MB/s network-disk and 60 MB/s network/tape for each T1, and CERN capable of supporting 1 GB/s for the transfers to disk and 400 MB/s for those to tape at the T1s. Some T1s also supported T2s, which may upload simulated data and download analysis data. In September–December 2005 (with a re-run in January 2006) a "service" phase started, focussed on stable operation, during which experiments were committed to carry out tests of their software and computing workflows, including additional software components, a Grid WMS, use of Grid catalogues, mass storage management services and a file transfer service (gLite FTS [5,6]).

The throughput tests in July 2005 did not meet targets (about 50% higher than SC2): on average, half of the targets were achieved, with scarce stability. A substantial improvement came from subsequent SC3 "throughput re-run" in January 2006: some sites (RAL, FNAL, TRIUMF, NIKHEF/SARA, Nordic DataGrid Facility) exceeded the disk-disk targets (reaching up to 250 MB/s) but not all of them, and daily average rates were still far for most sites. It was nevertheless an important step to gain experience with the services before the start of SC4 (see Sect. 4). Since then, many fixes were put into production:SRM [3] ready on each site, dCache [4] upgraded to version 1.6.6+, EGEE gLite FTS upgraded to version 1.4 (and a lot of effort was spent on debugging transfers by experiment people), bug fixes were applied to Castor [7] at CERN. All these upgrades were released and deployed before SC4.

4 Service Challenge (phase 4)

The SC4 phase aims to demonstrate that all of the offline data processing requirements expressed in the experiments' Computing Models, from raw data taking through to data access, can be handled within Grid at the full nominal data rate of the LHC (see Table 1). The SC4 time slot is April–September 2006: T0, all T1s and the majority of T2s participate. The ÒthroughputÓ phase in April 2006 aims to a throughput demonstration sustaining for 3 weeks the target data rates at each site. The target is a stable, reliable data transfer to T1s at target rates to any supported SRM implementation (dCache, Castor, etc), plus a factor of two for backlogs/peaks. The "service" phase (May–September 2006) gets the basic software components required for the initial LHC data processing service into the loop. The target is to show capability to support full Computing Models of each LHC exp, from simulation to end-user batch analysis at Tier-2's. This will become the initial production service for LHC and will be made available to the experiments for final testing, commissioning and processing of cosmic ray data.

Table 1. Nominal SC4 "throughput" target rates (in MB/s, from CERN to both disks and tapes at the Tier-1), and hosted LHC experiments at each WLCG Tier-1 centre

Tier-1 centre	ALICE	ATLAS	CMS	LHCb	disk-disk	disk-tape
ASGC, Taipei		X	X		100	75
CNAF, Italy	X	X	X	X	200	75
PIC, Spain		X	X	X	100	75
IN2P3, Lyon	X	X	X	X	200	75
GridKA, Germany	X	X	X	X	200	75
RAL, UK		X	X	X	150	75
BNL, USA		X			200	75
FNAL, USA			X		200	75
TRIUMF, Canada		X			50	50
NIKHEF/SARA, NL	X	X		X	150	75
Nordic Data Grid	X	X			50	50

5 Summary

WLCG is getting ready to provide Grid infrastructure/components for the LHC era, matching the MoU targets and the experiment requirements. Along this path, WLCG is approaching the 'regime' through several Service Challenges involving Tiers. The computing Tiers participate to such challenges to get prepared to operate WLCG-enabled services to realize the Computing Models of the experiments. In particular, the SC3 phase (recently closed) and the SC4 phase (now running) are experienced to be useful exercises to understand how to proficiently deploy and run a real, wide and reliable Grid service. The Tiers are continuing to collaborate with WLCG to ramp-up essential Grid services to target levels of reliability, availability, scalability and end-to-end performance, as requested for the LHC era.

References

1. LHC Computing Grid (LCG) project: `http://www.cern.ch/lcg/`
2. CERN-C-RRB-2005-01/Rev. 21 March 2006
3. Storage Resource Manager (SRM) Working Group: `http://sdm.lbl.gov/srm-wg/`
4. dCache: `http://www.dcache.org/`
5. EGEE Middleware Architecture, EGEE-DJRA1.1-594698-v1.0: `https://edms.cern.ch/document/594698/1.0/`
6. gLite, the EGEE Lightweight Middleware for Grid Computing: `http://glite.web.cern.ch/glite/`
7. CERN Advanced STORage Manager, (CASTOR): `http://castor.web.cern.ch/castor/`